Developmental Biology
of Cultured Nerve,
Muscle, and Glia

Developmental Biology of Cultured Nerve, Muscle, and Glia

David Schubert

The Salk Institute
San Diego, California

A Wiley-Interscience Publication
JOHN WILEY & SONS
New York · Chichester · Brisbane · Toronto · Singapore

Library of Congress Cataloging in Publication Data:

Schubert, David, 1943–
    Developmental biology of cultured nerve, muscle, and
glia.

· "A Wiley-Interscience publication."
    Includes bibliographical references and index.
    1. Developmental cytology.   2. Cell culture.
I. Title.   [DNLM: 1. Neurons—Cytology.   2. Muscles—
Cytology.   3. Neuroglia—Cytology.   4. Cells, Cultured.
5. Cell differentiation.   6. Neuromuscular junction.

WL 102.5 S384d]
QH491.S38   1984   591.1'88   83-26115
ISBN 0-471-86592-3

Printed in the United States of America

10 9 8 7 6 5 4 3 2 1

To
William M. Smith
of
Rapid City, Michigan

Preface

During the last decade there has been an exponential increase in the use of cultured cells to study the problems of developmental biology. This has been due largely to the increasing availability of clonal cell lines and of relatively homogeneous populations of explanted cells that exhibit differentiated characteristics in culture. One of the most extensively studied areas of developmental biology is the differentiation and interactions of nerve, muscle, and glia. The rapid advances that have occurred in these areas could only have been made through the use of cultured cells. This monograph attempts to summarize the contribution of cell culture to the cellular biology of nerve, muscle, and glia. The reason for collecting this information in a single volume is to make it readily available to students of developmental biology and the neurosciences, and to define the types of problems that can best be solved through the use of cultured cells. The information that has been gleaned from the study of this subset of cultured cells should also be of use to research scientists who are employing cultured cells to elucidate the developmental mechanisms of other cell types.

This book focuses primarily on those experiments that have sought to critically test alternative explanations for developmental phenomena. Some of the experiments are presented in detail and are analyzed in terms of possible outcomes, experimental results, and the interpretation of the data. Although this analytical approach may be somewhat tedious for some, it was adopted in order to focus on experimental and conceptual problems that need to be understood by all readers who use cultured cells. Many students are not aware of the exquisite sensitivity of the phenotypes of cultured cells to environmental changes and of the need to incorporate this knowledge into their experimental protocols.

Some of the experiments used in this text are taken from our own work. This is a decision for which I apologize, but the familiarity with the material will, I hope, enhance the clarity of its presentation. Although an extensive literature is reviewed, I have limited the citations to the first and the most recent publications in reviewed journals on each subject rather than burdening the text with extensive lists of references.

The introductory chapter briefly outlines the origins of cell culture and discusses both the advantages and limitations of this paradigm in studying cellular differentiation. The following four chapters describe the knowledge gained by studies of the differentiation of cultured muscle, nerve, and glia, and discuss the functional interactions between pairs of these cells. The final chapter is pertinent to all of the preceding material, for it outlines complex phenomena that can be experimentally approached only through the use of cultured cells—the secretion of extracellular macromolecules and cellular adhesion.

With the recent development of new anatomical, molecular biological, and electrophysiological techniques, the study of the nervous system is at a crossroads. It is quite likely that experiments using cultured cells will make even greater contributions to developmental neurobiology in the future than they have in the past. A great deal of information has, however, been acquired from the studies of cultured cells during the last decade, and a thorough understanding of this work must be acquired to move forward. The mistakes of the past need not be repeated. It is hoped that this monograph will help to bridge the gap between an experimental system largely developed during the past decade and the more enlightening science of the next.

DAVID SCHUBERT

La Jolla, California
March 1984

Acknowledgments

I wish to express my sincere appreciation to Drs. J. H. Steinbach, W. Stallcup, J. Levine, Y. Kidokoro, F. Collins, S. Halegoua, B. Hyman, B. K. Schrier, M. Cowan, A. Traynor, R. Bloch, and S. Sine for their encouragement and their helpful criticism of portions of this text. I also wish to thank Annie Steinbach for lending her editorial skills toward the preparation of this manuscript and Bonnie Harkins for her patience and expertise in translating my marginally legible script into typescript. The artistic talents of Professor Jamie T. Simon are expressed in all of the original and redrawn figures. Finally, without the dedication of Monique LaCorbiere, Helgi Tarikas-Jobe, F. George Klier, and William Carlisle none of the original work from our laboratory would have been possible.

The work from my laboratory was supported by grants from the Muscular Dystrophy Foundation of America and the National Institutes of Health.

D.S.

Contents

Developmental Biology
of Cultured Nerve,
Muscle, and Glia

—1—

Introduction

1.1 The Beginning

In 1907 Ross G. Harrison published a brief article describing an experiment designed to distinguish between two models for the *in vivo* development of nerve fibers. One hypothesis, proposed by H. Held, was that nerve fibers arose by the formation of protoplasmic bridges between cells. The other hypothesis, promulgated by Wilhelm His and Ramon y Cajal, was that the nerve fiber grew out from the body of the nerve cell. To rule out the cytoplasmic bridge hypothesis, Harrison removed a piece of frog embryo spinal cord and placed it in a drop of lymph on a cover slip. The cover slip was then inverted over a hollow slide and sealed with paraffin. The lymph clotted and it was possible to observe the transplanted tissue continually under a microscope. The nerve cells were initially round, but within a day many fibers were observed growing away from the explanted tissue. Actively moving, amoeboidlike tips of the nerve fibers were also observed—structures which have since been named growth cones. This experiment showed that a pair of cells was not required for the formation of nerve processes, and that neurites arose by elongation. It was also the first use of an experimental approach which has dominated the field of cell biology over the last three decades.

During the 30 years following Harrison's initial experiment little new information emerged from the application of the tissue culture technique. This period did, however, yield some improvements, such as the development of the culture flask and the use of embryo extract (fetuses homogenized in saline) as a nutrient mixture. The culture media were initially supplemented with bouillon. In the mid-1930's more defined nutrient mixtures containing amino acids, glucose, vitamins, and physiologically balanced salts were introduced. These mixtures were used in conjunction with animal sera to promote long-term cell viability. The major problem of lack of sterility was overcome with the introduction of antibiotics in the culture media. The stage was then set for the rapid developments in the use of cell culture that were initiated by the establishment of the first permanent cell line from mouse (the L cell line; Sanford et al., 1948). In 1951 George Gey established the first human line from neoplastic tissue (Harvey, 1975). This cell line, named HeLa, was established from a cervical cancer biopsy of Henrietta Lacks. The biopsied cells were placed in culture by Gey, and the cells continued dividing and were eventually cloned. The HeLa and L cell lines were then used to improve the culture methods for mammalian cells, and they are still widely used as hosts for animal viruses.

Since the establishment of HeLa, there has been an exponential increase in the experimental application of cultured cells. One area which has greatly benefited from tissue culture technology is the study of the nervous system. The period of use of cultured cells in the neurosciences can be divided into three eras (R. Bunge, 1975). The work by Harrison and others through the early 1950's was concerned mainly with axon growth and structure and the description of the supportive glial cells (P. Weiss, 1955). The second era evolved from the first demonstration that cultured neurons were electrically excitable (Crain, 1956), an observation that quelled the nonbelievers who felt that the cultured nerve cells would not be able to express the most characteristic property of neurons. The following decade yielded a clear description of the cellular arrangement in cultured tissue explants, showing that explanted tissues can retain a cellular organization similar to that observed *in vivo*. Chemical synapses were demonstrated within spinal cord explants (Crain, 1966), and a limited number of synaptic interactions were observed between explants from various parts of the nervous system. These data suggested that cultured cells could mimic some aspects of normal cell–cell interactions within the nervous system. Finally, the current era of cell culture neurobiology began in the late 1960's and has been devoted largely to the study of clonal cell lines and long-term dissociated cell cultures.

Although there was initially a preoccupation with maintaining the "cytoarchitectonic" properties of the tissue when it was placed in culture, the retention of the extremely complex *in vivo* structure in cultured pieces of tissue gave little advantage over studying the cells in the intact organism. Methods were therefore needed for the isolation of homogeneous cell populations in which it would be possible to study the biochemical properties of individual cell types. Since the initial attempts at separating viable nerve and glial cells from whole brain met with limited success, methods were developed for maintaining dissociated cells in culture for several months (Bray, 1970), and tumors of the nervous system were used as a source of dividing cells from which to adapt cells to grow in continuous clonal culture (Sato, 1974).

In addition to increased experimental accessibility over cells *in vivo,* cultured cells have two other uses. The cellular homogeneity of clonal lines and some dissociated cell culture systems makes it possible to study in detail the molecular mechanisms underlying phenomena previously described *in vivo.* An example is the mode of action of nerve growth factor (NGF), the chemistry of which is virtually impossible to study in the animal. Also, cultured cells can be employed to define new phenomena and to isolate macromolecules that had not previously been shown to exist in the animal. For example, the study of cultured cells led to the discovery of many cytoskeletal components of cells and opened up the whole field of cytoarchitecture. It cannot be assumed, however, that a process defined in culture necessarily performs a function *in vivo.* It is necessary to use the information gleaned from culture to verify its role *in vivo.*

1.2 Types of Cell Culture

There are three experimental paradigms for the *in vitro* maintenance of cells. The earliest studies simply explanted intact pieces of tissue into the culture environment—a technique called organ culture. If the tissue is dissociated into single cells and the entire population placed in culture, the result is appropriately called dissociated cell culture. Finally, if a population of cells is obtained from a single cell, the resultant cultures are termed clonal. A clonal population may divide forever to generate a permanent clonal cell line, or it may divide for a finite number of generations as in some dissociated cell cultures. Organ, dissociated cell, and continuous clonal cell cultures have all been used extensively to examine various aspects of the nervous system, and examples of the use of all three types of cultures will be presented. The choice of which system is employed is usually dictated by the nature of the questions that the experimenter is posing and the type of tissue he intends to study. At the present time only a limited number of cell types can be cultured by all three procedures. It is therefore important to understand the strengths and weaknesses of each. The following paragraphs discuss the three culture systems, using examples of peripheral nervous tissue and skeletal muscle. This group of cultured cells also forms the basis for the majority of the experimental work outlined in this monograph.

1.2.1. *Organ Culture.* When a piece of tissue is explanted from an animal directly into a cell culture environment it is said to be in organ culture. The tissue may be a thin slice, a chunk, or the complete organ, as in the case of sympathetic ganglia. There is generally a limitation on the size of the tissue, because larger masses of cells tend to become necrotic due to the poor exchange of nutrients in their core. One experimental advantage of organ culture is that it tends to maintain, at least for a limited period of time, the original tissue histology and cellular physiology. Thus, brain tissue slices have been used to examine the pharmacological properties of nerve cells in an environment where it is easier to visualize the cells and to vary the external milieu than in the intact animal. It cannot be assumed, however, that the properties of organ-cultured cells are identical to their counterparts *in vivo,* for the massive denervation and cell death caused by the explantation procedures, along with the new influence of the tissue culture environments, must change at least a subset of the cell's metabolism. If, however, the particular property of the cell which is being studied is the same in culture as *in vivo,* then the experimental advantages of the culture system can be exploited.

Another advantage of organ culture is the ease with which tissue samples can be prepared. This is particularly useful if the biological activity of a large number of fractions must be assayed. Perhaps the most elegant use made of organ culture was in the characterization and purification of nerve growth factor (NGF). It was initially shown that an activity from Sarcoma 180, a mouse tumor, increased the growth of sympathetic ganglia in intact chick embryos (Bueker, 1948). The chick embryo assay was, however, too laborious to allow the purification of the factor responsible for the effect. An easy assay was needed. This was achieved by explanting whole sympathetic ganglia into culture and assaying the effects of various cell homogenates on neurite outgrowth

(Levi-Montalcini et al., 1954). It was shown that tissues that promoted hypertrophy of sympathetic ganglia *in ovo* caused the rapid extension of a halo of neurites from explanted ganglia (Figure 1.1). On the basis of this cell culture assay, NGF was purified to homogeneity and shown to play a major role in the development of the autonomic nervous system (see Chapter 3).

There has been a more limited use of organ cultures of muscle. Since multinucleate muscle cells are exceedingly large compared with all other cell types, it is difficult to explant chunks or slices of tissue into culture and maintain cell viability. Greater success was achieved when rat hemidiaphragms were placed in organ culture (Ziskind and Harris, 1979). These preparations have been used to study reinnervation and to measure the rate of acetylcholine receptor (AChR) turnover. Although the viability of the hemidiaphragm preparations is limited to a few days, they have helped to solve several problems that were experimentally inaccessible in the animal.

1.2.2. *Dissociated Primary Cultures.* The most widely used culture system in neurobiology is the primary culture of dissociated cells. Typically, tissue is explanted from the animal and placed in a solution of proteolytic enzymes such as trypsin or collagenase to dissociate the cells. The practice of separating cells by treatment with proteolytic enzymes before placing them in culture was introduced by Rous and Jones in 1916. The ease and extent of dissociation is dependent on the age of the tissue; embryonic tissues are much easier to dissociate. The single-cell suspension is then washed to remove the enzyme and the cells are placed in culture dishes. After a few hours some of the input cells adhere to the substratum. In the case of anchorage-dependent cells such as myoblasts and central nervous system (CNS) nerve and glia, those cells that do not adhere usually die. The fraction of input cells which adheres and survives (the plating efficiency) varies greatly with the type of tissue. The plating efficiency of myoblasts from chick embryonic skeletal muscle is generally around 80%, while that of embryonic brain tissue may be as much as 10-fold lower. There is an initial selection for those cell types that are able to survive the dissociation procedure, followed by selection for those cells that are compatible with the culture environment. If the initial cell population is extremely heterogeneous, as in CNS tissue, it is currently impossible to obtain a group of cells in culture that is completely representative of those *in vivo*. However, starting with a tissue, such as skeletal muscle, that contains primarily muscle and fibroblasts, it is feasible to obtain a more representative population in culture.

The problem of cellular heterogeneity in dissociated cell cultures has

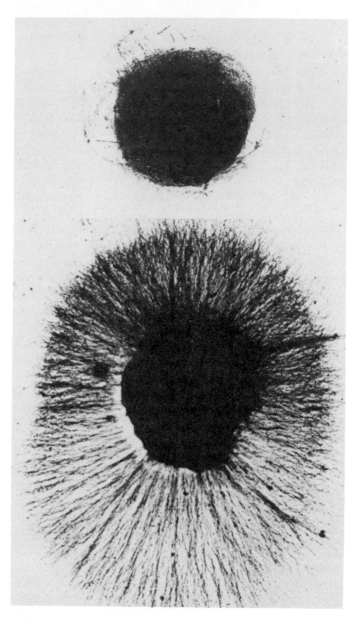

Figure 1.1. Explanted sensory ganglia of chick embryo. Sensory ganglia from 7-day chick embryos were cultured without NGF (top) or with NGF (bottom) for 24 hr and stained. From Levi-Montalcini (1964).

been reduced by two methods: (1) The desired cell type can be separated from the others before plating in culture. For example, glial and fibroblastlike cells are more adhesive to substrata than are nerve cells. By selectively adhering the nonneuronal cells to culture dishes or glass beads, a more pure population of nerve cells may be obtained (Hanson et al., 1982a). (2) The cell type of interest can be selected by a variety of methods from the mixture once the cells are growing in culture. For example, a cytotoxic antiserum may be used to kill one class of cells within the population, or the deletion of a required hormone from the culture medium can selectively eliminate another type of cell. In primary cultures of nerve and muscle, the undesired glia and fibroblasts are frequently eliminated by the use of DNA synthesis inhibitors that kill only dividing cells. Glia and fibroblasts are among the most prolific of cultured cells, but the nerve cells and muscle fibers do not divide. A relatively homogeneous dissociated cell culture is of significantly greater experimental usefulness than are organ cultures.

The primary advantage of dissociated cell cultures stems from the fact that they can be made up of a single cell type. Since the cells are alike, what is true for one should be true for the population. This presents the opportunity to make biochemical measurements on large numbers of similar cells rather than on a single cell. With some rapidly dividing cell types, such as glia and myoblasts, a large cell population can be derived from the progeny of a single cell. It is therefore possible to have clonal primary cultures. Cloning is the only way to assure that all of the cells in the culture are phenotypically identical. Clonal primary cells are, however, limited in the number of times that they can divide. The limited life span of cultured normal cells is apparently a reflection of in vivo aging (Orgel, 1973).

The cell type perhaps most frequently studied in primary culture is the skeletal muscle myoblast. When embryonic or neonatal skeletal muscle is dissociated with trypsin and plated onto collagen-coated substrata, a mixed population of cells attaches to the substratum and starts dividing. Eventually the myoblasts in the culture fuse to form multinucleate myotubes, while the fibroblasts continue dividing. Since the DNA synthesis in the myotubes is minimal, the rapidly dividing fibroblast population can be eliminated by introducing DNA synthesis inhibitors such as cytosine arabinoside, leaving a fairly homogeneous population of myotubes (Figure 1.2). Dissociated skeletal muscle cultures have been used extensively to study the mechanisms of myogenesis and the formation of synaptic junctions. Very little would be known about myogenesis if myoblast cultures had not been introduced about 20 years ago (see Chapter 2 for extensive discussion).

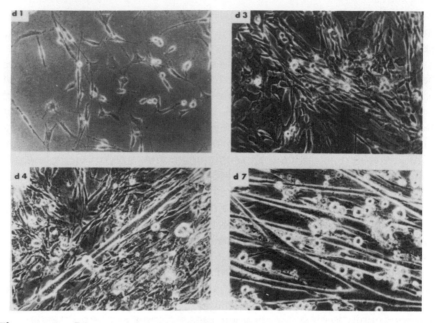

Figure 1.2. Primary culture of fused skeletal muscle myoblasts. Myoblasts were prepared from neonatal rat thigh muscle, placed in culture, and allowed to fuse to form multinucleate myotubes. The days in culture are indicated. Courtesy of Dr. Robert Bloch.

Cultures of dissociated peripheral neurons are a more recent innovation, primarily because they are more fastidious with respect to their culture media. The pioneering efforts of Dennis Bray (1970) showed that sympathetic ganglia could be dissociated with trypsin and placed into an air-buffered culture medium containing rat serum, methocel to increase the medium viscosity, and nerve growth factor (NGF), which is essential for the survival of the neurons. Under these culture conditions the growth of non-neuronal glia and fibroblasts was suppressed. The nerve cells did not divide, but they did send out long processes or neurites which covered the dish (Figure 1.3). These cultures and similar ones from sensory ganglia have been used extensively to study problems of nerve differentiation and electrophysiology which will be discussed in Chapter 3.

The experimental limitations of dissociated primary cultures are somewhat fewer than those of organ cultures. A major problem is that the growth conditions select for some phenotypes and against others. To obtain reproducible results with primary cultures it is therefore mandatory that the conditions of dissociation and culturing be identical each

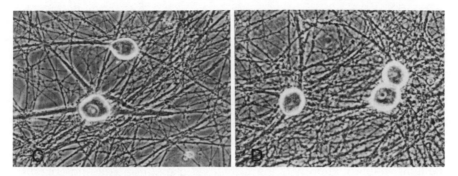

Figure 1.3. Dissociated sympathetic neurons. Sympathetic neurons from neonatal rat sympathetic ganglia were grown in culture for approximately 3 weeks. The cell body diameters are approximately 35 μm. From Hawrot (1980).

time the experiment is done. If the tissue is heterogeneous, it is unlikely that the final cell population will reflect the initial one. In addition, if some of the cells in the population are able to divide, the composition of the culture will change drastically with time because of the selective advantage of the more rapidly dividing cells. Even if it is possible to clone cells from the original cultures, these clones only divide a limited number of times. It is therefore necessary periodically to isolate new clones. It is never certain, however, that the phenotypes of the earlier and later clones are identical. Finally, it is not possible to isolate mutants from primary cultures. Clones of continuous cell lines are required for this.

1.2.3. *Continuous Cell Lines.*   The final class of culture systems is that of continuous cell lines. These are cells that will divide forever if maintained properly. The major attributes of continuous clonal cell lines are their homogeneity and the ease with which large quantities of cells can be grown. This situation allows for many more biochemical experiments than can be done with dissociated cell cultures. In addition, once the phenotype of the cell line is established it does not change. This high degree of reproducibility is not always possible with primary cultures, for there often are subtle variations in culture conditions among laboratories which lead to the selection of different phenotypes.

The homogeneity of clonal cells presents the opportunity to define new cellular determinants, such as cell surface antigens, and then use these markers to identify the phenotype *in vivo*. This approach to elucidating complex developmental sequences was initially used with a great deal of success in the immune system (Cohn, 1967) and has been extended to the nervous system (Stallcup and Cohn, 1976). Finally, mutants can be selected from clonal cell lines in much the same manner as with bacteria. In theory, one can use these mutants to define complex differentiated functions by isolating steps in the sequence of events which leads to phenotypic changes in much the same way as amino acid auxotrophs were used to outline the metabolic pathways. Although this potential has not been realized to the extent one would expect, given the enthusiasm of its early supporters, a number of significant biological insights have been obtained through the use of variant cell populations.

One of the first clonal cell lines that proved useful to the study of cellular differentiation was the clonal rat skeletal muscle myoblast L6 (Yaffe, 1968). This cell line was isolated from rat skeletal muscle, and has been used for numerous studies of the biochemistry and electrophysiology of muscle differentiation (see Chapter 2). The cells divide as mononucleate myoblasts, and once the cells reach confluency they fuse to form multinucleate myotubes (Figure 1.4). The biochemical and electrophysiological changes which accompany the fusion event are quite similar in the clonal L6 line and in primary cultures of skeletal muscle myoblasts.

The PC12 clonal sympathetic nerve cell line is to dissociated sympathetic ganglion cell cultures as Yaffe's L6 clone is to myoblast primary cultures. The PC12 line was isolated from a rat tumor of the adrenal medulla—a tumor type called a pheochromocytoma (Greene and Tischler, 1976). The cells were adapted to culture and cloned, and were shown to have many properties in common with primary cultures of dissociated sympathetic ganglia. Like sympathetic ganglia, the cells respond to exogenous NGF by the formation of long neurites (Figure 1.5). The PC12 line has been used extensively to examine the mechanisms underlying the NGF response and nerve differentiation (see Chapter 3).

There are at least four problems inherent to the use of clonal cell lines; some are more apparent than real.

1. Perhaps the greatest difficulty is the limited number of cell lines available that express well-characterized *in vivo* phenotypes. For example, clonal cells resembling sympathetic ganglion neurons exist, but no cell line of Purkinje cells has been isolated. This problem may diminish with time, but new methods are needed to establish continuous cell lines from selected cell types.

2. Since most continuous cell lines are derived from neoplastic tissue, it is frequently stated that the very origin of the cells makes them abnormal and detracts from their usefulness. It should be remembered, however, that all cells acquire "abnormal" characteristics when placed

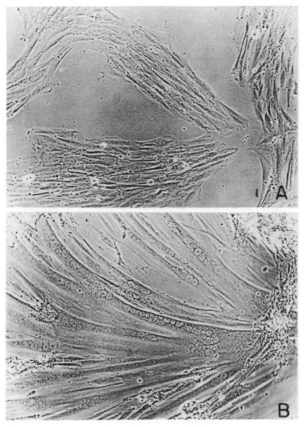

Figure 1.4. L6 skeletal muscle clone. The clonal L6 rat skeletal muscle myoblast cell line of David Yaffe was plated on tissue culture dishes and photographed as exponentially growing myoblasts (A) or multinucleate myotubes (B). From Schubert et al. (1973).

in a tissue culture environment. Although continuous lines are, by definition, lacking in the type of growth regulation observed *in vivo,* the critical point is whether or not they express the phenotype of the normal cell. If the differentiated characteristics of interest in the clonal cells are identical to those *in vivo,* then there is little reason to doubt that the biochemical mechanisms underlying them are the same. If, however, the characteristics of interest are lacking in the cultured cell, then there is little to do except try again, for the wrong cell type may have been selected or the culture environment may not have been sufficient to allow expression of the desired phenotype.

3. Another problem unique to clonal cell cultures is the incompatibil-

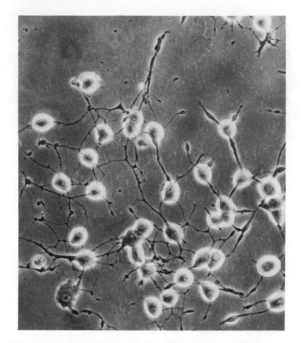

Figure 1.5. PC12 sympathetic nervelike cell line. The PC12 clonal rat cell was photographed 3 days after the addition of NGF. The cell body diameters are approximately 25 μm. Courtesy of Dr. William Stallcup.

ity of cell division and cellular differentiation. *In vivo* the end cell in a differentiation sequence usually does not divide. In a continuous clonal cell line, it frequently is difficult to inhibit cell division experimentally to obtain a stable population of differentiated cells. For example, although NGF inhibits the division of responsive sympathetic-ganglionlike cells and cellular fusion removes clonal myoblasts from the cell cycle, there is no effective way to induce the long-term differentiation of most nerve and glial lines.

4. Finally, there are some problems with the phenotypic stability of clonal cell lines. These problems usually are avoidable and frequently are helpful in understanding the original phenotype. They will be discussed in the next section.

1.3 The Genetic and Phenotypic Stability of Clonal Cell Lines

The exponential increase in the use of permanent cell lines which occurred during the 1960's was accompanied by many questions about the phenotypic stability of cells derived from "differentiated" tissues. For example, clonal liver cell lines frequently lost their ability to synthesize and secrete albumin after a few dozen passages in culture, and skeletal muscle myoblasts ceased differentiating after a few transfers. The reasons for the loss of specific traits in culture can broadly be attributed to two factors: (1) the genetic stability of any given phenotype, and (2) the influence of the tissue culture environment on the expression of that phenotype. The first problem involves the mechanism by which variant phenotypes of mammalian cells are produced, and the second involves our ability to recreate in culture the conditions that allow full expression of the cell's phenotype.

1.3.1. *Genetic Stability.* When a variant phenotype replaces the original clonal cell population it is usually due to a stable change in a single cell of the clone to a phenotype which gives a selective advantage over the wild type. For example, if 20% of a cell's protein synthesis is devoted to the synthesis of a product which is useless in cell in culture, such as the immunoglobulin produced by plasmacytomas, then cells which lose the ability to make immunoglobulin can expend more energy on reproducing; they eventually overgrow the original immunoglobulin-producing cells (Cohn, 1967; Schubert et al., 1968).

What is the nature of these and other variations in cultured cells? There are two alternatives—mutational or epigenetic changes. According to the criteria of Siminovitch (1976), mutation is defined as any heritable change in the primary structure of DNA. This includes nucleotide base changes or deletions, or chromosomal rearrangements or deletions. An epigenetic event is anything else that leads to a relatively stable change in phenotype. When dedifferentiated variants, selected drug-resistant variants, and auxotrophic variants were first examined, it was suggested that their frequency of occurrence was too high for a classical mutational mechanism to be involved. In addition, there appeared to be no correlation between frequency and ploidy. For example, the spontaneous mutation rate to 8-azoguanine resistance is the same in diploid, tetraploid, and octoploid cells (M. Harris, 1971). Since the expression of a recessive mutation would require the mutation of each copy of the gene before the variant phenotype was expressed, the frequency of mutation should be an exponential function of the chromosome number. The failure of this prediction requires that other mechanisms be operative in producing many of the variants. Nonetheless, there are examples in which variation in somatic cells apparently occurs by classical mutational mechanisms. For example, certain plant lectins bind to cell-surface carbohydrates and are lethal to cultured fibroblasts. Many lectin-resistant mutants have been isolated. The frequency of these mutations was increased by mutagens, the mutants were stable in the absence of the selecting agent, and a deficiency in a carbohydrate biosynthetic enzyme was demonstrated (Siminovitch, 1976). A variety of recessive mutations in diploid cells resulting in drug resistance and auxotrophy have also been described that meet the criteria expected from classical genetics.

Despite evidence for normal mutational mechanisms in some diploid cell lines, the frequency of recessive mutations is too high in many other cells, suggesting that they are genetically haploid. One of the best-studied cell lines is the Chinese hamster ovary line (CHO), which was first isolated in 1958. These cells, like the majority of permanent cell lines, are aneuploid, meaning that the chromosome count varies from the normal diploid number. An examination of the extensive data acquired with CHO cells led to the proposal that the apparent reason for the high rate of variant production in this line is that it is functionally hemizygous at many loci (Siminovich, 1976). The cell may have only one copy of some genes instead of the normal two found in diploid cells. This con-clusion was based on the data derived from chromosome banding pat-terns in the CHO cell line and normal hamster cells. Although the total DNA content of both cell types was diploid, there was extensive chromo-somal rearrangement in the CHO cells. These rearrangements took place during the establishment of the cell line in culture and gave no selective disadvantage. Changes in chromosome structure could isolate single gene copies and account for the high rate of variation and reces-sive mutations in permanent cell lines. Practical support for these argu-ments may be found in the considerable difference in frequency of spe-cific mutations among different cell lines (see, for example, Worton et al., 1980).

In addition to the examples of variant selection in clonal cultures that can be explained in terms of point mutations, deletions, or chromosomal rearrangements, studies using cytotoxic reagents and other growth-limit-ing factors have generated insights into other mechanisms of variation in cultured mammalian cells. When cells are exposed to such selective agents as enzyme inhibitors, they frequently respond by overproducing the target enzyme (see, for example, Wahl et al., 1982). Of the four cases in which cloned cDNA probes were available to assay both genomic DNA and mRNA, cells responded in three by making multiple copies of the genes coding for the target protein. The increase in gene number ranged from several- to 100-fold (Schimke, 1982). The fourth case of enzyme overproduction resulted from an increase in the mRNA without a change in gene number. These data strongly suggest that cells are able to respond in a variety of ways to strong selective agents, and that the wide range in the frequency of variant selection in cultured cells may reflect an interplay between two or more independent mechanisms for dealing with the problem.

Although the above arguments may explain the origins of some variants, they are not sufficient to explain all of the phenotypic changes in cultured cells. An example is an analysis of rat hepatoma variants (Deschatrette et al., 1980). Two variants of a cloned rat liver cell line were isolated which lacked up to seven liver-specific functions, including two gluconeogenic enzymes, two inducible aminotransferases, and the ability to synthesize serum albumin. The wild-type hepatoma cell was able to grow in glucose-free medium, while the variants, which lacked the gluconeogenic enzymes, were unable to survive without glucose. This permitted a direct selection for revertant cells making these tissue-specific enzymes. The variant phenotype was not due to the loss of an element necessary to maintain the differentiated state, for if either variant was fused with the wild-type cell line, all liver-specific traits were lost in the fused cells. If the variants arose by classical mutation, then treatment with mutagens should have increased the frequency of reversion from the variant phenotype to the wild-type cell. Since selection was based only on the requirement for the two gluconeogenic enzymes, it was asked if one event (e.g., mutation) could cause the reexpression of all seven liver-specific functions. By selection of cells in glucose-free medium, thirteen revertants of one variant clone were isolated and none of the second variant clone. The reversion frequency was not increased by mutagenesis and all of the revertants expressed all seven liver-specific functions. These data showed that all seven functions were controlled by a single mechanism and that there was only one pathway for reversion. Since a variety of mutagens had no effect on the reversion frequency, this mechanism was not a mutation of the type extensively studied in prokaryotes.

The above data suggest that the origin of variants in clonal cells can be explained in terms of classic mutational mechanisms in some cases, but in others epigenetic phenomena have not been ruled out. The apparent need to invoke epigenetic phenomena may, however, be due to inadequate selection procedures or to deficiencies in other aspects of the analysis. What is clear is that the stability of the differentiated phenotype of cultured cells is dependent on the tissue culture environment in which they are grown.

1.3.2.   *The Influence of Culture Conditions on the Phenotype.*   The growth and differentiation of cells *in vivo* is dependent on a complex set of environmental influences. These include nutritional and hormonal factors and the nature of the substratum to which the cells adhere. Since cells are exquisitely sensitive to their environment, to produce a cell in culture which is identical to its counterpart in the animal all external influences must be duplicated. Although it may never be possible to reconstruct completely the ideal situation in culture, the culture media and substrata have been improving steadily during the last decade, particularly with the introduction of chemically defined, serum-free media. The extensive amount of work with cells exhibiting tissue-specific or differentiated traits in less than perfect tissue culture environments has, however, provided a great deal of information on the effects of environmental changes on cellular metabolism. In the following paragraphs the influence of culture media, substrata, and culture density on cell physiology will be discussed.

What with the wide variety of culture media and media supplements now available, one must be aware of the extreme variations in cellular metabolism and gene expression that result when cells are grown in different conditions. For example, cells grown in a defined serum-free medium metabolized epidermal growth factor (EGF) differently from cells grown with serum (Wolfe et al., 1980). EGF was internalized by cells grown in serum, but remained on the surface when the cells were grown in a chemically defined medium. Another example is the effect of serum type on steroid secretion. Adrenal cortical cells grown in medium supplemented with horse serum secreted large amounts of corticosterone, while cells grown in fetal calf serum had a very different morphology and produced only 1% of the amount of steroid made by the cells in horse serum. This effect was completely reversible, for the phenotype was changed by varying the type of serum in the culture medium (Turley, 1980).

Other problems derived from inconsistent cell culture techniques originate from the routine changing of the culture medium, or "feeding," of the cells. Since the phenotype of some cells is altered by the addition of new culture medium, care must be taken in experimental designs that the desired experimental effects will be distinguishable from those caused by the addition of new medium. An example of a "fresh growth medium" effect is the decrease in endothelial cell plasminogen activator (PA) activity caused by the addition of new growth medium (Levin and Loskutoff, 1980). Endothelial cells line blood vessels and are thought to be involved in the removal of fibrin clots by virtue of their ability to synthesize plasminogen activator, a proteolytic enzyme that cleaves the inactive serum protein plasminogen into its active form, plasmin. Plasmin plays a major role in the degradation of fibrin. When endothelial cells were isolated from bovine aortae and grown in culture, the production of PA increased rapidly with cell density and confluent cultures maintained a high level of activity for at least one week. However, when the culture medium of confluent cells was replaced with fresh medium containing 10% fetal calf serum there was a transient 70% drop in the activity of cellular PA within 2 hr (Figure 1.6). The activity regained its previous level within 96 hr after the medium change. The fresh medium effect was due totally to the serum in the medium, with the maximum decrease being caused by 1% serum. Since the serum did not stimulate secretion of PA, it probably contained molecules that suppressed PA activity but were inactivated or metabolized by the endothelial cells, resulting in the kinetics of recovery shown in Figure 1.6. It follows that culture manipulation and the use of biologically complex serum can introduce uncontrolled variables into experimental protocols. This problem can be circumvented by the use of chemically defined, serum-free culture media.

Most cultured cells are grown in media containing serum plus a basic nutrient mixture of required sugars, salts, amino acids, and vitamins. The only way to determine the factors required for cell growth and differentiation is to grow cells in a completely defined medium. Serum contains a number of growth hormones as well as other growth and attachment factors. Attempts to purify these components have been largely unsuccessful because (1) two or more components may act synergistically in growth stimulation, (2) hormones are present in small amounts, and (3) many compounds are bound to carrier proteins. An alternative to growing the cells in a defined medium is to start with the basic nutrient mixture and add hormones, additional nutrients, and attachment factors

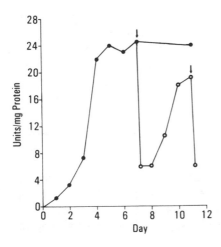

Figure 1.6.   Plasminogen activator activity changes with cell density and the addition of fresh serum. Endothelial cells were plated at a low cell density in media containing 20% fetal calf serum, and PA activity was measured daily. Cultures formed a confluent monolayer on the dish at day 4. On day 7 the old medium was removed from some of the cultures and new medium containing 10% serum was added (arrow). Some of the cultures were fed again on day 11 (second arrow). The enzyme activity per mg cell protein is presented. ●, unfed cultures; ○, cultures given fresh medium. From Levin and Loskutoff (1980).

to it until the cells are able to survive, express their tissue-specific traits, and divide. With this approach it may be possible to define quantitatively the growth and differentiation requirements for each of the numerous phenotypes expressed *in vivo* (see, for example, Barnes and Sato, 1980). Chemically defined serum-free media have been devised for many cell lines, including nerve (Bottenstein and Sato, 1979), glia (Morrison and DeVellis, 1981), and skeletal muscle (Dollenmeier et al., 1981). For example, a medium that supports the growth of CNS nerve cell lines consists of Eagle's nutrient mixture plus insulin (5 μg/ml), transferin (100 μg/ml), progesterone (20 nM), sodium selenite (30 nM), and putrescine (100 μM) (Bottenstein and Sato, 1980). The dishes are coated with fibronectin or poly-L-lysine to promote cellular adhesion. A protein-free medium for mouse neuroblastoma cells has also been developed (Agy et al., 1981). Defined media have been used for primary cultures as well as clonal cell lines, and there are several examples of cell types that maintain their differentiated state in serum-free medium but not in the presence of serum (Sato, 1981). Finally, since different cell types have distinct hormonal and nutritional requirements, it is possible to select for the survival or growth of a desired cell type in a mixture of cells in primary culture. This approach has proved useful in the study of the nervous system, where it is frequently difficult to separate nerve and glial populations (Bottenstein et al., 1980).

Like culture media, the substrata on which cells are grown play a major role in their division and metabolism. Substrata frequently used to culture cells include highly sulfonated tissue culture plastic, collagen, fibronectin, and polylysine. Some cells, such as fibroblasts, require an adhesive surface on which to adhere and flatten in order to proliferate. These cell types are described as being anchorage dependent for growth. Other cell types, such as most nerve cell lines, can proliferate either loosely attached to the substratum or in suspension culture. The affinity of the substratum for the cells can be altered in many ways. For example, certain basement-membrane-associated proteins, such as fibronectin and collagen, increase the cell–substratum adhesion of some cells, whereas the growth of cells over a layer of agar prevents cell–substratum adhesion. If cells are grown under conditions in which their state of adhesion is experimentally varied, many biochemical characteristics of the cells also change. Rat adrenal cortical cells grown on normal tissue culture dishes secrete large amounts of corticosterone, but when the cells are grown on tissue culture dishes exposed to sulfuric acid, which increases the negative charge density from 3 charges per 100 Å^2 to 200 charges per 100 Å^2, the amount of steroid synthesis decreases at least 200-fold (Turley, 1980). Polylysine is a reagent used to increase cell–substratum adhesion. It is frequently employed to assure that cells are not washed from the substratum during various kinds of uptake experiments and in experiments measuring the binding of isotopically labeled ligands to cell-surface receptors. Polylysine causes a number of phenotypic changes in cells, including changes in morphology, ultrastructure, and metabolism. The cell–polylysine interaction alters cAMP synthesis (Wolff and Hope-Cook, 1975), and represses the synthesis of glycosaminoglycans in fibroblasts (Gillard et al., 1979). It is likely that these effects are due to a direct and global action of the polycation on the cell membrane rather than to specific modifications of the particular phenotypic trait being examined.

Cells grown in suspension and cells which divide attached to a surface constitute the extremes of cellular adhesion. Here again very large differences in metabolism have been observed when the same clone is grown in the two conditions. Different amino acid transport systems are used in attached and suspended fibroblasts (Otsuka and Moskowitz, 1975), and fibroblasts grown on a surface contain five times as much tubulin as cells growing in suspension culture (Ostlund and Pastan, 1975). Substrate-attached fibroblasts are also much more sensitive than cells in suspension to cytolysis by antisera against the cell surface, although the number of cell-surface antigens is the same in both cases (Ishii et al., 1977).

The substratum also is able to regulate the division rate and the morphology of cells, two parameters which appear to be tightly coupled (Folkman and Moscona, 1978). If fibroblastlike cells are grown on tissue culture plastic they are very flat and elongated and divide rapidly. The same cells grown on hydrophobic bacteriological plastic do not attach, are round, and divide very slowly if at all. If there is a correlation between cell shape and division rate, then sequentially varying the shape of the cell should generate systematic alterations in the rate of cell division. The adhesiveness of tissue culture plastic can be quantitatively reduced by the application of poly(2-hydroxyethyl methacrylate) (poly HEMA) (Folkman and Moscona, 1978). When poly HEMA was used to decrease the adhesiveness (and make the cells less flat and more round), an inverse relationship between cell height (a measure of "roundness") and DNA synthesis was observed. This observation suggests that fibroblast-like cells can initiate DNA synthesis only when they are sufficiently flattened on the substratum.

Finally, cellular physiology can be altered drastically by the aggregation state of cells. Glutamine synthetase, an enzyme which ligates ammonia to glutamic acid, is normally induced by cortisol in neural retina tissue. This enzyme was not inducible, and the cytoplasmic cortisol receptors were decreased, when neural retina cells were maintained as single cells in suspension. When the cells were allowed to reaggregate, cortisol receptors were again detectable and cortisol induced glutamine synthetase to levels similar to those found in tissue (Saad et al., 1981). These results suggest that cell–cell contact is involved in the regulation of hormone receptor function.

In summary, the interactions between cells and between cells and their substrata can alter a wide variety of phenotypic traits of cultured cells. Subtle changes in the tissue culture media frequently result in dramatic changes in cellular biochemistry. All experimental designs which disturb the cell culture environment must have the proper controls or neutralize the tissue culture effects relative to the experimental variable.

1.4   Changes in Protein Synthesis throughout the Culture Growth Curve

As shown in Figure 1.6, the ability of cells to synthesize plasminogen activator increases throughout the growth of the culture. Low-density, exponentially dividing cells synthesize little material, but the enzyme activity per cell increases with the number of cells in the culture (culture density) until the stationary phase of culture growth is reached, at which point the culture is confluent and the total number of cells remains constant. This variation of protein synthesis with culture density is a general rule (Garrels, 1979a). A systematic examination of the rates of synthesis of approximately 300 proteins in a clonal nerve line and a clonal glial line showed that there were no detectable qualitative differences between dividing and stationary phase cells. However, both increases and decreases in the rates of synthesis of up to 15-fold were detected for about 40% of the 300 most abundant proteins of each cell line (Garrels, 1979a). The amounts of extracellular glycoproteins which are released into the culture medium by exponentially dividing and stationary-phase 3T3 cells are also different (Schubert, 1976). An examination of the release of specific proteins into the culture medium by CHO cells showed that Type I procollagen was secreted by stationary-phase cultures, but not by dividing cells; fibronectin was secreted throughout the growth curve (Aggeler et al., 1982). These data show that the nature of cells in confluent cultures is quite distinct from that in exponentially dividing cultures. It follows that experimental protocols involving reagents which alter the cell cycle (e.g., antimitotic drugs) must be designed to distinguish between the primary experimental effect and the secondary consequences of altering the growth cycles of the cultures.

1.5   Alterations of Cellular Behavior within the Cell Cycle

A great deal of information about animal cell mitosis had been acquired well before the elucidation of the structure of DNA in 1952. This was largely through the work of cytologists, who, working at the level of the light microscope, were able to describe the cytoplasmic and nuclear organization of the cell. In the early 1950's, with the realization of the importance of DNA and the rapidly expanding use of cultured cells, two questions remained—how is cell division controlled and what is the distribution of macromolecular synthesis within the cell cycle? To date these questions have been answered only partially; they are the subject of the remainder of this introductory chapter. No attempt will be made to summarize all of the detailed experimental and theoretical work that has gone into cell cycle analysis over the last decade, but an overview of the information most relevant to cell differentiation will be presented as a background for some of the experimental work on nerve and muscle. Excellent review articles and a monograph on the cell cycle have recently been published (Rozengurt, 1980; Edmunds and Adams, 1981; Gelfant, 1981; John, 1981).

1.5.1.   *The Cell Cycle.* Cytologists have divided the mitotic cell cycle into a number of sequential phases based largely on the staining behavior of chromosomes. After the physical separation of daughter cells, chromosomes are not visible in the nucleus for a period called the interphase. At the end of this phase chromosomes begin to appear, the mitotic spindle forms (prophase), the nuclear envelope disappears, and the duplicated chromosomes line up on the metaphase plate (metaphase). Anaphase begins when the chromosomes start to move towards opposite poles. When chromosomal movement ceases, telophase begins, in which the chromosomes elongate and disappear, the spindles break down, a new nuclear membrane is formed, and the cytoplasm is separated by a constriction of the plasma membrane (cytokinesis), generating interphase cells. It was recognized in the 1950's on the basis of microspectrophotometry and autoradiography data that the duplication of cellular DNA took place during a very limited period of interphase. This period of duplex DNA synthesis was called S, and the gaps preceding and following S were G_1 and G_2, respectively (Howard and Pelc, 1953). M was used to designate the period between G_2 and G_1 during which mitosis occurred, that is, the time needed to go through prophase, metaphase, anaphase, and to complete telophase. The sum of M, G_1, S, and G_2 is the cell generation time (Figure 1.7).

The phases of the cell cycle can be identified by labeling cells with a radioactive pyrimidine (tritium-labeled thymidine, ^{3}H-TdR), which is in-

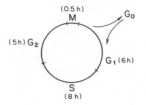

Figure 1.7. Schematic representation of the mammalian cell cycle. The durations of G_1, G_2, S, and M are approximate times for 3T3 fibroblastlike cells.

corporated into DNA and can be detected by autoradiography. Only cells synthesizing DNA incorporate large amounts of isotopically labeled precursors into their nuclei. After pulse-labeling for 10–15 min with ^{3}H-TdR, all cells which were in S for the duration of the isotopic labeling will progressively appear as labeled mitotic cells after going through G_2. This is followed by a decline in the number of labeled mitoses. This bell-shaped curve will then be repeated, with the isotope diluted by 50% as detected by autoradiography (Figure 1.8). The time between the addition of isotope and the lowest point on the curve corresponds to the generation time. The G_2 period is determined by the time interval between the pulse of isotope and the point at which 50% of the metaphase cells are labeled. The S period is calculated from the fraction of interphases which are labeled after a short period of isotopic labeling. For example, if 60% of the interphases are labeled after a 15-min pulse, then the S period comprises 60% of the generation time. Finally, the

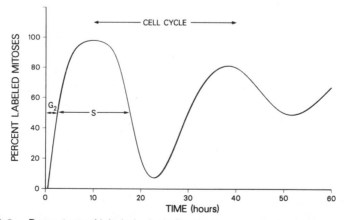

Figure 1.8. Percentage of labeled mitotic figures in the retina of newborn rats at different times after injection of ^{3}H-TdR. On the basis of these data it was calculated that the cell cycle was 28 hr long, and that S = 12.5 hr, M = 1 hr, G_1 = 13 hr, and G_2 = 1.5 hr. Redrawn from Denham (1967).

length of mitosis (M) can easily be determined by time lapse cinematography. The length of G_1 is then obtained by subtracting the other known times from the generation time. Cell cycle analysis of hundreds of cell lines has shown that DNA synthesis is, with one exception, periodic rather than continuous. The exception is a clone of Chinese hamster cells which has no G_2 (Liskay, 1977).

There is a large variation among the generation times of different mammalian cells, and most of the variation is in G_1. Both *in vivo* and in culture most cells which are not dividing have a diploid DNA content, showing that they are in the G_1 phase. *In vivo,* this quiescent state of the cell is usually associated with a differentiated phenotype. In culture, noncycling diploid cells are sometimes associated with cellular differentiation, but more frequently they are the result of inadequate nutritional, hormonal, or physical conditions in the culture. The quiescent state of cells, which usually exceeds the length of G_1, is defined as distinct from G_1. For example, if a population of dividing cells is deprived of an essential amino acid, the individual cells usually proceed through S, G_2, and M and are growth arrested at a point in G_1. The growth inhibitory response is not immediate—the population continues to divide for some time, and the nondividing cells are diploid. When the deleted amino acid is added to the cultures, cell division is not initiated immediately, but rather only after a period of time which exceeds the length of G_1. From these types of experiments it has been argued that the physiological state at which the cells are growth arrested lies outside of G_1. This quiescent state is called G_0 (Figure 1.7) (Lajtha, 1968; Smith and Martin, 1973). Cells are also growth arrested in the G_0 state when they reach a high cell density within the culture, that is, at the stationary phase of the culture growth cycle. This growth inhibition is due to depletion of some component in the serum, for when fresh serum is added to stationary-phase cultures DNA synthesis is reinitiated (but only after a delay longer than the normal length of G_1). Cells can be growth arrested at various points throughout G_1, showing that there are multiple restriction points in the cell cycle. It is thought that some of these restriction points are utilized by cells after they complete their final division cycle prior to terminal differentiation.

The end cell in a differentiation pathway almost always has a diploid DNA content, showing that it is growth arrested in G_1. Except for multinucleate skeletal muscle myotubes all differentiated cells probably have the normal DNA content. Frequently the terminal cell is unable to reinitiate DNA synthesis, but the concept that mammalian cell proliferation and differentiation are always mutually exclusive is incorrect. Clearly, skeletal muscle and most post-mitotic CNS neurons are unable to divide.

However, cells from cardiac muscle, liver, and the pancreas all have a high degree of cytological and biochemical differentiation but are capable of dividing when placed in the proper environment (Cameron and Jeter, 1971). It is likely that the number of differentiated mammalian "end cells" which can be coaxed to divide by the proper experimental manipulation will increase in the future.

1.5.2. *Phenotypic Changes Associated with the Cell Cycle.*   The phenotypes of cells vary not only between the different growth phases of the culture, but also throughout the cell cycle. For example, an analysis of cell-surface proteins on synchronously growing lymphoma cells indicated that at least two proteins were much more abundant on S-phase cells than on cells in the other phases of the cycle (Reyero and Waksman, 1980). Similarly, an examination of the cell–cell adhesive properties of CHO cells showed that G_1 cells adhere to each other twice as rapidly as S cells adhere to each other (Hellerquest, 1979). Finally, protein synthesis and hormone induction can vary within the cell cycle. Although most proteins are synthesized at a similar rate throughout the cycle, some are synthesized primarily during one growth period. The histones, a group of basic nuclear proteins which interact directly with DNA, are synthesized predominantly during the S period; their synthesis is very tightly coupled with that of DNA (Spalding et al., 1966). In contrast, tyrosine aminotransferase is synthesized by hepatoma cells throughout the cell cycle. The inducibility of this enzyme by adrenal steroid hormones is, however, limited to the last part of G_1 and the entire S phase (Martin et al., 1969).

1.6 Summary

Three culture systems have been employed extensively to study the differentiation and interaction of nerve and muscle. The choice of which to use depends on the experimental objective. Organ culture minimizes tissue disruption, while dissociated primary culture can provide relatively homogeneous cultures of some types of tissue. Clonal cell lines offer a tremendous advantage for biochemical studies, since large numbers of homogeneous cells can be obtained and mutant or variant cells can be isolated from the parental cell line. Although there are some problems with the genetic and phenotypic stability of clonal cells, they can be avoided if the proper culture conditions are employed; even minor variations in the tissue culture environment can lead to qualitative changes in the biochemical characteristics of the cells. Finally, the phenotype of the cells varies as a function of the growth curve of the culture and of the stage of the cell cycle. Under normal growth conditions, macromolecular synthesis in exponentially dividing cells is always quantitatively different from that in stationary-phase cells.

−2−

Muscle Differentiation

2.1 Embryonic Development of Skeletal Muscle

Skeletal muscle is derived primarily from masses of mesodermal cells called somites which are arranged radially around the central cavity of the early embryo. The somites divide to generate the cells which constitute the limb buds and other mesodermally derived tissues. The same area of the somites that gives rise to skeletal muscle also generates cells which form cartilage, bone, and skin. It is not clear at which point in development some cells of the limb bud mesenchyme become developmentally committed to the muscle lineage. However, immediately before the appearance of muscle fibers, their dividing precursor cells, called myoblasts, constitute a major fraction of the cells in the outer portion of the limb bud. Myoblasts can be identified as bipolar, spindle-shaped cells which eventually line up end to end and fuse to form multinucleate myotubes. The myotubes then start synthesizing and assembling myofibrillar proteins and acquire the morphology of adult muscle. The number of muscle fibers does not increase once embryonic differentiation is complete. The increase in fiber length of most muscle types during development is due to an increase in the number of sarcomeres. The increase in fiber diameter is due to an increase in the number of myofibrils.

2.2   Formation of the Multinucleate Myotube

The multinucleate nature of myotubes immediately presents two problems to developmental biologists; both were initially solved by the use of muscle cells grown in culture. The first concerns the mechanism by which myotubes become multinucleate; the second is whether nerve is required for the differentiation of myoblasts.

For multinucleation, there are two alternatives. Either there is a sequential replication of the single myoblast nucleus or multinucleate cells are formed by the fusion of many myoblasts. The initial studies addressing this question showed that DNA synthesis was neither required for myotube formation nor detected in cultured skeletal muscle fibers (Konigsberg et al., 1960; Stockdale and Holtzer, 1961). However, the most direct experiments showing that muscle fibers are formed by myoblast fusion were done by Yaffe and Feldman in 1965. They obtained primary cultures of mononucleate cells from rat thigh muscle by trypsinization of the tissue, which dissociated the cells. When plated in tissue culture dishes, the cells divided for several days, and then multinucleate myotubes appeared. To test the hypothesis that myotubes resulted from the fusion of mononucleate myoblasts, they labeled myoblast nuclei with ^{3}H-TdR so that all cells were labeled. Immediately prior to the formation of multinucleate fibers, they again dissociated the cells with trypsin and replated them in the absence of exogenous isotope under two conditions: (1) labeled cells only, or (2) labeled and unlabeled cells mixed in equal numbers. The cells in the cultures were then allowed to become multinucleate, and the cultures were autoradiographed to determine the distribution of the ^{3}H-TdR in the nuclei of the myotubes. If myotubes were formed by the fusion of mononucleate cells, then all of the nuclei in individual fibers of set (1) and 50% of those in set (2) should have been labeled. Alternatively, if multinucleation occurs by endomitosis in mononucleate cells, then all of the nuclei in fibers which contained isotope should have been labeled in both experimental conditions. The results showed that in group (1) 96% of the nuclei within fibers were labeled, and approximately half of the nuclei in fibers of set (2) contained detectable isotope. It was concluded that muscle fibers are formed by the fusion of mononucleate cells. Similar experiments established that cell–cell fusion is specific with respect to cell type, for myoblasts could not fuse with kidney cells. There was, however, little species specificity with respect to myoblast fusion, for skeletal muscle myoblasts from rat fused with those from chicken.

The observation that myotubes arise from the fusion of mononucleate myoblasts was later confirmed *in vivo* by the use of allophenic or tetra-parental mice (Mintz and Baker, 1967). Allophenic mice are formed by assembling blastomeres *in vitro* from mice of two different genotypes, and then reimplanting the hybrid embryo into a pseudopregnant animal to allow normal development. Mice were used which possessed different allelic variants (isozymes) of isocitrate dehydrogenase. Isozymes are enzymes which have the same activities, but different molecular forms. Isocitrate dehydrogenase, for example, has two allelic activities, which can be experimentally defined by electrophoretic separation followed by histochemical staining. Each produces a single band in homozygous animals. However, in cells in which both alleles are expressed, an intermediate band is found in addition to the two original bands. In the case of isocitrate dehydroxygenase twice as much intermediate or "hybrid" enzyme activity is produced as the two original forms, suggesting that the active enzyme is a dimer. Since these dimers do not form spontaneously outside of the cell, it can be concluded that tissues producing hybrid isozyme molecules are expressing both alleles in the same cell. When the skeletal muscle of allophenic mice was tested, it contained both of the electrophoretically distinct parental enzymes and the hybrid enzyme produced by the cytoplasmic mixing of the two parental enzymes. The tissues composed of mononucleate cells contained both of the parental enzymes, but no hybrid. Since no hybrid enzyme would be found if multinucleation in muscle arose by endomitosis, it was concluded that the myotube syncytium originated from the fusion of myoblasts.

A second major question in myogenesis concerned the role of nerve tissue in the multinucleation step of muscle development. Since motor nerves are present in limb buds before the time of myotube formation, it is possible that nerves are needed to initiate myogenesis. That this is not the case was initially shown by the establishment of clonal cell lines from rat skeletal muscle (Yaffe, 1968). These cells can be cloned repeatedly as permanent cell lines and still retain the ability to fuse to form spontaneously contracting, striated myotubes which synthesize muscle-specific proteins. More recent experiments have shown that the removal of a section of the chick spinal cord before the outgrowth of motor neurons does not prevent myotube formation in the areas lacking innervation

(Pockett, 1977). It follows that nerve is not required for at least the initial stages of myogenesis.

2.3 Origin of Cultured Myoblasts

Once multinucleate myotubes are formed in the developing embryo, they become surrounded by a basement membrane consisting of collagen, glycosaminoglycans, and additional glycoproteins. Lying between the plasma membrane of the multinucleate fiber and the basement membrane are mononucleate cells. These are called satellite cells, and their frequency in embryonic animals can be as high as 25% of the number of myotube nuclei. The relative number of satellite cell nuclei declines to about 5% in adult animals (see Reznik, 1976, for a review).

The number of satellite cells present also varies with muscle type. Muscles have been divided according to contraction speed into two groups—fast and slow. There are fewer satellite cells associated with fast muscle than with slow muscle, and this difference is apparent early in myogenesis (Kelly, 1978). Growth of slow muscle is largely due to the division of the satellite cells and their incorporation into existing multinucleate fibers. The growth of fast muscle fibers, which constitute the major fiber type in mammals, is due to an increase in muscle cytoplasm without an increase in the number of nuclei per fiber; there is little division of satellite cells.

The number of satellite cells increases several-fold following denervation; they are also the cells that reconstitute muscle tissue following injury. Adult mammalian skeletal muscles have the ability to regenerate, and following tissue damage satellite cells proliferate to attain numbers similar to those in embryonic tissues. They eventually fuse with the existing muscle fibers to reconstitute the original muscle tissue. Lipton and Schultz (1979) demonstrated this directly by inducing muscle degeneration in adult animals and then explanting the damaged tissue into tissue culture. Some of the mononucleate cells which divide in culture can form myotubes, and these they cloned. The clonal myoblast cells were exposed to ^{3}H-TdR and then reinjected into the muscle of the original animal. The ^{3}H-TdR was incorporated into nuclear DNA, and autoradiographic techniques were used to follow the isotopically labeled nuclei of the clonal myoblasts. It was shown that the cultured myoblasts could be incorporated into existing normal fibers or damaged fibers which were regenerating. Since labeled myoblast cells were found inside the basal lamina of the muscle, either they passed through this structure or the lamina was reconstituted around the myoblasts.

The above evidence indicates that the satellite cells are primarily responsible for the regeneration of skeletal muscle. They are also the source of myoblasts when skeletal muscle tissue is dissociated with trypsin and placed in tissue culture. To obtain primary cultures of myoblasts, pieces of tissue, usually from breasts of fowls or thighs of rodents, are dissociated into single cells with trypsin or a mixture of trypsin and collagenase. This procedure destroys the basement membrane and other materials which hold the tissue together, as well as the majority of the muscle fibers. Myogenic cells must be isolated by enzymes that break down the basement membrane, for proteolytic enzymes that do not destroy the basement membrane liberate only fibroblasts, macrophages, and fat cells (Bischoff, 1979). The dissociated cells are separated from the enzymes by centrifugation, resuspended in culture medium, and placed in culture dishes. Mononucleate cells, a mixture of "fibroblasts" and myoblasts, adhere to the dish along with a few multinucleate myotubes. The initial multinucleate cells usually degenerate within a few days. The mononucleate cells divide rapidly, and at least some of the myoblasts then fuse to form myotubes on top of a monolayer of fibroblastlike cells. Embryonic muscle tissue is routinely used to prepare primary cultures of skeletal muscle because the yield of myoblasts is exceedingly small when adult tissue is dissociated. The yield of viable myoblasts surviving in tissue culture from skeletal muscle seems to be proportional to the number of satellite cells associated with muscle fibers *in vivo*. If muscle fiber degeneration is induced in adult rats with a local anesthetic there is a massive proliferation of satellite cells. When the damaged tissue is placed in culture, the yield of myoblasts is similar to that obtained from embryonic tissue and much greater than that obtained from the normal adult skeletal muscle (Jones, 1977). It follows that the major source of myoblasts in primary cultures is the satellite cell.

2.4   The Morphological Differentiation of Skeletal Muscle

There are at least three well-defined developmental states for cells of the myogenic lineage: the dividing myoblast, the fusing myoblast, and multinucleate myotubes. These will be discussed in sequence, followed by an examination of the proteins which constitute the muscle contractile apparatus.

2.4.1. *Dividing Myoblast.*   Primary cultures established from embryonic skeletal muscle tissue contain at least two cell types. These are the myoblasts, operationally defined as the cells which fuse to form myotubes, and those cells which do not fuse. The latter consist of fibroblasts, various cartilaginous cell types, and some myoblasts that have lost their ability to fuse. Evidence for nonfusing myoblast derivatives has been obtained from two sources. Subcloning of clonal populations of chick myoblasts yielded cells which did not fuse and were morphologically indistinguishable from fibroblasts (Abbott et al., 1974). Similarly, clonal rat skeletal muscle myoblast lines lose their ability to fuse at between 150 and 200 generations in culture (Yaffe, 1968). Subcloning has shown that all of the cells within the original population contemporaneously lose their ability to fuse, ruling out the possibility that nonfusing stem lines were selected from the original population. Thus, the inability to fuse *per se* does not exclude a close developmental relationship between nonfusing mononucleate cells in primary cultures and myoblasts.

Myoblasts in primary cultures of rat and mouse cells are difficult to identify, because the myoblasts share most of their morphological features with the other mesodermally derived cells in the culture. However, in primary cultures of chick and quail muscle, myoblasts possess a characteristic bipolar shape. When cells from primary cultures of chick were cloned, 96% of the bipolar cells yielded cultures which fused, while an equal percentage of the more flattened, pleiomorphic cells found in the cultures did not fuse (Konigsberg, 1963).

At the ultrastructural level, quail myoblasts have a high density of free ribosomes with very little rough endoplasmic reticulum (RER), while fibroblastlike cells have extensive RER and a higher concentration of 10-nm filaments than the myoblasts (Lipton, 1977a). It should be stressed, however, that both the light-microscopic and ultrastructural criteria used to distinguish myoblasts from fibroblastlike cells work only under optimal culture conditions, for these features can be extensively altered by variations of the culture medium and the surface on which the cells are grown.

After an indeterminate number of divisions, embryonic myoblasts begin fusing to form myotubes. In the developing embryo fusion may be initiated by cell density, the local environment, or a genetically programmed timing mechanism. Since the fusion of myoblasts inhibits mitosis, the question has been asked whether division is reversible up to the moment of fusion. Is there a time prior to fusion at which myoblasts cease dividing and become irreversibly committed to fusion? In developing chick wing, pulse-labeling studies showed that 92% of the mononucleate cells incorporated ^{3}H-TdR over a labeling period of 21 hr. Therefore, they must have passed through the DNA synthetic (S) period of their division cycle (Buckley and Konigsberg, 1977). On the basis of calculations from these and similar data it was concluded that only a very small fraction of the fusing myoblast population could have been withdrawn from the division cycle prior to fusion.

Studies with myoblast cultures are essentially in agreement with the *in vivo* work, and extend it with a more detailed analysis of the relationship between the cell cycle and fusion. Cultured myoblasts usually begin fusion after the cells form a confluent monolayer. The following evidence has shown clearly that the fusion of myoblasts occurs when the cells are in G_1. (1) Myoblasts do not fuse in M, for cells mitotically inhibited with colchicine still fuse, and time lapse studies have shown no nuclear division in myotubes. (2) Since there is no nuclear division after fusion, if the cells fused in G_2, each myotube nucleus would have a $4n$ complement of DNA. Direct spectroscopic measurement of nuclear DNA has shown it to be $2n$. (3) The interval between a ^{3}H-TdR pulse and the appearance of a labeled nucleus in myotubes is too long for the cells to have fused in the S phase of their cycle. The remaining part of the cell cycle is G_1, and the remaining question is how long the cells must be in G_1 before they are capable of fusion. Alternatively, are the cells irreversibly removed from G_1 to a post-mitotic prefusion state? Studies with quail primary cultures suggest that myoblasts can spend between 4 and 16 hr in G_1 prior to fusion, and that they can either fuse or reenter S at any time during this period (Konigsberg et al., 1978).

In contrast to avian primary cultures, clones of Yaffe's L6 rat myoblast line may become irreversibly committed to fusion in G_1, with a concomitant loss of proliferative capacity (Nadal-Ginard, 1978). Cells were recloned from growth conditions in which they proliferate (in medium containing fetal calf serum) or differentiate (in medium containing horse serum). If the cells become irreversibly committed to fusion prior to fusion, then the cloning efficiency of the cells in the differentiation-promoting medium should decrease as they approach the differentiated state. The results showed that the cloning efficiency of the myoblasts in the differentiating medium did indeed decline with time as the cells reached the point of fusion. Although these results were suggestive of an irreversible withdrawal from the division cycle prior to fusion, they did not formally show that the cells that did not divide were those committed to fusion. In addition, there was the technical problem of comparing the cloning efficiency of two sets of cells exposed to different growth conditions, one of which could have adversely affected the myoblasts, thus reducing their cloning efficiency. When making pairwise comparisons with cultured cells, it is always preferable to employ identical media.

Another hypothesis relative to the mitotic events prior to fusion was put forth by Holtzer and his colleagues. It is that the transition from an exponentially dividing myoblast to a post-mitotic cell capable of fusing requires a unique mitotic event—a "quantal" mitosis (Bischoff and Holtzer, 1969). This was an appealing idea and led to a great many experiments, most of which ruled it out as a true description of the events prior to fusion. The simplest experiment consisted of plating cells under conditions in which they could not divide and then determining whether fusion took place. Inhibiting DNA synthesis by serine starvation or with β-D-arabinofuranosylcytosine (AraC) did not inhibit fusion (Doering and Fischman, 1974; Nadal-Ginard, 1978). Myotube formation also occurred in the absence of DNA replication following X-irradiation (Friedlander et al., 1978) or when cells were plated at a high density (O'Neill and Stockdale, 1972). Although some aspects of the relationship between the myoblast cell cycle and fusion are still controversial, it can be concluded that myoblasts fuse during the G_1 phase of their growth cycle and that a unique DNA replication prior to fusion is not required.

2.4.2.　*Fusing Myoblast.* Following entrance into G_1, myoblasts which are in close physical contact with each other can initiate fusion. The first ultrastructural change in cultures of embryonic chick muscle is the appearance of 0.1-μm-diameter tubules between juxtaposed cells. These structures occur in regions distinct from those occupied by gap junctions. The tubules produce small regions of cytoplasmic continuity, which eventually open up completely to form the fused cells (Shimada, 1971). The resultant myotubes remain attached to the substratum at their ends by flattened pseudopodia. This initial ultrastructural work was extended using freeze–fracture techniques which can make visible the internal structure of membranes. Although it is difficult to temporally order electron-microscopic observations, a plausible model for the process of myoblast fusion has been constructed on the basis of transmission and freeze–fracture electron microscopy of fusing chick myoblast cultures (Kalderon and Gilula, 1979). The primary observation was that intracellular unilamellar cytoplasmic vesicles, 0.1–0.2 μm in diameter, were associated with cells in the process of fusing. Freeze–fracture images showed that the membranes of these vesicles contained no particles in the fracture planes. Since membrane particles observed with this technique are thought to be proteins, it follows that the particle-free vesicles were lipid enriched. In fusing cells, these vesicles were observed in close association with the myoblast plasma membrane and near areas of membrane between paired cells which consisted of a single shared bilayer. This bilayer was also particle-free, while the majority of the plasma membranes contained particles. Similar observations were made on embryonic muscle *in vivo*. It is therefore likely that the following events occur. (1) The membranes of two myoblasts become closely associated with each other. (2) Intracellular vesicles from each cell fuse with their respective plasma membranes on opposing sides of the fusion area, creating two particle-free bilayers. (3) These then fuse to form a single particle-free bilayer, which dissolves to form the cytoplasmic continuity between the two cells. Gap junctions are assumed not to be directly involved in the fusion process because none were seen in freeze–fracture images of the fusion regions. Finally, a correlation was made between the appearance of the phospholipid vesicles and fusion: when myoblast fusion was inhibited by bromodeoxyuridine (BUdR), cycloheximide, or phospholipase C, no vesicles were observed in confluent myoblast cultures, whereas gap junctions were present.

The network of cytoskeletal filaments which underlies the plasma membrane also undergoes major changes in organization during myogenesis (see Section 3.4 for discussion of the cytoskeleton). The highly interconnected network of filaments in dividing myoblasts is apparently lost just prior to fusion, leaving a cytoplasm devoid of cytoskeletal material except for a few filaments that terminate at the cell periphery (Fulton et al., 1981). After fusion, the cytoskeleton reforms, but within a cytoplasm dominated by the contractile proteins of the myofibril.

The sequence of membrane changes during myoblast fusion outlined above has been supported by studies relating membrane fluidity to myogenesis. Since fluid lipid bilayers fuse more readily than their more rigid counterparts, an increase in myoblast lipid fluidity should favor fusion. By experimentally altering the fatty acid composition of myoblast membranes Horwitz et al. (1979) showed that enrichment by fatty acids, which makes the membrane lipids more fluid, increases the rate of cellular fusion, whereas experimentally decreasing membrane lipid fluidity inhibited fusion. In addition, an increase in membrane fluidity was detected immediately preceding myoblast fusion (Prives and Shinitzky, 1977; Herman and Fernandez, 1978). Both groups of studies suggest that myoblast fusion is analogous to the fusion of lipid vesicles in that lipids participate directly in the fusion process.

Myoblast fusion was more precisely characterized by using an experimental design in which cells fused in suspension (Knudsen and Horwitz, 1978). Chick myoblasts were grown on collagen, detached with a chelating agent, and then placed in suspension culture in a low-calcium medium to prevent aggregation. When calcium was added, the cells aggregated and fused in suspension. The aggregation step was separated from membrane fusion by imposing a variety of conditions. Inhibitors of energy metabolism, protein synthesis, and cholesterol synthesis all prevented calcium-induced aggregation, while alterations in membrane lipid composition effected fusion but not aggregation. In particular, enrichment of cell lipids for the fatty acid elaidate (*trans*-9-octadecenoic acid), which reduces membrane lipid fluidity, inhibited fusion but not aggregation. On the basis of these results it was argued that myoblast fusion is a two-step process. The first is cell–cell recognition and adhesion, and the second is membrane union. Membrane lipid appears to play a direct role in the latter process.

The cellular and/or environmental events which trigger myoblast fusion are unknown. Conditions that restrict myoblast division tend to promote fusion, although only after the cells have reached a fairly high density. There is also a transient 10-fold increase in intracellular cAMP which occurs approximately 5 hr before fusion starts (Zalin and Montague, 1974). When fusion was blocked with BUdR, there was no increase in intracellular cAMP; the converse experiment, in which cAMP was increased in myoblasts with exogenous prostaglandin E produced precocious fusion. Because inhibitors of prostaglandin synthesis also blocked myogenesis, it was suggested that prostaglandins mediate the cAMP rise, which in turn triggers fusion (Zalin, 1979). Since inhibition of division in some cultured cells is correlated with high cAMP levels, the modulation of intracellular cAMP could also account for precocious fusion in culture conditions which inhibit growth.

After myoblasts fuse to form myotubes there is no DNA synthesis as defined by the uptake of ^{3}H-TdR into myotube nuclei. In addition, there is much less DNA polymerase activity in fused cultures than in myoblasts (O'Neill and Strohman, 1969). Since myogenesis is a frequently used example of "terminal" differentiation, it may be asked whether it is possible to reinitiate DNA synthesis in myotube nuclei. The answer is yes; when replicating embryonic fibroblasts were fused with myotubes by Sendai virus, the myotube nuclei were able to incorporate ^{3}H-TdR (Schwab and Luger, 1980). DNA synthesis in muscle nuclei was dependent on the ratio of fibroblast to muscle nuclei—the higher the ratio, the more muscle nuclei incorporated ^{3}H-TdR. It can be concluded that although myotube nuclei normally do not synthesize DNA, they are not irreversibly blocked in this function.

2.4.3. Myotubes.   Under normal growth conditions, the fusion of myoblasts is associated with the synthesis of myofibrillar proteins. Fusion is not, however, a requirement for myofibrillar protein synthesis and assembly. Experiments in which myoblast fusion is uncoupled from muscle-specific protein synthesis will be detailed after the normal myogenic sequence is described. Before that, however, a discussion of muscle proteins and ultrastructure is required.

The major myofibrillar proteins are listed in Table 2.1 and the organization of a typical skeletal muscle fiber is shown in Figure 2.1. The mature fibrils are cylindrical structures composed of repeating units called sarcomeres. The sarcomeres are delineated by darkly staining Z lines which run perpendicular to the fibril; the distance between the Z lines is the sarcomere length. The area immediately surrounding each Z line is the I band and between each pair of I bands is an A band. The terms A and I are derived from the appearance of the sarcomere under the polarization microscope. The A band is anisotropic and the I band is isotropic. In the center of the A band is the H zone, and in the center of

Table 2.1.   Major Myofibrillar Proteins

Protein	Approximate Percentage of Total Myofibrillar Protein	Approximate Molecular Weight (Daltons)	Location in Sarcomere	Reference
Myosin heavy chain	50	220,000	Thick filaments of A band	Szent-Gyorgyi (1948)
Myosin light chain	10	15,000–30,000	Thick filaments of A band	Szent-Gyorgyi (1948)
Actin	20	42,000	Thin filaments of I band	Szent-Gyorgyi (1948)
Tropomyosin	10		Thin filaments of I band	Cummins and Perry (1974)
α Subunit		32,000		
β Subunit		36,000		
Troponins	3	30,000–40,000	Thin filaments of I band	Wilkinson (1978)
α-Actinin	3	100,000	Z line	Ebashi and Kodama (1965)
Desmin	0.1	55,000	Z line	Granger and Lazarides (1978)
Filamin	0.1	250,000	Z line	Gomer and Lazarides (1981)
Vimentin	0.1	55,000	Z line	Granger and Lazarides (1978)
85K Amorphin	—	85,000	Z line	Chowrashi and Pepe (1982)

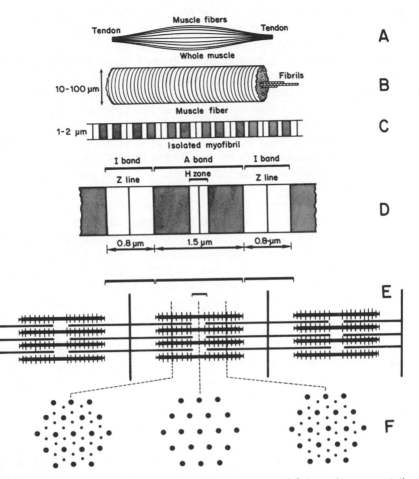

Figure 2.1. Organization of mammalian skeletal muscle. (A) Schematic representation of complete muscle. (B) Muscle fiber. (C) Individual myofibril. (D) Nomenclature of banding pattern. (E) Organization of thick myosin and thin actin filaments. The small lines perpendicular to the thick filaments represent the cross bridges responsible for muscle contraction. (F) The hexagonal array of thin filaments around each thick filament in the overlapping areas of the A band. From Karp (1979).

that the M band. The apparent banding of the sarcomere is due to the partial overlap of two filament types—the thick and thin filaments. The thick filaments contain myosin and the thin filaments actin. In relaxed skeletal muscle the thin filament extends from the Z line toward but not into the H zone. The thick myosin filaments are in the center of each sarcomere and run the length of the A band. Thus the I band contains only thin filaments and the H zone only thick filaments; they overlap in the darker areas on either side of the H zone. Electron micrographs of cross sections through this area show that thin filaments are arranged in a hexagonal array around each thick filament. The contractile force generated by muscle is thought to be created by the sliding of the thick and thin filaments over each other, resulting in a shortening of the sarcomere length (see Huxley, 1982, for a review of the sliding-filament model of muscle contraction). Finally, there are two elaborate membrane systems in skeletal muscle, the transverse tubular system (T system) and the sarcoplasmic reticulum, whose primary functions are, respectively, to transmit the muscle action potential into the central portion of the fibers and to store intracellular calcium. The action potential triggers the release of calcium which, in turn, initiates muscle contraction. The T system consists of folds in the plasma membrane which penetrate deeply into the muscle fibers, transverse to their length. The sarcoplasmic reticulum is derived from the endoplasmic reticulum and consists of a meshwork of flattened tubules which surround the myofibril and extend along its length. The sarcoplasmic reticulum is associated with the tubules of the T system at regular sites along the myofibril called triads.

With this brief description of muscle ultrastructure, it is now possible to outline the changes associated with developing myotubes in culture. The initial evidence of myofibrillar assembly is the aggregation of thick and thin filaments parallel to each other near the plasma membrane of the myotube. The hexagonal arrays of thick and thin filaments then appear, well before the appearance of the M and Z bands. It follows that neither M nor Z bands are required for orientation of the thick and thin filaments in the hexagonal lattice. The myofibrils increase in thickness by the addition of new filaments at the edges nearest to the plasma membrane, and longitudinal growth is accomplished by the addition of new sarcomeric units. Once myotubes begin spontaneously to contract, cross striations and Z lines appear. The Z lines are points of attachment for the thin filaments at the ends of the sarcomeres.

α-Actinin comprises more than half of the protein in the Z line. Immunohistochemical data show that α-actinin is found uniformly distributed in myoblasts; after the cells fuse it is associated with longitudinal filaments in the myotubes (Jockusch and Jockusch, 1980). By the time the muscle has begun to contract most of the α-actinin has localized to the Z band. If spontaneous muscle contraction is inhibited by tetrodotoxin, the Z line pattern does not form and well-defined sarcomere development does not take place, suggesting that contractile activity is required for Z band organization. Ultrastructural maturation of the sarcomere beyond A and I band development with Z lines is infrequent in cultured avian cells, where M bands are rarely observed. M bands were, however, seen in the initial fiber bundles of a clonal rat cell line well before sarcomere development (Klier et al., 1977). The intermediate filament proteins vimentin and desmin are primarily located in and at the periphery of the Z lines, where they are thought to maintain the lateral registry of the myofibrils as well as to stabilize the muscle fibers in the cytoplasm (Granger and Lazarides, 1978). Filamin becomes associated with the Z line slightly before desmin and vimentin do, and may therefore be a prerequisite for the association of the two intermediate filament proteins with actin in the Z line (Gomer and Lazarides, 1981). Finally, the sarcotubular system of cultured myotubes develops in conjunction with the myofibrils. The T tubules develop from invaginations of the plasma membrane and the cisternae of the sarcoplasmic reticulum bud off from the endoplasmic reticulum. The development of the sarcotubule systems is not as complete in cultured cells as in mature muscle, although extensive coupling between the T system and sarcoplasmic reticulum is observed (Klier et al., 1977).

The final morphological aspect of myogenesis to be described is the synthesis of the cell's basal lamina and basement membrane. The plasma membrane of adult muscle fibers is ensheathed by an electron-dense layer, 10–15 nm thick, which is somewhat fuzzy in appearance. Lying immediately outside of this layer is a more darkly staining matrix material containing collagen fibrils. The material lying between the plasma membrane and the collagen matrix is the basal lamina, and the complete extracellular structure (basal lamina plus collagen matrix) is called the basement membrane. The terms basement membrane and basal lamina are, however, frequently used interchangeably to describe the immediate extracellular matrix. This structure has also been called the glycocalyx, the greater membrane, and the cellular exoskeleton.

The basement membrane appears to play a major role in both tissue differentiation and synaptogenesis within the peripheral nervous system. Muscle basement membrane is composed primarily of a group of complex carbohydrates (the glycosaminoglycans), a specific kind of collagen (Type IV), and a basement-membrane-specific glycoprotein (laminin) (Ahrens et al., 1977; Kuhl et al., 1982). Acetylcholine (ACh) esterase is also associated with the matrix (Inestrosa et al., 1982). Basement membrane serves as a substratum for embryonic cell migration and delineates tissue boundaries. In skeletal muscle, a basement membrane surrounds the individual fibers, even interposing between nerve endings and muscle at the synapse. The basal lamina associated with the synapse is thought to be involved in a number of synaptic functions. These include (1) causing nerve–muscle adhesion (Betz and Sakmann,

1973), (2) providing the structural matrix in which ACh esterase is embedded (Hall and Kelly, 1971), (3) maintaining and inducing the clustering of ACh receptors in the synaptic area (Burden et al., 1979), and (4) providing the structural clue to the exact topographical site of the original synapse for muscle reinnervation (Sanes et al., 1978). At least some of these functions require that the basal lamina at the synapse be distinguishable from that of the rest of the cell in some respect; evidence for an immunological difference between synaptic and extrasynaptic basal lamina has been obtained (Sanes and Hall, 1979).

For many years it was not known if the muscle basement membrane was synthesized by the muscle or by one or more of the connective tissue cell types that surround the differentiating fibers. The problem was solved by examination of a clonal cell line. Electron microscopy studies of a rat skeletal muscle line showed that myoblasts had no extracellular matrix material. Immediately after myoblasts fused, a fuzzy material began to appear on the surface of the myotubes. As differentiation proceeded, a basal lamina of electron-dense material approximately 40 nm thick was observed, with clearly defined collagen fibrils lying external to it (Schubert et al., 1973). The basement membrane of the cultured cells was always discontinuous, even when the cells become well striated. A clonal mouse skeletal muscle line also makes basal lamina material (Silberstein et al., 1982). Although muscle cells are capable of synthesizing their own basement membrane, it is likely that other cell types contribute to muscle basal lamina *in vivo*. For example, nerve may deposit a "marker" in the synaptic basal lamina that allows it to recognize the previous site of innervation.

2.5   Acetylcholine Receptors

Once myoblasts have fused to form myotubes, a variety of biochemical and electrophysiological changes take place which accompany the ultrastructural differentiation of the cells. These include the appearance of ACh receptors, ACh esterase production, increased electrical excitability, and numerous changes in total protein synthesis. These dramatic changes, many of which reflect qualitative changes in gene expression, will be dealt with sequentially.

The isolation, characterization, and deployment of two proteins from snake venoms has led to the exponential increase in our knowledge of the ACh receptor over the last 10 years. These proteins are the α-neurotoxin from the Indian cobra, *Naja naja,* and the α-bungarotoxin from *Bungarus multicinctus.* The former binds reversibly to the nicotinic ACh receptor and is used both to identify cell surface ACh receptors and to purify them on neurotoxin affinity columns. α-Bungarotoxin binds irreversibly to nicotinic ACh receptors and is widely used to estimate their number. This is readily accomplished by coupling ^{125}I to the toxin and measuring the binding of toxin to muscle cells with and without a known ACh receptor antagonist such as d-tubocurarine. The difference between the amounts of toxin bound in the two cases represents the specific binding to ACh receptors (Figure 2.2). Two criteria must be met to establish that the toxin binding is specific for the ACh receptor. The first is that the rate of loss of the physiological receptor activity parallels that of the rate of loss of toxin-binding sites. Figure 2.3 shows that the rates were the same when the inactivation of the cellular response was measured as the inhibition of the response to iontophoretically applied ACh. The second criterion for specificity is that the affinities of receptor agonists and antagonists for the toxin-binding sites are similar to the affinities for ACh receptors in other systems. The nicotinic ACh receptors on cultured muscle cells have the same pharmacology as those *in vivo*. Once a binding assay was developed for cultured cells, it was possible to ask when ACh receptors appeared on the surface of cells during myogenesis. Figure 2.4 shows that nicotinic ACh receptors are absent on myoblasts of the clonal L6 rat muscle line and normally appear only after the cells fuse. Similar observations were made with cultured chick embryo skeletal muscle cells (Sytkowski et al., 1973).

The ACh receptor, along with its associated ion channel, is an oligomeric protein made up of four different polypeptide chains of approximately 40,000, 50,000, 60,000 and 65,000 daltons each and of known amino acid sequence (Noda et al., 1983). The biosynthesis, assembly,

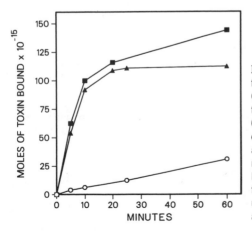

Figure 2.2. Binding of ^{125}I-labeled toxin to fused L6 skeletal muscle cells. Cells were incubated in 15 nM toxin in the presence of (O) or absence of (■) 0.2 mM d-tubocurarine, and the amount of toxin bound per plate was determined at the indicated times. Triangles (▲) represent the specific binding of α-bungarotoxin as determined by the difference between the toxin binding values in the presence of and in the absence of the antagonist.

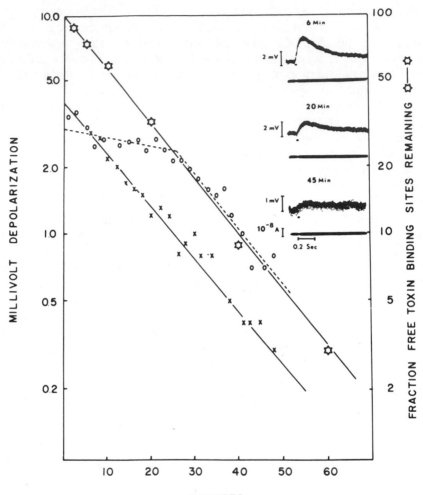

Figure 2.3. A comparison of the rates of α-bungarotoxin (αBT) binding and ACh channel inactivation in response to iontophoretically applied ACh. Toxin was added at 9 nM and the depolarizing response to iontophoretically applied ACh determined as a function of time. Crosses and circles represent responses with two different cells. Stars represent the loss of αBT-binding sites using parallel cultures and the same concentration of toxin. The insert in the upper right-hand corner shows traces of the electrophysiological responses of cells at various times.

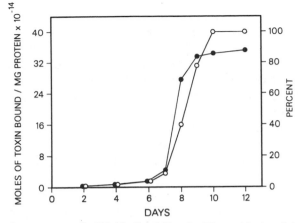

Figure 2.4. Appearance of αBT binding sites in L6 myoblast cultures. Clonal L6 myoblasts were plated at 5×10^4 cells per 60-mm culture dish, and at the indicated days after plating the amount of toxin bound per mg protein was determined along with the percentage of total nuclei residing in cells with more than one nucleus (percent fusion). ○, toxin bound; ●, percent fusion. Redrawn from Schubert et al. (1974a).

and insertion into the plasma membrane of ACh receptors appears to be similar to those of other membrane glycoproteins (see Fambrough, 1979, for a review). Neurotoxins which bind to functional receptor were used to identify structurally complete receptor molecules. These require less than 15 minutes to be synthesized and assembled. The newly synthesized ACh receptors are transported via the rough endoplasmic reticulum and the Golgi apparatus to the cellular surface within about 3 hr, and all of the receptors synthesized make it to the surface. Once at the surface of the myotubes, the ACh receptors are generally uniformly distributed over the plasma membrane. In some cases, however, aggregates or clusters of ACh receptors occur on the surface of cultured muscle cells. The possible role of these ACh receptor clusters in the trophic interactions between cells has been studied extensively, and will be discussed in detail in Chapter 5.

Ultimately, the ACh receptor is degraded or "turned over." In cultured cells the average lifetime for ACh receptors on the cell surface is approximately 20 hr (Fambrough, 1979). This can be determined either by directly labeling ACh receptors with radioactive amino acids or by measuring the degradation rate of ^{125}I-labeled α-bungarotoxin after it has bound irreversibly to cell-surface receptors. Although the results from both methods agree, the latter is more widely used. Since the ^{125}I-labeled toxin and the ACh receptor to which it is bound are degraded at the same rate, one simply has to measure the release of isotope into the culture medium to determine the ACh-receptor turnover rate. The mechanism by which ACh receptors are degraded is similar to that of other membrane and bulk cellular proteins (Libby and Goldberg, 1981). Electron-microscopic autoradiography of cultured myotubes labeled with [^{125}I]α-bungarotoxin indicated that the surface-receptor–toxin complex was internalized and transported to lysosomes, where it was proteolytically degraded. The [^{125}I]tyrosine from the degraded toxin–receptor complex is then released from the cell.

There are two types of electrophysiological responses in myogenic cells to the iontophoretic application of ACh, but only one is correlated with the appearance of nicotinic ACh receptors during myogenesis. In myotubes there is a rapid depolarizing response to ACh similar to that seen in adult skeletal muscle (Figure 2.3). This response is blocked by α-neurotoxin or β-bungarotoxin, and has the pharmacological properties typical of a nicotinic ACh receptor. Rat myoblasts also respond to ACh, but the nature of the response is quite different from that of myotubes. Myoblasts have a slow hyperpolarizing response to ACh which is not blocked by curare, neurotoxin, or α-bungarotoxin (Steinbach, 1975). A similar response has been observed in fibroblastlike cells (Nelson and Peacock, 1972). Although the function of the hyperpolarizing ACh response and the nature of the ACh receptor which mediates it are not known, this response may be a trait of relatively undifferentiated cells of mesodermal origin which is lost as the cells differentiate further.

2.6   Electrophysiological Differentiation

Three aspects of the electrophysiological differentiation of myogenic cells have been studied extensively. The first is the electrical coupling between myoblasts, the second is the "development" of the cells' resting potential, and the third is the differentiation of the action-potential mechanism.

Electrotonic coupling between cells is operationally defined as follows. When a current is passed into a cell by means of a micropipette, and the change in the membrane potential of a neighboring cell is monitored by a recording pipette, the cells are said to be electrically coupled if the applied current passes from the first to the second cell with minimal attenuation. An examination of embryonic tissues has shown that adjacent cells are generally physiologically linked in this manner during early stages of development (Furshpan and Potter, 1968). The ultrastructural basis for the electrotonic junction in most cell types is the gap junction (see Loewenstein, 1981, for a review). In electron micrographs of thin sections, gap junctions appear as closely opposed plasma membranes. At higher magnification the junction shows a pentalamellar structure (three dense and two light bands) and is about 15 nm thick (Figure 2.5). The central dense stratum is about 5 nm wide. Gap junctions appear as small plaques of tightly packed particles in the inner half of the fractured membrane in electron micrographs of freeze–fractured material (Figure 2.5). On the basis of ultrastructural, chemical, and X-ray diffraction a detailed structure for the gap function has been proposed (Unwin and Zampighi, 1980). Both ultrastructural and electrophysiological evidence have demonstrated electrotonic junctions between embryonic myoblasts and between myoblasts grown in culture.

Taking advantage of the fact that myoblasts are the major class of mononucleate cells within the developing basal laminae of embryonic rat muscles, gap junctions were demonstrated between pairs of myogenic cells by freeze–fracture techniques (Rash and Staehelin, 1974). In cell culture, electrophysiological and ultrastructural data demonstrate similar junctions between myoblasts in the clonal rat L6 line (Klier et al., 1977; Schubert et al., 1974a) and in primary cultures (Shimada, 1971; Rash and Fambrough, 1973; Kalderon et al., 1977). When myoblast fusion was inhibited by the removal of calcium from the culture medium or by the addition of 5-bromodeoxyuridine or cycloheximide to the medium, the gap junctions and electrotonic junctions were still present (Kalderon et al., 1977). Therefore, gap junctions may be necessary but are not sufficient to produce fusion.

Gap junctions mediate the flow of current and also the exchange of metabolites such as amino acids and nucleotides between cells. This type of metabolic coupling was demonstrated between myoblasts by isotopic labeling of "donor" cells with [^{3}H]uridine and "recipient" cells with ^{3}H-TdR. The two groups of cells were mixed, allowed to interact, and then autoradiographed to detect the distribution of isotope within the cells. Since [^{3}H]uridine labels RNA and nucleotide pools that are found primarily in the cytoplasm, and ^{3}H-TdR is incorporated into nuclear DNA, cells which are metabolically coupled contain isotope in both their cytoplasms and their nuclei. By this assay, chick myoblasts were shown to be metabolically coupled (Kalderon et al., 1977).

Gap junctions, electrotonic coupling, and metabolic interactions between pairs of cultured cells are not unique to myoblasts. Most types of anchorage-dependent cultured cells interact in this way, including fibroblastlike cells (Gilula et al., 1972). Since gap junctions exist *in vivo* between myogenic cells, the interactions in culture may, however, reflect

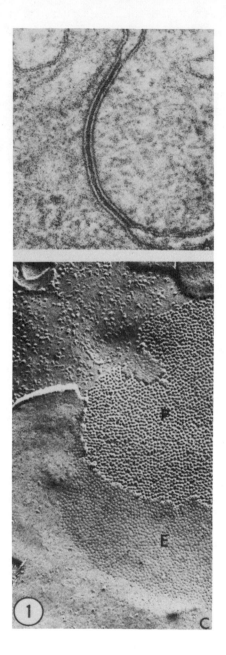

Figure 2.5. Ultrastructure of the mammalian gap junction. The upper panel shows a section through the rat liver, demonstrating both the tightly opposed membranes of two cells characteristic of the gap junction and its pentalamellar structure; 192,500× magnification. The lower panel is a freeze–fracture image of a gap junction in the guinea pig pancreas. There is a polygonal arrangement of 8–9-nm intramembrane particles which are preferentially associated with the inner membrane half (P-fracture face). The complementary arrangement of depressions is located in the outer half of the membrane (E-fracture face); 100,000× magnification. From Epstein and Gilula (1977).

only those occurring during development. There appears to be no electrotonic coupling between mature muscle fibers *in vivo* or in culture.

With respect to resting potential, two apparently contradictory sets of results have been reported. Kidokoro (1975a) showed that in the clonal L6 myoblast cell line there was no difference between the resting membrane potentials in exponentially dividing myoblasts and in myotubes. Both had membrane potentials of approximately −70 mV. In contrast, other investigators have found that in primary cultures of rat and chick skeletal muscle there was an increase in membrane potential from approximately −10 mV in myoblasts to −60 mV in older myotubes (Ritchie and Fambrough, 1975; Spector and Prives, 1977). For the electrophysiologist, the finding of a small (−10 mV) resting potential usually presents a number of alternatives; the least likely is that the measured value is a valid representation of the cell's actual membrane potential. The usual explanation is that some cells, particularly small ones, are easily damaged by the penetrating electrode, resulting in an electrical shunt and the recording of only a fraction of the true membrane potential. As cells become larger (e.g., myotubes vs myoblasts) they are less easily damaged on penetration and the recorded membrane potentials are higher. This problem can usually be overcome by improving one's technique and using better electrodes. However, the discrepancies in the myogenesis data can apparently be explained without invoking differences in technique, for at least some of the experiments were done in different external potassium concentrations (K_o^+). K_o^+ has a dramatic effect on the developing resting potentials of cultured myoblasts (Engelhardt et al., 1980). When chick myotubes were cultured in 2.7 mM K_o^+, their resting potentials increased from −40 mV to −85 mV over a period of 4 days, whereas those cultured in 5.4 mM K_o^+ maintained a constant resting potential at about −70 mV. Some of the experiments which described a change in resting potential development during myogenesis used lower K_o^+ than were employed in Kidokoro's experiments. The reason for the resting potential change in the low K_o^+ medium is not known, but it may be related to the preferential leakage of potassium from young myotubes (Engelhardt et al., 1980). That it is of developmental significance is doubtful.

There are several noteworthy aspects of the differentiation of electrical excitability in cultured muscle cells (for a review see Kidokoro, 1981). In Yaffe's clonal L6 rat myoblast line, the exponentially dividing myoblast is weakly electrically excitable, as a regenerative action potential is found at the end of the current pulse (anode-break stimulation) when the cells are hyperpolarized to −100 mV (Kidokoro, 1975a,b). The myoblast action potential was apparently carried by sodium ions, although a small calcium component would have escaped notice. Once the L6 myoblasts fuse, the magnitude of the action potential evoked by anode-break excitation increases. The action potential of myotubes is carried by both calcium and sodium ions (Kidokoro, 1975a,b). Since adult rat skeletal muscle has a pure sodium action potential, it is likely that the mixed sodium and calcium action potentials in L6 myotubes represent an intermediate stage of muscle development and that the calcium component of the action potential is lost in the adult tissue. Similar observations have been made in tunicate muscle, in which the developing muscle cells have mixed sodium and calcium action potentials. By maturity the sodium component is lost, leaving a pure calcium spike (Takahashi et al., 1971). The results with L6 cells demonstrate, however, that the differentiation in cultured rat muscle cells is not as complete as that of adult skeletal muscle. This is also true of the electro-physiological differentiation of chick muscle primary cultures (Fukuda et al., 1976). There are two alternative explanations for these results. Either the culture medium and substratum limit differentiation or functional in-nervation is required for maximum differentiation. Neither has been ruled out.

Finally, the electrical excitability, defined in terms of rate of rise of the action potential and the ability of the cells to fire repetitively, increases with myotube age in chick skeletal muscle cultures (Spector and Prives, 1977). Figure 2.6 summarizes the relationship between myoblast fusion, the increase in membrane potential, the action potential rise time, and the production of ACh receptor and ACh esterase. In chick cells ACh receptors are first detectable at the onset of fusion, increase in number for several days, and then decrease. This decrease coincides with the onset of spontaneous electrical activity and contraction, and may be related to the ability of electrical activity in muscle to regulate ACh receptor synthesis (see Chapter 5).

2.7 Acetylcholinesterase

As shown in Figure 2.6, production of acetylcholinesterase increases as fusion progresses. There are two classes of this enzyme, each of which exists in multiple molecular forms. They are the so-called "non-specific" cholinesterases (Ch esterase) and the "true" acetylcholinester-ases (ACh esterase). Both classes hydrolyse choline esters; they are differentiated by the more rapid rate of ACh hydrolysis by ACh esterase and by being inhibited by different esterase inhibitors. When applied to a sucrose gradient, these activities separate into components of different

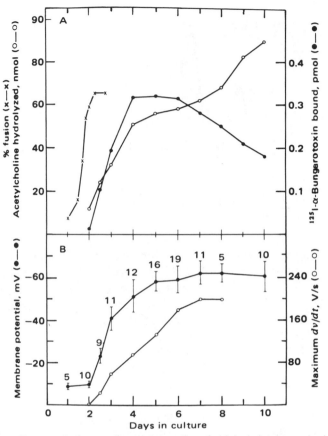

Figure 2.6. Changes in the membrane properties of chick skeletal muscle during their differentiation in culture. The various parameters were assayed in replicate cultures of chick myoblasts plated at an initial density of 2×10^6 cells per 60-mm culture dish. (A) ×-×-×, cell fusion; ●, appearance of ACh receptor assayed by αBT binding; O, ACh esterase activity. (B) ●, resting potential; O-O-O, maximum rate of rise of action potential (maximum dv/dt). The resting potentials as shown as the mean ± standard deviation, with the number of experimental cells shown. From Spector and Prives (1977).

sedimentation values (Fig. 2.7), with major peaks at 16S, 10S, and 6S for the ACh esterase activity; the corresponding Ch esterase activities have slightly greater S values (Vigny et al., 1978). In addition, antibodies prepared against ACh esterase do not react with Ch esterase. The thermal stabilities of the two activities are also quite different. As the data for rat skeletal muscle in Figure 2.7 show, the molecular sizes and, with the exception of the 16S form, relative abundances of the components within

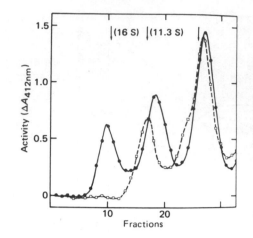

Figure 2.7. Sedimentation profile of two forms of esterase from skeletal muscle. Extracts were prepared from the nerve endplate region of a skeletal muscle and sedimented toward the left in a 5 to 20% sucrose gradient. The activities of ACh esterase and Ch esterase were determined in each fraction and plotted on an arbitrary scale. The arrow on fraction 10 represents the sedimentation velocity of β-galactosidase (16S). ●, ACh esterase; □, Ch esterase. From Vigny, et al. (1978).

the two classes of esterases are very similar; yet they appear to be completely different molecules. Finally, not all of the esterase activity is found with the cells, as a large amount of enzyme is released into the culture medium (Wilson et al., 1973) and some becomes associated with the extracellular matrix (Silberstein et al., 1982).

In skeletal muscle the majority of the esterase activity is of the ACh type. The 16S ACh esterase activity is of particular interest since it is localized to the endplate region of muscle (Hall, 1973). As shown in Figure 2.6, the activity of ACh esterase increases after the cells fuse. Exponentially dividing rat myoblasts do not contain the 16S form of ACh esterase (Rieger et al., 1980). The 16S material appears soon after the first spontaneous contractions are observed, and increases in amount for several days thereafter (Figure 2.8). If, however, the spontaneous contractile activity is inhibited by tetrodotoxin, which blocks sodium channels and therefore the action-potential mechanism, the 16S ACh esterase disappears after one day. This effect is reversible, as removal of the tetrodotoxin led to the reappearance of the 16S form after 2 days (Rieger et al., 1980). These findings indicate that the appearance of the 16S ACh esterase in skeletal muscle is not dependent on the presence of nerve, but is related to muscle electrical activity.

2.8 Changes in Protein Synthesis during Skeletal Muscle Differentiation

As outlined above, there is an increase in the ultrastructural complexity of mesodermal cells during the development of skeletal muscle. The

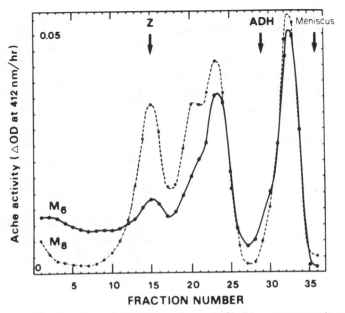

Figure 2.8.   The induction of 16S ACh esterase (AChe) in spontaneously contracting myotubes. Primary cultures of rat skeletal muscle were homogenized on days 6 (M6) and 8 (M8) after plating and the distribution of ACh esterase assayed after sedimentation into a sucrose gradient. Z marks the sedimentation of β-galactosidase (16S) and ADH that of alcohol dehydrogenase (4.8S). The amount of the 16S form of ACh esterase was greatly increased in the older, spontaneously contracting myotube cultures. From Rieger et al. (1980).

most striking change is the appearance of the contractile apparatus soon after fusion. The major structural components of these myofibers are the proteins myosin and actin. Although they were among the first muscle proteins characterized, it was not until recently that the full extent of their diversity was appreciated. In addition, the development of new techniques for separating proteins and quantitating their synthesis has enabled researchers to assay changes in overall protein synthesis during myogenesis. In 1975, O'Farrell described a procedure that exponentially increased the number of proteins that could be resolved on acrylamide gels. This procedure, termed two-dimensional (2-D) gel electrophoresis, separates proteins according to their isoelectric points and molecular weights. Proteins are first electrophoresed in acrylamide tube gels containing ampholytes until they reach positions in the pH gradient corresponding to their isoelectric points. They are then electrophoresed in a second dimension on a rectangular slab gel containing sodium dodecyl sulfate (SDS) to separate the proteins according to their molecular weights. Two-dimensional gel electrophoresis can resolve about 1000 proteins from mammalian cells, and has a combined resolution of 0.1 charge unit and 500 daltons for a 50,000-dalton protein. In addition, this technique offers the advantage that a large number of proteins can be assayed without prior identification of the proteins of potential interest. This is of particular importance when pairwise comparisons are being made between cells at different stages of differentiation or between variants and their parental cell lines. Finally, new computerized techniques have been developed that accurately measure the relative rates of synthesis of proteins resolved on 2-D gels (Garrels, 1979a).

There are two types of changes in protein synthesis during cell differentiation. Qualitative changes result in the synthesis of a protein not detected in the precursor cell, or in the elimination of a protein found in the precursor. Quantitative changes are alterations in the amount of any protein. Both classes of changes are observed during skeletal muscle myogenesis. When the clonal L6 myoblast cell line was labeled with [^{35}S]methionine for 2 hr and the proteins synthesized by myoblasts and purified myotubes quantitated by 2-D gel electrophoresis, the synthesis of approximately 10% of the proteins was qualitatively altered (half of these were induced during myogenesis, half were repressed). Quantitative changes in protein synthesis were more frequent, for the synthetic rates of about 40% of the proteins were altered by more than a factor of two (Garrels, 1979b). Figure 2.9 shows 2-D gel profiles of L6 myoblast and purified myotube proteins which isoelectrically focus between pHs 4.5 and 6. An examination of these gels reveals that numerous proteins are present either exclusively in myotubes or exclusively in myoblasts. When similar experiments were done with primary cultures, which always contain fibroblasts and unfused myoblasts, no proteins were missing in "differentiated" myotube cultures relative to dividing myoblast cultures, although many new proteins did appear after fusion began. The inability to detect protein repression was due to contamination of the fused cultures by myoblasts and fibroblasts. Thus, a major aspect of protein synthetic changes during cell differentiation can be missed if clonal cell lines or homogeneous cultures are not employed. The following sections concentrate on the synthesis of the multiple forms of actin, myosin, and creatine kinase during myogenesis. It will be shown that their synthesis is regulated during development by changes in gene expression.

2.8.1.   Actin.   The actins are major components of the contractile systems of all cell types. Since it was first observed that actin isolated from

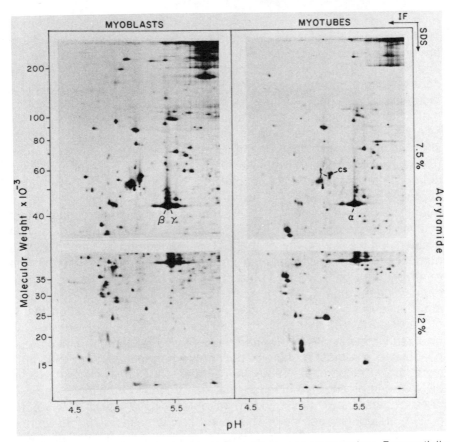

Figure 2.9. Proteins synthesized by L6 myoblasts and L6 myotubes. Exponentially dividing myoblasts and a homogeneous population of L6 myotubes were labeled for 2 hr with [³⁵S]methionine, and the proteins were separated initially in a pH 5 to 7 ampholyte gradient, followed by electrophoresis in sodium dodecyl sulfate (SDS) on 7.5% and 12% acrylamide gels to maximally resolve the high- and low-molecular-weight proteins, respectively. The marked proteins are α-actin (α), β-actin (β), γ-actin (γ), the tubulin (t) subunits, and an intermediate filament protein (CS). From Garrels (1979b).

brain and actin isolated from muscle have different peptide maps (Gruenstein and Rich, 1975), at least five actin proteins have been characterized. Actins from skeletal and cardiac muscle (α-actins) have indistinguishable mobilities on 2-D gels, but they have different amino acid sequences (Vandekerckhove and Weber, 1978). Smooth-muscle actin has a different primary structure from the α-actins, and also a different mobility on 2-D gels. There are two forms of actin (β and γ) in nonmuscle cells; they have different mobilities on 2-D gels and are the products of different genes (Hunter and Garrels 1977; Vandekerckhove and Weber, 1978). Two additional forms of β- and γ-actins have been described that are one charge unit more basic than the normal β and γ forms (Garrels and Hunter, 1979). These two actins, termed delta and epsilon, are the nonacetylated precursors to β- and γ-actins. The post-translational modification of the β- and γ-actin gene products is due to the acetylation of their N-termini. Muscle α-actins also have acetylated N-termini.

Adult skeletal muscle contains only α-actin. Both primary and clonal skeletal muscle myoblasts contain only the β and γ forms of actin (Garrels and Gibson, 1976; Whalen et al., 1976); the α form appears in the cultures after the initiation of cellular fusion. An example of this change in protein synthesis during L6 myogenesis is shown in Figure 2.9. When quantitated, the increase in α-actin synthesis in myotubes relative to that in myoblasts was more than 300-fold (Garrels, 1979b). It may actually be greater, for the sensitivity of the 2-D gel technique is limited by the number of proteins in the gel area surrounding the protein of interest. The relatively high background in the actin area prohibited setting a minimum value for the myoblast α-actin. Further support for the idea that α-actin is found only in differentiated muscle is the observation that a cloned DNA probe that hybridized specifically with α-actin mRNA hybridized well with mRNA prepared from *in vivo* rat skeletal muscle and postfusion myoblast cultures, but only marginally to mRNA from myoblasts (Katcoff et al., 1980). In a similar study employing an α-actin cDNA recombinant DNA clone to analyze mRNA in chick myogenic cultures, it was shown that dividing myoblasts have about 130 α-actin mRNA molecules per cell. This number increases 270-fold during myogenesis, while the amounts of β- and γ-actin mRNA decrease (Schwartz and Rothblum, 1981).

Although the chromosomal arrangement of mammalian actin genes is unknown, the genomic distribution of actin genes in the fruit fly, *Drosophila,* has been described (Fryberg et al., 1980). There are six genes per haploid genome, and they are found at six different sites on polytene chromosomes. Since only three actin mRNA species have been found, some of the actin genes may code for identical mRNAs, or some genes may not be transcribed, or there may be additional actin mRNAs in low abundance.

Finally, the functional significance of the multiple forms of actin and of other polymorphic muscle proteins is not known.

2.8.2. *Myosin.*   Like actin, both myosin heavy and light chains exist in several forms whose expression changes during skeletal muscle development. At least four distinct myosin heavy chains have been identified by peptide mapping and isoelectric focusing procedures, and the existence of an even greater number has been suggested by immunological techniques.

Adult skeletal muscle has been divided into two types, fast and slow, on the basis of contraction speed, contractile protein synthesis, and energy metabolism. Embryonic skeletal muscle is physiologically similar to adult slow muscle. The predominant form of myosin heavy chain in bulk muscle, which is mostly of the fast type, is fast-type heavy chain (MHC_F). Slow muscle and cardiac muscle contain heavy chains which are similar, but not identical, to each other according to peptide mapping procedures. Two other forms of heavy chain, MHC_{emb} and MHC_{neo}, have been found in embryonic and neonatal fast muscle. MHC_{emb} occurs in primary cultures of rat skeletal muscle and in the clonal L6 myoblast cell line (Whalen et al., 1979). In both primary and clonal cell cultures only MHC_{emb} heavy chain has been observed, whereas during the *in vivo* development of fast skeletal muscle there is sequential expression of MHC_{emb}, MHC_{neo}, and MHC_F (Whalen et al., 1981; Bandman et al., 1982). The transition to the adult heavy chain may require extensive electrical activity in the muscle or innervation.

Myosin light chains also exist in multiple molecular forms, and the amounts and types of these isozymes change over development. The 2-D gel pattern of myofibrillar proteins of a mixture of rat fast twitch and slow twitch muscles is shown in Figure 2.10. The light chains, tropomyosins, and actins are marked. *In vivo*, myosin light chains of adult fast muscle (LC_{1f}, LC_{2f}, and LC_{3f}) and adult slow muscle (LC_{1s}, $LC_{1's}$, and LC_{2s}) are all found in embryonic chick fast and slow muscle (Obinata et al., 1980). As development proceeds, there is an increase in the relative amounts of the light chains associated with the adult fiber type. It follows that there is a coordinate regulation of myosin light-chain synthesis as the fiber type becomes defined; this regulation is possibly by factors such as innervation or activity patterns. At the time of innervation either the slow or fast myosins are suppressed (Stockdale et al., 1981). These influences are, however, reversible, for if two fiber types are cross-innervated, or if a new activity pattern is imposed on a muscle by electrical stimulation, the contractile properties and the light-chain synthesis change to reflect the new innervation or electrical activity (Weeds et al., 1974; Salmons and Sreter, 1976).

The regulatory proteins of the myofibril are the troponins and the tropomyosins. They couple the myofibrillar ATPase activity to the rise in intracellular calcium during muscle stimulation, and thus their properties are important in determining the contractile parameters of the muscle. At least two forms of tropomyosin, α and β, are present in skeletal

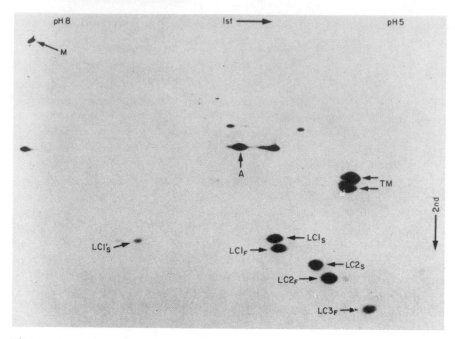

Figure 2.10. Two-dimensional gel electrophoresis of rat myofibrillar proteins. Myofibrils were prepared from both fast and slow twitch muscle. The samples were mixed and analyzed by 2-D gel electrophoresis. The positions of myosin heavy chain (M, 200,000 MW), actin (A, 43,000 MW), two tropomyosins (TM, 34,000 and 36,000 MW), and the six light chains (LC) are indicated. From Whalen et al. (1978).

muscle. They differ slightly in amino acid sequence. Embryonic mammalian fast skeletal muscle synthesizes primarily the β-subunit and adult fast muscle synthesizes primarily α-tropomyosin. Adult cardiac muscle contains exclusively α-tropomyosin (Roy et al., 1979).

The forms of troponin also differ in slow, fast, and heart muscle. The appropriate form of troponin is uniquely associated with fast or slow skeletal muscle in adult mouse tissues, but embryonic muscle fibers contain both slow and fast isozymes (Dhoot and Perry, 1979). Since the *in vivo* changes in myofibrillar protein synthesis described above take place in nondividing myotubes, it follows that they are due to changes in gene expression and not to changes in muscle-fiber populations. These conclusions are supported and extended by studies with cultured cells.

The regulation of tropomyosin and myosin light-chain synthesis in cultured cells has been studied most extensively in birds and rats, and the results are qualitatively different. Avian myogenesis in culture will be discussed first. When cloned primary cultures from quail embryonic slow muscle (anterior latissimus dorsi) and embryonic fast muscle (pectoral) were examined by 2-D gel electrophoresis, no adult light chains were found in dividing myoblast cultures. However, after myoblast fusion both the fast and the slow muscle synthesized all five avian adult myosin light-chain types of fast and slow muscle (Keller and Emerson, 1980). LC_{1s}, LC_{2s}, LC_{1f}, LC_{2f}, and LC_{3f} were all synthesized by myotubes. These findings support the results obtained *in vivo* and suggest that the synthesis of a particular type of light chain is regulated by such extrinsic factors as innervation or muscle activity, which are imposed on the cell after myogenesis. Thus, in avian species the synthesis of the entire set of fast and slow muscle light chains is activated on myoblast fusion, and the synthesis of subsets of the light chains is turned off during the functional activation of the muscle fiber.

When the synthesis of several contractile proteins from embryonic quail breast muscle was assayed by 2-D gel electrophoresis, there was a coordinate appearance of all of them coincident with myoblast fusion (Devlin and Emerson, 1978). These proteins included myosin heavy chain, the large subunit of troponin, α-actin, all myosin light chains except LC_3, and α- and β-tropomyosin. Synchronously fusing cultures were pulse-labeled during myoblast fusion and the appearance of these proteins was quantitated. Their synthesis was initiated nearly simultaneously (Figure 2.11). The rate of incorporation of $[^{35}S]$methionine into most of these proteins increased at least 500-fold. The rate of synthesis of myosin heavy chain increased at least 2000-fold, from fewer than 20 molecules per nucleus per minute in exponentially dividing myoblasts to 40,000 per nucleus per minute in fused fibers (Emerson, 1977). These were again minimum increases, due to the limitations of the 2-D gel technique. Finally, when the stoichiometry of contractile protein synthesis was determined by correcting the incorporation of $[^{35}S]$methionine into the proteins for their methionine content, all had similar rates except actin, which had three times the molar synthetic rate of the other contractile proteins (Devlin and Emerson, 1978).

In a continuation of the above studies on protein synthesis during myogenesis, Devlin and Emerson (1979) asked whether the initiation of contractile protein synthesis was regulated at the level of mRNA synthesis (transcription) or of protein synthesis from preexisting mRNA (translation). To distinguish between these two alternatives total cellular RNAs were isolated from primary cultures of embryonic skeletal muscle throughout the culture cycle, from exponentially dividing myoblasts to fused myotubes. The RNAs were then translated *in vitro* in the presence of $[^{35}S]$methionine, and the contractile protein translation products were identified by 2-D gel analysis and quantitated. The mRNAs for myosin

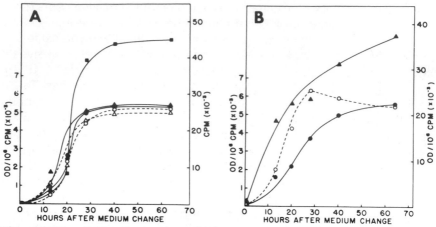

Figure 2.11.   Contractile protein synthesis during myogenesis. Exponentially dividing myoblasts were pulse-labeled for 2 hr at various times after the cells had been shifted to a culture medium that promoted cell fusion. Fusion was initiated 12 hr after the shift and was complete at 39 hr. The labeled proteins were separated by 2-D acrylamide gel electrophoresis and subjected to autoradiography, and the optical density in each spot corresponding to the known myofibrillar proteins was determined. The optical density is, under optimal conditions, proportional to the amount of isotope incorporated into each protein. (A) ●, LC_1; ○, LC_2; △, the large troponin subunit; ▲, the small troponin subunit; ■, myosin heavy chain. The ordinate for myosin heavy chain is on the right and the ordinate for the other proteins on the left. (B) ○, one tropomyosin; ●, second tropomyosin; ▲, α-actin. The actin ordinate is on the right and the tropomyosin ordinate on the left. From Devlin and Emerson (1978).

heavy chain, α-actin, three myosin light chains, α- and β-tropomyosin, and the heavy troponin subunit accumulated with similar kinetics at the onset of fusion, corresponding to the beginning of contractile protein synthesis. The increase in terms of mRNAs per nucleus was at least 200-fold. These results were confirmed through the use of gene-specific hybridization probes for the mRNAs of each muscle protein (Hastings and Emerson, 1982). The data therefore rule out the alternative that contractile proteins are synthesized from mRNAs existing prior to fusion, and show that during avian myogenesis there is a coordinate activation of genes coding for contractile proteins.

In contrast to primary cultures of quail skeletal muscle, neither clonal nor primary cultures of rat muscle synthesize a complete set of adult slow and fast myosin light chains. Myosin from fused rat muscle primary cultures contained LC_{1f}, LC_{2f}, and a small amount of LC_{3f}; no slow light chains are detectable (Yablonka and Yaffe, 1977; Whalen et al., 1978).

An additional light chain has been found on 2-D acrylamide gels that is slightly more acidic than, but has the same molecular weight (26,000) as, LC_{1f} (Whalen et al., 1978). This protein was characterized as an embryonic light chain (LC_{lemb}) by four criteria. (1) It was found in highly purified myosin. (2) It was found *in vivo* in embryonic muscle of all mammals examined, but not in adult tissue. (3) It was localized to the globular head region of the myosin molecule. (4) In the clonal L6 myoblast cell line, LC_{lemb} replaced LC_{1f} in myosin in stoichiometric amounts. LC_{lemb} is not a light chain of nonmuscle myosin, as nonmuscle light chains have molecular weights of 20,000 and 16,000 (Burridge and Bray, 1975). The nonmuscle myosin light chains are, however, found in dividing myoblast cultures before the onset of fusion (Yablonka and Yaffe, 1977). No function has been assigned to the transient appearance of LC_{lemb} during the *in vivo* differentiation of skeletal muscle, but its presence in primary cultures is further evidence that these cultures represent a normal intermediate in muscle development. The clonal L6 line, which synthesizes only LC_{lemb} and LC_{2f}, may be developmentally frozen at a stage prior to the expression of LC_{1f}.

When the contractile proteins in L6 myoblast and purified L6 myotubes were quantitated by the computer analysis of 2-D acrylamide gels, there were increases of greater than 1000-fold for α- and β-tropomyosin and LC_{2f}, greater than 300-fold for actin, and about 600-fold for LC_{lemb} in myotubes relative to myoblasts (Garrels, 1979b). In contrast, there were at least 100-fold decreases in the rate of synthesis of three nonmuscle tropomyosins.

The synthesis of rat muscle proteins is regulated at the transcriptional level in a manner similar to that of avian muscle. When RNA was extracted from rat myoblast cultures and translated *in vitro*, there was a rapid accumulation of mRNA for light chains and α-actin following the onset of myoblast fusion (Hunter and Garrels, 1977; Yablonka and Yaffe, 1977). The synthesis of muscle-specific myosin heavy chain, LC_{2f}, and α-actin RNAs was assayed by using cDNA copies of purified protein mRNAs as probes for protein mRNAs during myogenesis. It was shown that the amounts of the RNAs for all three proteins increased during myogenesis at rates similar to the rates of synthesis of the corresponding proteins (Benoff and Nadal-Ginard, 1979). In a study of tropomyosin synthesis, β-tropomyosin was separated into two components by isoelectric focusing. The more acidic form, β-1, was found in both myoblasts and myotubes, whereas the β-2 component was specific to differentiated cultures (Carmon et al., 1978). Like the other myofibrillar protein mRNAs, translatable mRNA for β-2 tropomyosin greatly increased in amount following the onset of fusion.

2.8.3.   *Creatine Kinase.* Numerous enzymatic changes are associated with the differentiation of skeletal muscle *in vivo*. These include increases in the specific activities (enzyme activities per unit cellular protein) of creatine kinase, fructose diphosphate aldolase, myokinase, pyruvate kinase, and glycogen phosphorylase. There is a decrease in DNA polymerase. Most of these enzymes are involved in the regulation of the elevated energy metabolism required for muscle contraction, and they are frequently used as markers for muscle differentiation. Although these enzymatic activities are found in many tissues, there are specific gene products responsible for their activities which are uniquely associated with adult muscle. For example, there are two types (or isozymes) of creatine kinase (CK), B and M, which are separable by electrophoresis. If both types of subunit are synthesized at the same time and in the same cell, a third band of intermediate electrophoretic mobility, the BM dimer, appears. During myogenesis *in vivo*, there is a gradual transition from the embryonic BB form, through an intermediate BM form, to the MM isozyme of adult muscle (Eppenberger et al., 1964). When the other enzymes listed above are examined, similar qualitative changes in isozyme forms are observed.

Embryonic muscle grown in primary cultures undergoes changes in enzyme activities and isozyme shifts which closely mimic those observed *in vivo*. Although these transitions have been studied for CK, pyruvate kinase, and adolase in cultured cells, we will discuss only observations made with CK isozymes. In all cases the nonmuscle (B) form is predominant in myoblasts and the increase in activity during myogenesis is due to the appearance of the adult muscle isozyme in myotubes.

Figure 2.12 shows the kinetics of appearance of the total CK specific activity and the contribution of each of the three CK dimers to the total enzyme activity during myogenesis in primary cultures of chick breast muscle. Myoblasts from 11-day embryos were plated at a relatively high density, and myotube formation was essentially complete by day 3. There was no MM isozyme detectable at day 1, but 70% of the CK activity was due to the MM dimer after 1 week in culture. In day-1 culture 84% of the activity consisted of BB, with the BM form accounting for the remainder. The contribution of BB dropped to 10% by day 6. The total amount of BB activity was relatively constant throughout the time of culture; the large increase in CK activity during myogenesis was due to the appearance of a different gene product, the M isozyme of CK.

When mRNA was isolated from prefusion chick myoblast cultures, only minute amounts of mRNA for the M isozyme were detected by translation in reticulocyte lysates and analysis of the products by specific

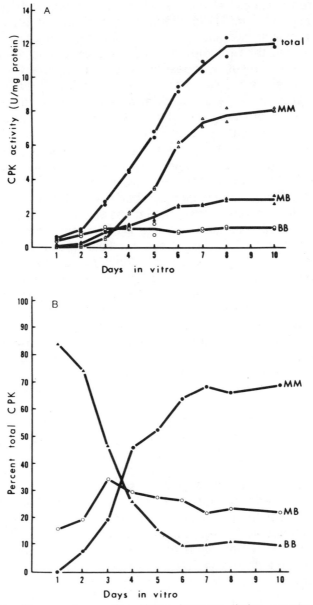

Figure 2.12. Changes in creatine kinase isozymes during myogenesis. Chick myoblasts were plated on day 0 and the activity and isozyme pattern of CK assayed for the following 10 days. (A) Specific activity of CK normalized to soluble protein. ●, total activity; △, MM isozyme; ▲, MB; ○, BB. (B) Activity of each isozyme expressed as a percentage of total activity. ●, MM; ○, MB; ▲, BB. From Lough and Bischoff (1977).

immunoprecipitation (Perriard, 1979). Following fusion, there was a 25-fold increase in the mRNA for the M isozyme. Translatable mRNA for B CK was present in myoblasts, increased fourfold immediately following cell fusion, and then returned to levels approximating those found in myoblasts. These results suggest that the increase in M CK associated with myogenesis is regulated at the level of mRNA production. Since primary cultures of skeletal muscle always contain contaminating fibroblasts, it is possible that some of the CK activities were due to this cell type. This problem was overcome by the use of clonal cell lines.

David Yaffe and his colleagues have isolated a number of clonal myoblast lines from embryonic and newborn rat skeletal muscle (Richler and Yaffe, 1970). These lines have generated our most detailed biochemical knowledge of myogenesis, because fibroblast contamination is not a problem, myoblast-free populations of fused myotubes are easily obtained, and variants can be isolated. When four clonal rat myoblast lines were examined for CK isozymes, two sets of results were obtained. Exponentially dividing myoblasts of clone M41 contained exclusively the BB isozymes, and in clone M58 only BB and trace amounts of MB were found (Figure 2.13). After fusion MM was the predominant dimer in both cell lines, with lower levels of MB and BB; the total CK activity was 20 times higher than in myoblasts. However, when myoblast and myotube cultures from another pair of clones, L8 and L84, were examined, only the MM isozyme was detected in both myoblasts and fused cultures. The data show that isozyme transitions characteristic of *in vivo* skeletal muscles can be found in clonal cell lines, but also demonstrate that aberrant results can be obtained. The inability of lines L8 and L84 to synthesize the BB isozyme probably reflects a mutation in these clones. These results do, however, establish that the B isozyme is not a necessary precursor to the synthesis of the M subunit and that the expression of the B gene is not required for the otherwise normal progression of myotube formation and CK enzyme increase. Since the normal sequence of CK isozyme transitions occurs in primary and clonal skeletal muscle cultures, innervation must not be required for this transition.

Similar studies of the isozyme changes of pyruvate kinase show that tetrodotoxin does not block the transition, suggesting that muscle contraction is not needed for the sequential expression of the different isozymes (Cardenas et al., 1979).

2.9   The Role of Myoblast Fusion in the Synthesis of Muscle-Specific Proteins

As outlined above, numerous changes in myoblasts are temporally associated with their fusion to form myotubes. A major question has been to

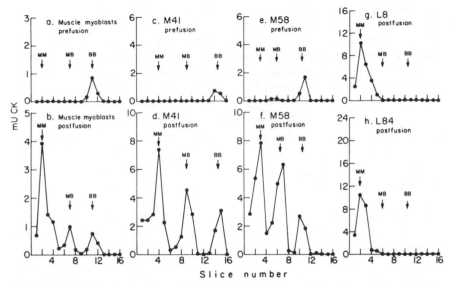

Figure 2.13.   Gel electrophoresis of CK isozymes of myogenic cultures. Extracts of the indicated cells were electrophoresed on polyacrylamide gels and the enzyme was extracted and assayed for activity. (a) Culture of rat primary myoblasts before cell fusion. (b) Primary myoblast culture after fusion. (c) Prefusion M41 clonal cells. (d) Postfusion M41 cells. (e) Prefusion clonal M58 cells. (f) Postfusion M58 cells. (g) Postfusion clonal L8 cells. (h) Postfusion L84 cells. From Dym and Yaffe (1979).

what extent cellular fusion is a prerequisite for the expression of those functions normally associated with differentiated skeletal muscle. There are two basic approaches to this problem. One is to block myoblast fusion by manipulation of the tissue culture environment, and the other is to select variant myoblast cell lines which are unable to fuse. One then sees to what extent the sequence of developmental events in nonfusing myoblasts follows that normally associated with myogenesis. The data show that myoblast fusion is not required for the expression of most muscle-specific activities.

The fusion of cultured myoblasts is dependent on the presence of calcium in the culture medium (Shainberg et al., 1969). If calcium is not present in the culture medium the rate of cell division is normal, but the cells fail to fuse when they reach confluency. The inhibition of fusion depends on the concentration of external calcium ions, and this dependency is apparently specific to the tissue and species from which the myoblasts are derived. Early experiments showed that the inhibition of fusion by calcium deprivation prevented the sequence of enzyme induction normally associated with myogenesis, but more recent experiments indicate that some low-calcium conditions inhibit myoblast fusion without inhibiting muscle-specific protein synthesis. These observations have been made in cultures of chick, quail, calf, and rat myoblasts, and are thus not species specific. These contradictory results were apparently resolved when dose–response curves were constructed which related external calcium concentration to myoblast fusion and CK activity (Morris and Cole, 1979). Using chick primary cultures, it was shown that the absence of calcium in the culture medium prevents cellular fusion but allows the development of the characteristic high CK activities in confluent cultures. However, if low concentrations (30–50 μM) of calcium were added to the medium fusion was still inhibited, while the specific activity of CK was reduced to 75% of its maximum value. A calcium concentration of 200 μM allowed cell fusion but continued to suppress CK activity, whereas calcium concentrations in the range of those found in normal culture medium (2–3 mM) permitted maximum fusion and enzyme activity. These data show that there is a defined range of exogenous calcium ion concentration which is inhibitory to CK activity. This acute calcium dependence of CK activity is not unique to myogenesis, for external calcium concentrations regulate the type of collagen synthesized by cultured chondrocytes (Deshmukh and Sawyer, 1977). Although the mechanism responsible for the inhibition of CK activity in myoblasts by low calcium is unknown, these results emphasize the necessity for careful titration of any variables introduced into a tissue culture regime.

In addition to calcium deprivation, there are numerous other conditions which inhibit skeletal muscle myoblast fusion. These include phospholipase C and 5-bromo-2'-deoxyuridine (BUdR). Phospholipase C is an enzyme which cleaves phospholipids such as phosphatidylcholine at the phosphodiester bond. Exogenous phospholipase C inhibits fusion of chick muscle myoblasts, but does not prevent the increase in CK activity that normally accompanies myotube formation (Keller and Nameroff, 1974). When fusion was inhibited by phospholipase C, the mononucleate cells organized myofibrils and apparently had normal T-tubule and sarcoplasmic reticulum differentiation. Mononucleate cells with striations and ACh receptors were also observed in primary cultures of rat muscle in which fusion had been inhibited by calcium deprivation (Fambrough and Rash, 1971).

Like calcium ion deficiency, 5-bromodeoxyuridine (BUdR) inhibits fusion but not the increase in CK. BUdR is a thymidine analogue which can be phosphorylated and incorporated into DNA. It is thought that BUdR, once it is incorporated into DNA, alters the binding of RNA-synthetic enzymes and regulatory proteins to DNA and therefore interferes with normal gene expression. In 1964 it was shown that the fusion of myoblasts was inhibited by BUdR (Stockdale et al., 1964). Since that time a great deal of data has accumulated which shows that BUdR alters a wide variety of cellular functions. For example, BUdR inhibits induced erythroid differentiation and hemoglobin synthesis in Friend leukemia cells (Preisler et al., 1973), suppresses pigmentation in mouse melanoma cells (Wrathall et al., 1973), causes the morphological differentiation of C1300 mouse neuroblastoma cells (Schubert and Jacob, 1970), and induces prolactin synthesis in a clone of rat pituitary cells (Biswas et al., 1979). BUdR thus tends to alter in fairly specific ways characteristics normally associated with more highly differentiated cells. Although it has repeatedly been assumed that BUdR must be incorporated into DNA to exert its effects, there are to date no conclusive experiments which prove this. There are, however, data which show that the effects of BUdR can be reversed without altering its accumulation in cellular DNA. For example, deoxycytidine reverses suppression of myogenesis and pigmentation by BUdR without changing the amount or distribution of BUdR in the DNA (Reff and Davidson, 1979; Rogers et al., 1975). These results show that the incorporation of BUdR into DNA is not sufficient to inhibit pigment production or myoblast fusion, and that extrachromosomal effects must be involved in at least some of the phenotypic alterations caused by BUdR. One possibility is that BUdR has a direct effect on the synthesis of plasma membrane constituents, perhaps through the inhibition of glycosyltransferases involved in the synthesis of glycoproteins and glycolipids required for myoblast fusion or pigment production (Schubert and Jacob, 1970; Lage-Davila et al., 1979; Kinoshita et al., 1982).

An alternative to inhibiting myoblast fusion by manipulation of culture conditions is selection of nonfusing variants from clonal cell lines. At least three methods have been used to obtain nonfusing cell lines from skeletal muscle myoblasts:

1. Selection by repeated passage of a clonal cell line after fusion (Kaufman and Parks, 1977). The cells which do not fuse were thus selected for after each passage, and after several postfusion transfers nonfusing variants were obtained.
2. Selection for growth over agar (Tarikas and Schubert, 1974). Skeletal muscle myoblast lines are anchorage dependent for growth, and variants were selected for growth in suspension culture by forcing the cells to grow over an agar surface to which the parent could not attach.
3. Selection by viral transformation. When chick myoblasts were trans-

formed with a temperature-sensitive Rouse sarcoma virus (RSV), they did not fuse at the permissive temperature at which the viral transforming gene is expressed, but differentiated normally at the nonpermissive temperature (Fiszman and Fuchs, 1975; Holtzer et al., 1975).

In addition to their inability to fuse, the variant cell lines selected by all three procedures share a number of other characteristics. All are less adhesive to the substratum of the culture dish than their parental lines, and all produce tumors in athymic *nude* mice. Therefore, the variants have a transformed phenotype, whereas the myoblast lines from which they were derived are not tumorigenic.

Rat myoblasts from Yaffe's L8 line were infected with avian sarcoma viruses, and the clones were shown to contain virus by virus rescue into permissive chick cells. None of the clones fused, nor was there the increase in myosin synthesis normally associated with fusion (Hynes et al., 1976). Similar experiments using temperature-sensitive RSV and chick myoblasts showed that at the permissive temperature virus transformation leads to the inability of cells to fuse and to express several biochemical markers for differentiation. The cells did form myotubes at the nonpermissive temperature, at which the viral transformation genes are not expressed (Fiszman and Fuchs, 1975; Moss et al., 1979). It follows that the expression of the transforming viral SRC gene inhibits myogenesis. The fact that temperature-sensitive mutants of the SRC gene reversibly inhibit myogenesis rules out the hypothesis that the integration of the viral genome into the myoblast DNA directly alters the activity of genes involved in normal myogenesis.

A large number of nonfusing variants have been selected from clonal cell lines of skeletal muscle myoblasts. Some lines express some of the biochemical markers associated with myogenesis; many do not. Since essentially all of these variants have altered chromosomes and are tumorigenic, it is impossible to make any conclusions about normal functions that are not expressed and regulated as in the wild type cells. For example, if a nonfusing myoblast variant synthesizes a low level of the muscle CK isozyme and this does not increase when the cells become confluent, it cannot be concluded that fusion is required for the maximum expression of this enzyme. This is because the massive genetic changes in the variant relative to the parent may alter the regulation of CK. An example is the fu-1 variant selected from postfusion cultures (Kaufman and Parks, 1977). The activity of creatine kinase remains constant throughout the growth curve of fu-1, although myokinase (MK) activity increases with the same kinetics and to the same extent as it does in the parental L8 line. If, however, CK and other biochemical markers increase in a nonfusing variant as they do in normal cells, then it is likely that fusion is not required for the maximum enzyme activity.

The M3A variant of L6 does not fuse and its regulation of CK and MK is very similar to that of the parental L6 line (Tarikas and Schubert, 1974). The properties of M3A will be discussed in detail, for it has some characteristics which are quite general to cultured cells. These are the regula-

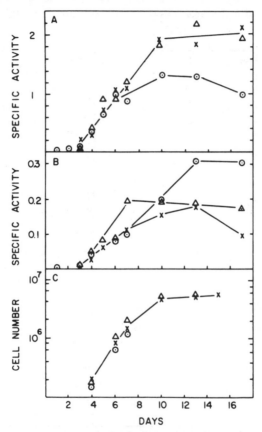

Figure 2.14.   Changes in the specific activities of MK and CK in L6 and M3A cells. Cells were plated at 5 × 10⁴ cells per 60-mm tissue culture or Petri dish, and cell number and the specific activities of MK and CK were followed as a function of the growth cycle. L6 cultures began cell fusion on day 7, making accurate cell counts impossible on the subsequent days. More than 90% of the L6 nuclei were in fused myotubes on day 10. The enzymes' specific activities are expressed as ΔA_{412}/min/mg protein. (A) Specific activity of MK. (B) Specific activity of CK. (C) Viable cell number per 60-mm culture dish. ☉, L6 grown on tissue culture dishes; △, M3A grown on tissue culture dishes; ×, M3A grown in suspension culture on Petri dishes. From Tarikas and Schubert (1974).

tion of enzyme specific activity by cell–cell contact, growth-conditioned medium, and cell volume. When the clonal myogenic line L6 (Yaffe, 1968) and its nonfusing M3A derivative are plated at low cell densities and the specific activities of CK and MK followed as a function of time, both enzymes increase with cell number in both cell lines (Figure 2.14). Thus, a normal increase in CK and MK can occur in the absence of cell fusion. At least two mechanisms could account for the enzymatic increase in the M3A line: (1) Enzyme activity may be a function of the number of cellular divisions in the culture dish after plating. (2) Enzyme

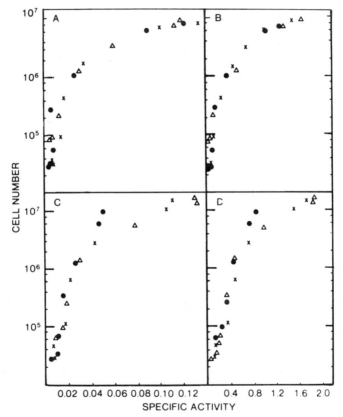

Figure 2.15. Variation of enzyme specific activities as a function of cell density. Cells from exponentially growing M3A and L6 cultures were plated at initial densities between 2×10^4 and 1×10^7 cells per 60-mm tissue culture dish. Enzyme activities were determined on days 1, 2, and 3 after plating and expressed as a function of the cell number per dish at the time of assay. (A) Specific activity of L6 CK. (B) Specific activity of L6 MK. (C) Specific activity of M3A CK. (D) Specific activity of M3A MK. ●, Day 1; ×, day 2; △, day 3. From Tarikas and Schubert (1974).

levels may be controlled by cell density, which in turn alters cell contact and medium conditioning. It was possible to distinguish between these two alternatives by plating L6 and M3A cells from low-density exponentially dividing cultures at densities between 2×10^4 and 1×10^7 cells per 60-mm culture dish. The cells were then assayed for CK and MK on days 1, 2, and 3 after plating, and the specific activity of each enzyme was plotted as a function of the number of cells per dish at the time of assay (Figure 2.15). There was no increase in cell number during the first 24 hr after plating, but there was an approximate doubling of cell number on each subsequent day for all but the two highest plating densities. The data show that the specific activities of the two enzymes changed as a function of cell density and were independent of the number of division cycles, thus ruling out the first possibility.

The specific activities shown in Figure 2.15 were further analyzed in terms of enzyme activity per cell and total protein per cell. Table 2.2 shows that changes in both protein per cell and enzyme activity per cell combined to generate the observed specific activities. The protein per cell decreased 4.5-fold and the enzyme activities per cell increased approximately 1.6-fold, together accounting for the observed sevenfold increase in specific activities. Mammalian fibroblasts also lose cellular protein during the stationary phase of growth, whereas other cell lines, such as the C1300 mouse neuroblastoma, increase in volume and cellular protein as they progress through their growth curve.

Table 2.2. Enzyme Activities as a Function of Cell Density[a]

Cells/Dish $\times 10^4$	mg Protein/ 10^6 Cells	Myokinase		Creatine Kinase	
		Specific Activity	Activity/Cell	Specific Activity	Activity/Cell
3.1	1.3	0.18	0.23	0.011	0.014
4.9	1.1	0.24	0.26	0.009	0.010
9.7	1.2	0.24	0.29	0.010	0.013
22	1.4	0.23	0.32	0.014	0.019
475	0.37	0.92	0.34	0.063	0.023
1100	0.29	1.3	0.36	0.068	0.020

[a] Exponentially growing M3A cells were plated at the densities indicated, and the enzymatic activities, cell number, and protein concentrations were determined 2 days after plating. The cell number doubled after 2 days for all but the highest plating density, which remained constant. The specific activity is defined at ΔA_{412}/min/mg protein. Activity per cell is defined as ΔA_{412}/min/10^6 cells. From Tarikas and Schubert (1974).

Table 2.3. Effect of Conditioned Medium on Enzyme Activities[a]

	Day 2			Day 3		
	Control	Conditioned Medium	High Cell Density	Control	Conditioned Medium	High Cell Density
Relative cell number	2.4	1.15	1.27	7	1.40	1.28
Protein (mg/10^4 cells)	1.24	1.36	0.307	1.19	1.29	0.323
Myokinase (specific activity)	0.203	0.331	1.44	0.248	0.366	1.62
Myokinase (activity/cell)	0.253	0.449	0.439	0.295	0.471	0.523
Creatine kinase (specific activity)	0.009	0.021	0.111	0.014	0.019	0.097
Creatine kinase (activity/cell)	0.011	0.029	0.034	0.015	0.025	0.032

[a] Exponentially growing cells were plated at 2×10^5 cells per 60-mm dish in conditioned medium from early stationary-phase cells or in fresh medium (control). In both cases the medium was changed daily. As positive control, cells were also plated at 1.2×10^7 cells per dish in fresh medium (high cell density). The cell number relative to day 1 after plating, enzyme activities, and protein/10^6 cells were determined. Activity is defined as in Table 2.2. From Tarikas and Schubert (1974).

Both enzyme activity per cell and protein per cell changed during the growth of M3A cells. The changes may have resulted from intercellular contact or from conditioned medium. If cells condition medium at a constant rate this conditioning may be a function of cell density and thus may be confused with other types of density-dependent phenomena, such as cell contact. To distinguish between these alternatives, M3A cells were plated at low density in conditioned medium from stationary-phase cells. Cells were also plated at the same density in fresh medium and at high density in fresh medium. Protein per cell, enzyme per cell, and the specific activities of CK and MK were determined on days 2 and 3 after plating. The data in Table 2.3 show that growth-conditioned medium did not alter the amount of protein per cell, but did affect the specific activities of CK and MK.

The experiments with the M3A variant are in agreement with the majority of studies that used other techniques to inhibit myoblast fusion. Myotube formation is not required for the expression of biochemical and ultrastructural changes associated with myogenesis.

—3—

Nerve Cells

3.1 Introduction

The basic structure of the mammalian nervous system is relatively well defined regarding the organization of neuronal sets and the major pathways of communication between them. The developmental pathways employed to generate this ultimate cellular assemblage can also be described in broad terms. However, essentially nothing is known about the molecular mechanisms that orchestrate the developmental program, which in humans involves over 10^{12} neurons and 10^{13} glial cells. Since this complexity makes the ontogeny of the whole nervous system virtually incomprehensible, it is necessary to study the differentiation and interactions of individual or small numbers of its components. There are many approaches to this problem. They range from describing the genesis of all cells in organisms such as nematodes whose nervous systems contain relatively small numbers of cells to the study of single explanted cells in microcultures. The cell culture technique was, in fact, developed by Ross Harrison in 1907 to study the problem of axon elongation. The bulk of this chapter describes what has been learned about the cell biology of neurons and the characteristics which distinguish them from other types of cells. This work tends to be more descriptive than that presented in the previous chapter on myogenesis, perhaps because the most interesting (and difficult) problems in neurogenesis have not yet been solved. Before discussing the contributions of cultured cells to neurobiology, we will give a very brief overview of the development of the nervous system (see Jacobson, 1978, for a more detailed intro-duction).

3.2 The Development of the Nervous System

The elaboration of the nervous system can be described within the framework of six discernible events (Hamburger and Levi-Montalcini, 1949; Cowan, 1979):

1. The *proliferation* of the precursor cells to nerve and glia. Nerve and glia are derived from the same anatomical locations—the neural tube, the neuroepithelium lining the ventricular system, the neural crest, and thickenings of the epidermis called placodes.
2. The *migration* of the presumptive nerve and glial cells away from the proliferative zones to their final anatomical locations. With the excep-tion of nerve cells derived from the neural crest and from the olfactory placode, the immediate nerve cell precursors, the neuroblasts, do not divide after they initiate migration. In contrast, some glial cells are able to proliferate once they reach their final locations.
3. Once the migrating cells reach their destinations, there is frequently an *aggregation* of like cells to form the discrete nuclei of nerve cells within the brain and the spinal ganglia.
4. *Differentiation* of the nerve cells occurs once the cells reach their final destination. Differentiation includes neurite elongation, increased ul-trastructural complexity, and probably maturation of the cell's ability to generate action potentials, synthesize and release neurotransmit-ters, and express surface receptors for specific transmitters. In most cases it is not yet clear precisely when the expression of the cell's ultimate phenotype begins.
5. During or immediately following cytodifferentiation, *synaptogenesis* occurs where nerve cells make specific contacts with each other or with their appropriate end organs.

6. Finally, there is *selective cell death* of some of the nerve cells that have been produced. It is assumed that cell death eliminates cells that have formed incorrect or superfluous synaptic connections, thus fine-tuning the system for maximum efficiency.

The nervous system is derived from an oval-shaped area on the dorsal side of the embryo called the neural plate. The neural plate is surrounded on its margin by presumptive epidermis and is underlain in part by the endodermal lining of the gut. In its earliest stages the neural plate is morphologically indistinguishable from epidermis, but soon the plate thickens and begins to fold and separate from the overlying epidermis to form the neural tube. In the area where the two sides of the folding neural plate unite, cells immediately migrate away from the closure into the area between the neural tube and the epidermis. This group of cells, called the neural crest, eventually segments on either side of the neural axis, forming the sensory and sympathetic ganglia, as well as a wide variety of other cell types and tissues. As the embryo elongates, part of the neural tube gives rise to the spinal cord nerve and glia, and the anterior portion undergoes more extensive proliferation and regional specialization, giving rise to the various brain regions. The neural tube is a columnar epithelium in which there is a stratification of cells according to their stage in the cell cycle. The nuclei of cells migrate to the inside or ventricular surface before mitosis and then away after mitosis is complete (for an example, see Sidman et al., 1959). Mitosis occurs only at the free surface. The daughter cells then migrate out of the "ventricular zone" along pathways of undefined nature to their final locations in the developing nervous system. Since the cells in the ventricular zone generate all of the neurons and most of the glia within the CNS, proliferation is very rapid, with generation times frequently less than 10 hr. The timetables for cell proliferation and migration within the various brain regions are very precise, and waves of division are associated with the layered cytoarchitecture of the various cortical areas.

3.2.1. *Central Nervous System.* Although glial cells usually are produced later in development than nerve cells, it is not clear whether one stem cell in the ventricular zone can give rise to both nerve and glia, or whether the cells arise from two separate cell lineages. It is known, however, that mitotic presumptive glial cells, defined by an intermediate filament protein [glial fibrillary acidic protein (GFAP), which is thought to be a specific marker for some glial cells], are intermixed with mitotic GFAP-negative cells in the ventricular areas at the same time during development (Levitt et al., 1981). This datum is compatible with the coexistence of distinct nerve and glial precursors in the ventricular zone (B and C, Figure 3.1), but does not solve the problem of an "uncommitted" germinal cell that gives rise to nerve and glial precursor cells (A, Figure 3.1). To address this problem cultured cells have been employed.

A clonal stem-cell line capable of producing both nerve and glial phenotypes has been described (Tomozawa and Sueoka, 1978). The clonal stem-cell line was derived from a chemically induced tumor and synthesizes the glial-specific proteins GFAP and S100; it is also weakly electrically excitable. This line is able to generate spontaneously morphologically distinct clonable cells which are either S100- and GFAP-positive and electrically passive (glialike) or S100-negative and electrically active (nervelike). The rate of cell-type conversion (Figure 3.1, steps 1 and 2) in this system is about one in 10^6 cells (Sueoka et al., 1982). Another clonal precursor-cell line, F7, was derived from the mouse hypothalamus by viral transformation. By the proper manipulation of the tissue culture environment it is also possible to convert F7 to phenotypes with exclusively neuronal or glial markers (DeVitry et al., 1983).

Stable cell lines from the CNS have been isolated which have properties of both nerve and glia, and are therefore potential stem cells. For example, some clonal lines have both Na^+ and K^+ channels characteristic of neurons, but also possess surface antigens normally associated with nerve or glial cells (Arner and Stallcup, 1981; Wilson et al., 1981). These cell lines also synthesize both S100 and the nerve-specific 14-3-2 protein (Schubert et al., 1974b). The clonal cell lines with "intermediate" phenotypes have been used to produce antisera which are specific for

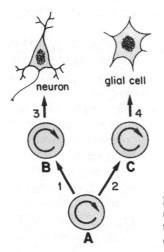

neuron

glial cell

Figure 3.1. The generation of nerve and glial cells. (A) represents a dividing precursor cell to both developmentally committed dividing nerve cell precursors (B) and glial cell precursors (C).

putative nerve and glial precursor populations *in vivo* and for precursor populations in clonal cultures of nerve-producing teratoma cell lines (Stallcup, 1981; Darmon et al., 1982). It is likely that these cell lines represent a phenotype of cells that are arrested in their normal developmental lineage *in vivo*.

Finally, cells containing both a nerve-cell marker protein (14-3-2) and a glial-specific protein (S100) have been identified in the ventricular surfaces of brain. These cells appear to give rise to cells containing either S100 (glial) or 14-3-2 (nerve) (DeVitry et al., 1980). This *in vivo* observation and the unambiguous evidence for cell-type conversion in clonal nerve and glial precursor cell lines suggest that at some point in the development of the nervous system precursor cells exist with properties of both nerve and glia. The characteristics of one phenotype are then greatly reduced during the terminal differentiation of the cell, concomitantly with the amplification of the differentiated phenotype.

3.2.2. *Peripheral Nervous System.*   The line of descent between neuronal and glial precursors and their differentiated derivatives in the CNS presents a very difficult problem to approach *in vivo,* primarily because of the density of the tissue. There is, however, a clearer understanding of the development of cells derived from the neural crest. The crest gives rise to the sympathetic and parasympathetic ganglia, which comprise the major nerve populations of the peripheral nervous system (see Le Douarin et al., 1981, for a review). The neural crest also produces melanocytes, the catecholamine-synthesizing adrenal chromaffin cells, and a variety of other cell types found in the head. Since the derivatives of the neural crest are spread throughout the animal and are readily distinguishable, this system has been a popular research subject for developmental biologists.

The neural crest is divided into several regions along the length of the spinal cord. The upper level gives rise to the enteric ganglia of the gut; other areas give rise to sympathetic ganglia and adrenal glands. By nuclear marking it is possible to follow the fates of chick cells explanted to various regions of quail embryos, and vice versa. Structural differences in the interphase nuclei of these two species allow recognition of cells of chick origin within the developing quail embryo. When cells were taken from a region of the chick neural crest that normally gives rise to catecholamine-synthesizing chromaffin cells and grafted into the region of the neural crest of early quail embryos that normally gives rise to the cholinergic enteric ganglia, they migrated to the gut, where they synthesized ACh characteristic of the enteric neurons (LeDouarin et al., 1981). These data showed that the migration of neural crest cells was determined by the pathways normally available to the cells at any position along the neural axis, rather than by a specificity imposed on the cells by their sites of origin. In addition, they showed that the phenotype of the neural crest cell is determined either by cues received during migration or by environmental factors at its final location. To distinguish between these alternatives, the migratory phase was eliminated by placing neural crest cells directly into the gut. The cells differentiated into Auerbach's and Meissner's plexuses and synthesized ACh. Since catecholamine-containing neurons were never observed, it is likely that cholinergic-nerve differentiation is independent of migration and specified by the terminal environment.

The chick–quail chimeras were also used to investigate the stability of the autonomic-cell phenotype once it was expressed. Cholinergic cells from quail ciliary ganglia were placed in the area of the chick neural crest that normally gives rise to the adrenergic cells of the adrenal medulla and sympathetic ganglia. When the chick embryos were examined 10 days later, the ciliary ganglion cells, which never synthesize catecholamines in their normal environment, had migrated into the host sympathetic ganglia and adrenal glands and were synthesizing the catecholamines characteristic of these tissues (LeDouarin et al., 1981). It follows that phenotype of normal neural crest derivatives is not irreversibly fixed during development, for the cells retain their ability to change their neurotransmitter metabolism and probably other phenotypic markers as well.

The *in vivo* experiments outlined above show that the microenvironment of the neural crest derivative has a major influence on the phenotype. There are two possible explanations for this phenomenon, which are difficult to address *in vivo:* (1) The environment may *instruct* the uncommitted cells to choose a specific neurotransmitter. (2) The local environment may *select* for survival a preexisting subpopulation from the neural crest derivatives that already has acquired the desired phenotype.

Cell culture systems employing both dissociated primary cultures and a clonal cell line derived from the neural crest have been used to show that the cell's environment can directly regulate the cell's choice of neurotransmitter and to rule out a selection process. The major breakthrough in the *in vitro* study of peripheral neurons was the development of a cell culture system in which dissociated neonatal rat sympathetic cells from the superior cervical ganglion (SCG) could be grown for extended periods of time in the presence or absence of "supportive" (fibroblast and glial) cells (Bray, 1970). This culture system was used to demonstrate that the environment has a major influence on whether perinatal sympathetic ganglion cells synthesize predominantly catechol-amines (CA) or ACh. If the SCG neurons are grown in the absence of non-neuronal cells, they synthesize predominantly CA, although some ACh synthesis is detected (see Bunge et al., 1978; Patterson, 1978; and Furshpan et al., 1981, for reviews). However, if SCG neurons are co-cultured with non-neuronal cells or with growth-conditioned medium from non-neuronal cells, the rates of ACh synthesis and choline ace-tyltransferase (CAT) activity per neuron increase approximately 1000-fold over cultures containing only neurons. There is also a reduction, but not an elimination, of the adrenergic properties of the cells. Individual cells with both cholinergic and adrenergic properties are observed in the normal cultures, as well as in cultures containing only one cell. A careful study of the biochemical and ultrastructural properties of these post-mitotic cultures ruled out the possibility that the selection of a sub-population of cells occurred over the relatively long culture periods (Johnson et al., 1980a).

These experiments clearly show that dissociated neonatal SCG neu-rons are developmentally pliable with respect to the predominant type of neurotransmitter that they synthesize. However, the ability of culture con-ditions to cause an increase in ACh synthesis is lost with advancing age of the animal from which the cells are isolated, for cells from animals older than a few weeks cannot be induced to synthesize ACh (Johnson et al., 1980b). In addition, other conditions, such as the composition of the culture medium, the sera, and the substratum, dramatically alter neuro-transmitter metabolism in the SCG cultures. This sensitivity to relatively nonspecific environmental influences is characteristic of most differenti-ated phenotypes expressed in cultured cells (see Chapter 1), and em-phasizes the caution required in extrapolating from this type of cell culture phenomenon to the *in vivo* situation.

The alternative to studying single cells in primary cultures is the use of clonal cell lines. Since a clonal population is derived from a single cell, if the modulation of a trait in the parental clone and all subclones derived from it is the same, then that phenotype is a valid reflection of the properties of a single cell. In addition, developmental changes induced by extrinsic factors frequently occur much more rapidly in clonal cells. This fact rules out the possibility of cell selection.

The clonal PC12 cell line synthesizes both ACh and CA and has essentially all of the phenotypic characteristics of dissociated rat SCG cultures, including the ability to respond to growth-conditioned media with an increase in the specific activity of CAT (Schubert et al., 1977). When the culture medium from low-density, exponentially growing PC12 cells is removed and replaced with growth-conditioned media from a variety of cell types, most conditioned media increase the specific activity of CAT between two- and ninefold; these media do not alter CA synthesis. Since conditioned media from chondrocytes, plasmacytes, and several types of nerve and muscle cells all stimulated CAT activity, there appeared to be little specificity in the stimulation of ACh synthesis. If this phenomenon reflects the *in vivo* situation, then some specificity must be demonstrated. The quantitation of the effects of conditioned media from a large number of different cell types on CAT induction in primary SCG cultures has not, however, been described.

The neural crest also generates the chromaffin cells of the adrenal medulla. These cells synthesize, store, and release CA, and are bathed in high concentrations of glucocorticoids released from the adjacent adrenal cortex. Like sympathetic neurons, neonatal rat chromaffin cells accumulate exogenous [^{125}I]NGF and respond to NGF with the induction of tyrosine hydroxylase and dopamine β-hydroxylase. Unlike sympathetic neurons, chromaffin cells are not killed by anti-NGF sera injected into neonates. In addition, the size of the CA-storage granule is larger (270 nm) in chromaffin cells than in sympathetic neurons (80 nm). A study of explanted adrenal medullary cells showed that glucocorticoids play a role in determining whether cells acquire the phenotype of chromaffin cells or of sympathetic neurons (Unsicker et al., 1978). Cultured neonatal sympathetic nerve and chromaffin cells respond to NGF with neurite outgrowth. However, when chromaffin cells are exposed to the steroids normally supplied by the adrenal cortex, NGF does not elicit neurite outgrowth. When sympathetic ganglion cells are cultured in the presence of glucocorticoids such as dexamethasone, they partially lose their responsiveness to NGF.

To gain more insight into the effect of steroids on sympathetic nerve cells, Schubert et al. (1980) exposed clonal PC12 sympathetic nerve cells to the glucocorticoids, hydrocortisone and dexamethasone (Dex). Although the glucocorticoids did not inhibit NGF-induced neurite outgrowth, they stimulated CA synthesis fivefold and inhibited 90% of the normal ACh synthesis. In addition, the steroids altered the structure of CA-storage granules. The diameters of dense-core vesicles in exponentially dividing cells which had been grown in $1 \times 10^{-6} M$ Dex for 1 month were compared with those in exponentially dividing control cultures. The population of dense-core vesicles in cells treated with Dex had a larger mean diameter [213 $\pm$ 2 nm (mean $\pm$ SEM.), n = 1099] than did the control cultures (158 $\pm$ 1 nm, n = 1248); the entire vesicle-diameter distribution was shifted toward larger diameters (Figure 3.2). This result is more dramatically seen when the dense-core vesicle diameters of PC12 cells grown in Dex are compared with those of cells treated with

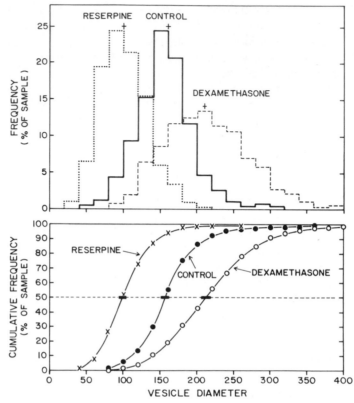

Figure 3.2. The effect of steroids on the diameter (d) of PC12 dense-core vesicles. The top panel shows frequency histograms for vesicle diameters with the mean ±1 SEM indicated (+). The solid line shows data for vesicles in control cells, the dotted line data for cells treated with $2 \times 10^6 M$ reserpine for 18 hr (d = 101 ± 1.0 nm, n = 949), and the dashed line data for cells treated with $1 \times 10^6 M$ dexamethasone for 30 days. At least 50 cells were examined in each example. The mean values are significantly different (p < 0.001). The bottom panel shows cumulative frequency histograms of the same data. In this plot, the ordinate value gives the percentage of the vesicle sample population which has diameters less than or equal to the abscissal value. The intersections of the dashed horizontal line at 50% with the interpolated cumulative curves give the approximate population medians. A 0.95 confidence interval for the median can be estimated. These confidence intervals are indicated in the lower panel as the short horizontal bars around the intersections of the cumulative frequency plots and the horizontal line at 50%. Note that the data medians are separated by several multiples of the estimated confidence intervals. From Schubert et al. (1980).

reserpine, which eliminates the CA-containing vesicles and leaves the smaller (80 nm) ACh-storage vesicles (Figure 3.2). These data show that the steroid hormones which normally bathe chromaffin cells are able to switch the phenotype of clonal PC12 cells from a "dual function" cell synthesizing both CA and ACh to a cell more closely resembling an adrenal chromaffin cell which synthesizes predominantly CA and contains large-diameter dense-core storage vesicles. Since these data were obtained with a clonal cell line over relatively short periods of time, and since the phenotypic response was reversible, the changes must have been due to alterations in the function of the cells caused by environmental "instruction" and not to a selection of subpopulations by the culture system.

The results showing the plasticity of neurotransmitter synthesis also point to a need to reexamine the widely held view that the terminal differentiation of most animal cells is irreversible. The plasticity of some differentiated cells is supported by results obtained in several culture systems in which the phenotype can be altered by extrinsic (environmental) conditions. For example, chick neural retina cells differentiate into lens or pigment cells in monolayer cultures (Okada, 1976), and fibroblasts can be converted to skeletal muscle (Constantinides et al., 1977), myoblasts to chondrocytes, and fibroblasts to bone (Lipton, 1977*b*; Schubert and LaCorbiere, 1977). In addition, there are perhaps hundreds of examples (some of which were discussed in Chapter 1) in which subtle changes in the cell culture environment drastically alter the ability of cells to synthesize specific molecules. The synthesis of CA and ACh by neural crest derivatives is one of these.

3.3 Clonal Nerve Cell Lines

As outlined in the introductory chapter, three types of culture systems have been used to study the biology of nerve cells—organ cultures, dissociated or primary cultures, and clonal cell cultures. Each has played a major role in one or more of the subjects covered in this monograph. As the relative merits of the different culture systems have been discussed (Chapter 1), they will not be reiterated. However, because of their unique properties and increasing usefulness in the neurosciences, a more detailed introduction to clonal nerve cell lines is warranted.

If it is accepted that clonal nerve cell lines are useful in delineating the development and functioning of the nervous system, then the kind and extent of the information which can be obtained is limited by the phenotypes of the available cell lines. At present the number of clonal nerve lines is relatively small, but during the past decade the methodology for making the cell lines has improved, and each year a few more cell lines become available. To establish a cell line in culture, a source of dividing cells is required. Nerve cells *in vivo* normally do not undergo mitosis, but there are three alternatives for obtaining the required cell population. The first is to rely on spontaneous tumors. The second is to chemically transform nerve cell precursors *in vivo* at an early stage of embryogenesis. The final method is to explant embryonic tissue to culture at a stage at which neurogenesis is still taking place and to transform the cells chemically or virally to generate a permanently dividing line. In the last technique strong selection against non-neuronal cells in the population is required since the nerve cells are a minor proportion of the total.

The first clonal cell line of neuronal origin was derived from a spontaneous mouse tumor that arose in 1940. The tumor, designated C1300, was maintained in animals for nearly 20 years, and was adapted to clonal cell culture in 1969 (Augusti-Tocco and Sato, 1969; Schubert et al., 1969). The C1300 mouse neuroblastoma was the only nerve cell line available for several years, and a great deal of information has been accumulated about it. However, perhaps the greatest contributions of the C1300 line were to teach us how to handle continuous nerve cell populations experimentally and to determine the limits of the useful information that can be extracted from this experimental paradigm.

The first group of experimentally induced nerve cell tumors from which clonal nerve cell lines were established were generated by transplacental induction of brain tumors with nitrosoethylurea (Druckrey et al., 1967). In this method, nitrosoethylurea is injected intravenously into rats at 15 days' gestation. If sufficient carcinogen is used to reduce litter size by about 50%, the majority of the offspring develop brain tumors between 150 and 250 days after birth. When permanent cell lines had been established from tumor explants and subpopulations of cells selected on the basis of their nervelike morphology, electrically excitable clonal nerve cells were isolated from about 5% of the induced tumors (Schubert et al., 1974b; West et al., 1977). To date nitrosoethylurea is the only chemical carcinogen that has been shown to transform nerve cells *in vivo*. It must be emphasized, however, that very few investigators have systematically examined brain tumors resulting from chemical carcinogenesis programs by methods other than histology. It is difficult to distinguish neuronal from glial tumors on the basis of histology alone.

Another source of cell lines which has not been exploited to the extent warranted is human neoplasms. Approximately a dozen human neuroblastomas have been adapted to culture (see Bottenstein, 1981, for a review), but there has been no systematic attempt to establish in culture and identify clones from large numbers of human brain tumors. Of the human neuroblastomas which do exist, many have remained uncloned and are therefore of minimal experimental use. Of the cloned neuroblastomas, some respond to NGF and cAMP and should be of use in defining the mode of action of NGF.

There are several severe limitations to depending on spontaneous and chemically induced neoplasms as sources of dividing nerve cells. The first problem is frequency; spontaneous rodent nerve cell tumors are extremely rare, and only a small percentage of chemically induced tumors are neuronal. Therefore, the absolute number of potential nerve cell lines is small. The second is that chemical carcinogenesis may select for only one cell type, thus limiting the diversity of cells obtained. The third problem is that we are not being able to transform specifically cells from selected brain regions. For example, transformed motor neurons or Purkinje cells would be of tremendous use in neurobiology. Several attempts have been made to circumvent these limitations, and they have met with varying success.

Since nerve cell precursors do divide during the earlier stages of embryogenesis, it should in theory be possible to adapt the precursors directly to cell culture. The problem is selecting the precursor cells out of the huge excess of rapidly dividing glial and fibroblastlike cells. This approach has been employed successfully in one laboratory. Bulloch et al. (1977) dissociated specific areas of embryonic brain and exposed the resultant cell population to an antiserum prepared against a clonal nitrosoethylurea-induced glial cell line. This antiserum, in the presence of complement, killed the glial cells in the population. By subculturing only cells which had nervelike morphologies, they isolated approximately a dozen cell lines which were electrically excitable and which shared nerve-specific antigens with other cultured nerve cells. Similar nerve cell lines were also obtained using thimerosal, an ethyl mercurial compound, which by an unknown mechanism selects against fibroblastlike cells and glial cells in the cultures (Bulloch et al., 1978). Although many of these cell lines were derived from the cerebellum, it was not possible to identify the exact cell type (i.e., Purkinje cell, granule cell, etc.) from which the cells originated. These results do, however, establish that it is possible to select nervelike cells from normal embryonic tissue, but they point to the difficulty of both selecting and identifying particular nerve cell types. They also raise the question of what cell types would be selected from any procedure dependent on extracting dividing cells from nerve precursor populations. If the interaction of the CNS nerve cell with its (post-migratory) environment is as critical as it is for the neural crest derivatives, then it is possible that only a limited number of mature nervelike phenotypes could be recovered as dividing cells. Precursor cells could predominate.

The final alternative for procuring clonal cell lines relies on the use of transforming viruses. This technique offers great promise, but has been used to obtain nerve cell lines in only two laboratories. The first virus-transformed nerve cell line was isolated by DeVitry et al. (1974). They dissociated embryonic mouse hypothalamic cells with trypsin, grew them in culture for 6 days, and infected them with SV40. Transformed foci of cells were selected on the basis of nervelike morphology, and eventually a clone of neurosecretory cells was isolated which made both neurophysin and vasopressin.

The other virus that has been used in transforming nerve cells in primary culture is a temperature-sensitive derivative of Rous sarcoma virus (ts-RSV). Cells infected with ts-RSV usually exhibit a phenotype which is under the control of the *ts* transforming gene. At permissive temperatures the viral transforming gene is expressed and the cells exhibit a transformed phenotype. At the higher nonpermissive temperatures the *ts* mutation is not expressed and the cells assume a normal, more differentiated phenotype. Transformation of cultures of rat cerebellum with ts-RSV has made possible the isolation of many cell lines, and at least one clonal cell line was identified as nerve on the basis of its electrical excitability and cell-surface antigens (Giotta et al., 1980*b*). Since the transformed nerve cells appear relatively normal at the nonpermissive temperature, this technique shows some promise of generating permanently dividing cells from defined brain areas.

Most of the known clonal nerve cell lines from rodents are listed in Table 3.1 along with some of their properties. Not included in this list is a large number of cell lines with nervelike properties which were generated by the fusion of fibroblasts or glial cells with nerve cells (Minna et al., 1972; Heumann et al., 1979). These cells have proven useful for studying nerve differentiation and are still of great value in examining interactions of some pharmacological reagents (e.g., opiates) with their membrane receptors (Collier, 1980). They will not, however, be discussed further in this monograph.

3.4   The Cytoskeleton and Motility of Fibroblastlike Cells

A major aspect of nerve differentiation is the elaboration of long processes, or neurites, from the cell body. Human axons can grow to several feet in length, and the dendrites of most nerve cells are extensively

Table 3.1.   Rodent Nerve Cell Lines

Cell Lines	Animal	Origin	Neurotransmitter	Reference
C1300	Mouse	PNS	Many	Augusti-Tocco and Sato, 1969; Schubert et al., 1969
B35, B50, B65, B103, B104	Rat	CNS	Primarily GABA	Schubert, et al., 1974*b*
A1, A9, C6, C9	Rat	CNS	Unreported	West, et al., 1977
RT4	Rat	PNS	Unreported	Tomozawa and Sueoka, 1978
F7	Mouse	Hypothalamus	Neurophysin, vasopressin	DeVitry et al., 1974
SC9	Rat	Cerebellum	Unreported	Giotta et al., 1980*b*
Many	Rat and mouse	Cerebellum	Unreported	Bulloch et al., 1977, 1978
PC12	Rat	PNS	ACh and dopamine	Greene and Tischler, 1976

Abbreviations: GABA, γ-aminobutyric acid; PNS, peripheral nervous system.

arborized. To describe the biochemical and biomechanical mechanisms that generate and maintain this most morphologically complex of all cell types, it is first necessary to understand the basic components of the cell that are involved in the maintenance of shape and the generation of motion. Due to their extensive use in the well-funded area of cancer research, most of the work on cell structure and movement has been carried out with fibroblastlike cells. This work will be outlined before discussing the morphological aspects of nerve differentiation.

The elegant but highly variable morphology of cultured eukaryotic cells is generated by a cytoarchitecture of great complexity. If cells are treated with nonionic detergents such as Triton X-100 the membrane lipid and the majority of the soluble cytoplasmic proteins are extracted. The intracellular material left behind is the framework of interconnected filaments called the cytoskeleton which was first described by the electron microscopists. On the outer edge of the cytoskeleton of Triton X-100-extracted cells is a lamina of proteins derived from the plasma membrane, which appear to be associated intimately with the cytoskeleton (Ben Ze'ev et al., 1979). The cytoskeletal filaments are grouped into three classes by electron microscopy—microtubules (25 nm in diameter), microfilaments (6 nm), and intermediate filaments (10 nm). The more recent application of biochemical techniques and high-voltage electron microscopy has permitted not only the detailed description of these three filament types, but also the definition of a new group of cytoplasmic filaments, the microtrabeculae (2–3 nm in diameter). The four cytoskeletal filament systems are probably present in all higher eukaryotic cells, and they have been implicated in numerous cellular activities, including movement, intracellular transport, modulation of shape, exocytosis and endocytosis, and regulation of molecular movement on the cell surface. We will describe each filament class and outline its role in cellular morphology and movement.

3.4.1. Microtubules. Microtubules are 25-nm-diameter hollow cylinders of variable length, with a wall thickness of 5 nm (see Weatherbee, 1981, for a review). The filaments are composed of the globular protein tubulin, a dimer of 110,000 MW. The subunits of tubulin, termed α and β, are similar but nonidentical proteins of 55,000 MW. As in the case of actin, there are several "isoforms" of α- and β-tubulin which can be separated from each other by isoelectric focusing (Edde et al., 1981). The polymerization of tubulin requires the hydrolysis of GTP, and is rigorously regulated by calcium ions and cAMP via a group of proteins which copurify with tubulin—the microtubule-associated proteins (MAPs). Two groups of MAPs have been purified, those of high (greater than 250,000) and those of low (55,000–65,000) molecular weight. MAPs increase microtubule nucleation and increase the net rate of tubulin polymerization through stabilization of the filament. In vitro studies of the assembly of microtubules from tubulin and MAPs showed that the phosphorylation of one high-molecular-weight MAP inhibits the rate of microtubule assembly but causes the formation of severalfold-longer microtubule filaments (Jameson and Caplow, 1981).

Microtubules are depolymerized by the alkaloid colchicine, and also by exposure to cold. Although it is frequently argued that microtubules play a role in a given cellular function if that function is inhibited by exposure to exogenous colchicine, it must be stressed that colchicine has a variety of other effects on cells. These include alteration of cyclic nucleotide levels and inhibition of protein synthesis. It follows that colchicine inhibition of a cellular function may be necessary, but it certainly is not sufficient, to establish a role for microtubules in the process.

When cytoplasmic microtubules of fibroblastlike cells are revealed by immunofluorescence staining with antibodies against tubulin, a net of fibers is seen, frequently running parallel to the long axis of the cell and extending to the cell's edges (Figure 3.3A). Similarly to centrioles in the formation of mitotic spindles, certain cytoplasmic structures nucleate the assembly of microtubules in the cytoplasm. Experiments designed to determine the number of microtubule-organizing centers in cells have suggested that the primary organizing region is again the centriole (for a review, see Solomon, 1980). The question of whether or not microtubules can *only* be assembled at centrioles is unresolved, for some cells appear to have up to 10 microtubule-organizing centers. It is generally believed that cells do not have that many centrioles.

Microtubules are frequently associated with membranes or membrane-bound vesicles, such as synaptic vesicles. Both immunofluorescence and electron microscopy show that they terminate on or in the plasma membrane, and some rather weak evidence suggests that microtubules bind specifically to purified plasma membranes and various types of vesicles *in vitro* (Weatherbee, 1981). Finally, tubulin may be an integral component of some membranes. Tubulin is found in purified membranes from many sources, including brain (Bhattacharyya and Wolff, 1975). It is therefore possible that both cytoplasmic organizing centers and membrane tubulin nucleate the assembly of microtubules.

3.4.2. *Intermediate Filaments.* Intermediate filaments (IFs), so named because their diameter of approximately 10 nm lies between that of microfilaments (4–7 nm) and microtubules (25 nm), are the most heterogeneous of the cell's cytoskeletal components. There are at least five classes of IF proteins: (1) Tonofilaments are made up of a group of proteins between 40,000 and 65,000 MW which share antigenic determi-

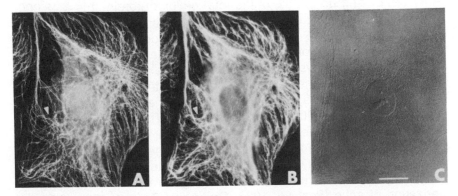

Figure 3.3. Fibroblast microtubules and intermediate filaments. Double immunofluorescent labeling of a fibroblast cell with (A) anti-tubulin and (B) anti-vimentin in the same focal plane. (C) is the same cell viewed with Nomarski optics. Bar in C represents 20 μm. From Ball and Singer (1981).

nants with prekeratin; they are found in most epithelial cells. (2) Desmin from chick cells is a 50,000-MW protein found in all three muscle types—skeletal, smooth, and heart. Desmin is a main component of Z lines in skeletal muscle, along with α-actinin (see Chapter 2). (3) Vimentin is a 57,000-MW protein and is found in many cell types. It is, however, the only IF in fibroblasts (Figure 3.3B). Essentially all dividing cultured cells contain vimentin in addition to the cell-type-specific IF, but nondividing differentiated cells in culture usually contain only the cell-type-specific intermediate filament. It is therefore possible that vimentin is a protein associated with dividing cells (Virtanen et al., 1981). (4) Mammalian neurofilament proteins are of 68,000, 145,000, and 200,000 MW and are thought to be localized exclusively to nerve cells. They are major structural elements of axons and dendrites, and, like all intermediate filament proteins, are phosphorylated. (5) The glial IF protein is 51,000 daltons in MW and is known as glial fibrillary acidic protein (GFAP). It is thought to be localized exclusively to astrocytes. The five groups of filament proteins are not mutually exclusive within the same cell, as two will frequently be detected within a cell. For example, desmin and vimentin are found together in some cell types, and they can be subunits of the same IF (Lazarides, 1981).

Intermediate filaments are thought to be responsible for the maintenance of cell shape, but there is no evidence that they are involved in cell movement. That role has been bestowed upon the actin-containing microfilaments.

3.4.3. Microfilaments.   Actin is one of the most abundant cellular proteins, comprising up to 20% of the total cellular protein. It is a 42,000-MW molecule which can exist as a free monomer (G-actin) or as a subunit of microfilaments (F-actin). The F-actin can occur as stress fibers or as a gelatinous meshwork. These double-stranded filaments are ~6 nm in diameter, and they usually terminate near the membrane or at adhesion plaques where the cell is in contact with the substratum (see Weatherbee, 1981, for a review). The actin filaments are therefore most frequently observed on the undersides of cells. They tend to run parallel to the long axis of the cell (Figure 3.4). A number of proteins are associated with the actin filaments in more or less defined patterns. These include myosin, tropomyosin, α-actin, calmodulin, filamin, vinculin, and a number of proteins which modulate sol–gel transitions and the polymerization of actin (Schliwa, 1981; Craig and Pollard, 1982). Myosin, tropomyosin, and α-actinin apparently serve the same function in fibroblastlike cells as they do in muscle (see Chapter 2). Calmodulin is a 16,500-MW calcium-binding protein which activates a variety of enzyme

Figure 3.4. Fibroblast Microfilaments. Staining was as in Figure 3.3, except with anti-actin. Courtesy of Dr. Ilan Spector.

systems (see Stocklet, 1981, for a review), filamin is a 250,000-MW protein which crosslinks F-actin to form a gel (Schloss and Goldman, 1979), and vinculin is a 130,000-MW protein which appears to be localized to sites of contact between microfilament bundles and the cytoplasmic membrane at areas of focal contact with the substratum (Geiger et al., 1980).

The organization of microfilament networks within the cell can change within a second in response to subtle physiological stimuli. It follows that the assembly and disassembly of this filament system must be exquisitely regulated. At least 15 proteins have been implicated in this process and purified to homogeneity (Schliwa, 1981). Some mediate the lateral association of actin filaments to form gels, some influence the rigidity of the F-actin gels once they are formed, and other influence the equilibrium between F-actin and G-actin. The activity of some of these factors is sensitive to calcium ions, making the actin filament system very responsive to intracellular free calcium. Studies using immuno-fluorescence techniques have demonstrated that while actin is continuously distributed along microfilament bundles, there is a regular arrangement of myosin, tropomyosin, and α-actinin. It is likely that microfilaments are made of sarcomeric units analogous to those of skeletal muscle, and it has been demonstrated that the microfilament sarcomeres are contractile (Kreis and Birchmeier, 1980). The force generated by the contraction of actin filaments is thought to generate the movement of cells, and it is possible to determine the direction of the force by "decorating" actin. In 1969 a technique for identifying actin filaments in whole cells was introduced (Ishikawa et al., 1969). Cells were made permeable by treatment with glycerin and were then exposed to muscle heavy meromyosin, which bound specifically to the 6-μm microfilaments, forming arrowhead-shaped complexes. It had previously been demonstrated in muscle that the force generated during contraction pulls the actin filament in the direction pointed by the arrowheads (Huxley, 1963).

3.4.4. Microtrabecular Lattice.   As outlined above, when cultured cells are extracted with nonionic detergents, a detergent-insoluble residual cytoskeleton remains which is composed of a variety of filament types, including microtubules, microfilaments, and intermediate filaments. The composition of the cytoskeleton varies with the amount and kind of detergent and the length of the extraction (Schliwa et al., 1981). To circumvent these extraction variables, Porter and his collaborators used the technique of stereoscopic high-voltage electron microscopy to examine the filamentous substructure of whole cells (Wolosewick and Porter, 1979). These experiments revealed a highly interconnected three-dimensional network of filaments between 2 and 25 nm in diameter which has been termed the microtrabecular lattice. A new system of filaments, 2–3 nm in diameter, was identified. Similar structures were observed by high-voltage electron microscopy following mild detergent extraction (Figure 3.5). The 2–3 nm fibers seem to serve as linkers between the other fiber types (Schliwa et al., 1981; Schliwa and van Blerkom, 1981). Actin filaments also appeared to contact microtubules and IFs. The highly interconnected nature of all of the filament systems in the overall lattice indicates that any experiment employing one reagent to eliminate a class of fibers (e.g., colchicine-induced microtubule breakdown) alters the organization and probably the functioning of all of the filaments. The very high density of the microtrabecular lattice also presents the possibility that all macromolecular cytoplasmic constituents are bound to this structure, with only the low-molecular-weight metabolites free in solution. This would produce a two-phase system within the cell and limit the spectrum of possible interactions between individual cytoplasmic protein molecules.

3.4.5. Cell Motility.   The *raison d'être* for at least some of the cytoskeletal components is to generate the superstructure and the force re-

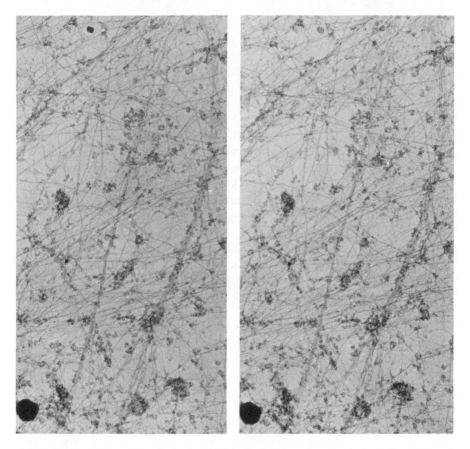

Figure 3.5. Microtrabecular lattice. Chick fibroblasts were extracted with the detergent Triton X-100 for 40 seconds and then examined by electron microscopy. Four types of filaments form a three-dimensional lattice within the cell: 25-nm microtubules; 9–11-nm intermediate filaments; 6–8-nm microfilaments; and 2–3-nm filaments. This figure shows a stereoscopic pair at 23,000× magnification. Courtesy of Dr. Manfred Schliwa.

quired for cellular movement. Cellular movement is manifested in many forms; two of these will be described here. One is change in cell shape, usually through the extension and retraction of small processes called pseudopodia. The extreme example of this type of movement is the extension of axons and dendrites from nerve cells. The technique of tissue culture was, in fact, first successfully used to study the movement of neurites (Harrison, 1907). The second is translational movement. Most cells are motile *in vivo,* particularly during certain stages of embryogenesis when mass cell migrations take place. Translational motility occurs in cultured cells; it has been studied intensely in fibroblastlike cells.

Since it is quite likely that similar mechanisms are responsible for both the movement of the growth cone during neurite extension and the translational movement of fibroblasts, the latter will be discussed first.

Time-lapse motion pictures of cultured cells have shown that fibroblasts move forward by initially extending a broad, thin region of surface membrane called the lamellipodium. During their extension the lamellipodia may lift off the substratum and form ruffled membranes. The lamellipodia usually extend 3–10 μm from the cell and are 0.1–0.4 μm thick; they contain no cytoplasmic organelles except a meshwork of actin-containing microfilaments. They have small protrusions called filopodia along their edges. The filopodia are the structures that make direct contact with the substratum. The membrane required for the formation of lamellipodia is apparently derived by surface flow from another part of the cell (Trinkaus, 1980).

To produce translational movement, the lamellipodial membrane adheres tightly to the substratum and stretches the cell along the axis of movement. The cell's posterior contacts with the substratum are then broken, and the cell moves forward (for a review, see Lazarides and Revel, 1980). Frequently long and exceedingly straight extensions, called retraction fibers, are formed between the substratum and the cell margin at the posterior end of the cell. These are sometimes broken off and remain associated with the substratum as the cell advances.

In muscle cells, contractile movement is due to the interaction of actin and myosin, a process which requires energy and is regulated by Ca^{2+} ions (see Chapter 2). The same mechanism may mediate the movement of nonmuscle cells, as the actin concentration of most motile cells is between 100 and 250 μM—up to 20% of the total protein. Actin filaments are more prominent in nonmoving cells, but more motile cells have a diffuse distribution of actin. Although one would assume, given the chemical similarities of actin and myosin from muscle and nonmuscle cells, that the actomyosin system of nonmuscle cells is responsible for the generation of cellular movements, there is no direct evidence for this. Two classes of models have been used to explain the translational movement of mammalian cells; neither is very compelling (Weatherbee, 1981). The first is that the contraction of cytoplasmic actin gels (meshworks) somehow squeezes the cytoplasm forward to form the lamellipodium, which then adheres to the substratum. The contractile mechanism then relaxes, enabling the cell to stretch along its axis of movement, and the cell moves forward after the rear adhesion sites are broken. The second hypothesis is that actin filaments are anchored at the leading edge of the cell and pull the trailing edge forward to generate motility. This model does not, however, explain the mechanism responsible for the protrusion of the leading edge. Whatever the mechanism, it is generally agreed that the regulation of the contractile process and movement involves calcium ions.

As outlined above, calcium regulates both sol–gel transformations of actin via interactions with cross-linking proteins and the contraction of actomyosin via its interaction with tropomyosin and troponin. In nonmuscle cells, however, a third form of calcium regulation has been described which involves the calcium-binding protein calmodulin and the phosphorylation of myosin light chains (see Pollard, 1981, for a review). If the concentration of intracellular calcium is sufficiently high, Ca^{2+} binds to and activates calmodulin, which then activates the light-chain kinase. The kinase phosphorylates the light chain and activates the myosin, resulting in contraction. This contraction can take place in filaments or in the more dispersed meshwork of actomyosin.

3.4.6. *Contact Inhibition of Cellular Movement.* The behavior of animal cells is frequently complex. They change shape, move around, divide, and interact with each other to form more elaborate structures. A fundamental aspect of all of these behaviors is the relative movement of cells. The movement of nontransformed fibroblastlike cells in monolayer culture is inhibited when they come into contact with each other (Abercrombie and Heaysman, 1954). When two migrating fibroblasts collide, the forward movement of both cells ceases, the wavelike movement (ruffling) of the leading edges (lamellipodia) of the cells usually stops, the lamellipodia retract, and the cells initiate movement in another direction. This process has been named the contact inhibition of movement and should not be confused with the erroneously labeled contact inhibition of cell growth whereby cultured cells cease net cell division at high cell density. Although cell–cell interactions may be involved in this type of growth inhibition, other factors such as nutrient depletion are also of major importance.

There are two types of contact inhibition of movement. The first type results in the inhibition of ruffling of the leading lamellipodium of the cell and soon thereafter in the lamellipodium's retraction. The second type of contact inhibition results in the cessation of forward cellular movement, but the ruffling of the leading lamellipodium is not inhibited, and its retraction is not induced. Although ruffling of the leading edge is frequently associated with movement, it is clearly not necessary for movement or for contact inhibition of movement. Cells migrating under agar have no ruffles (Abercombie et al., 1970), and when cells move to a less adhesive substratum, forward movement is stopped but ruffling continues with contraction of the leading lamellipodium (Carter, 1967).

The mechanisms which produce contact inhibition are not known. Forward movement may be inhibited by the inability of the cell to make functional adhesion with the opposing substratum, such as the surface of another cell. If homologous cells are less adhesive to each other than to the substratum on which they are growing, then contact inhibition of movement could be explained in terms of variation in adhesive strength. The cell is "driven" to move to the substratum of relatively greater adhesiveness. A second alternative is that cell–cell contact causes a direct inhibition of the locomotive apparatus within the cell, perhaps through the rapid sequestration of free calcium.

If a collision between cells inhibits one cell from moving across another, a monolayer will eventually form, and there will be no cellular overlap. In practice, there is a wide range of overlapping among various types of cultured cells, with the tumorigenic or transformed cell lines exhibiting the greatest tendency to pile up. The substratum and culture medium also influence the degree of cellular overlap (Harris, 1973).

Whether contact inhibition of movement is due to cell–cell and cell–substratum adhesive differences, direct inhibition of the machinery for locomotion, or a combination of both remains to be determined. There is, however, a direct analogy between the problem of cellular locomotion and one of the important problems of neurobiology—neurite outgrowth and the function of the growth cone.

3.5   Morphological Aspects of Nerve Differentiation

The first studies using cultured cells were done by Harrison (1907) to determine whether nerve fibers formed via cytoplasmic bridges or through fiber outgrowth from single cells. His work demonstrated that neurons can extend long processes (also called nerve fibers or neurites) from the cell body, and that at the tip of the extending neurite is an actively motile flattened structure called a growth cone. The growth cones of the cultured nerves were similar to those described in 1890 by Cajal in chick embryos. Growth cones are usually flattened, fan-shaped expansions of the nerve ending from which extend long thin (0.1–0.2 μm) projections called filopodia or microspikes (Figure 3.6). In addition, broad and very thin sheets of cellular material are found at the leading edge of growth cones. These are similar to the leading edges of motile fibroblasts and are therefore called lamellipodia. Harrison observed that the growth cone is in continuous motion, and that both the filopodia and lamellipodia are transient structures which move about and are continually being retracted and reformed. Plasma membrane is inserted at the

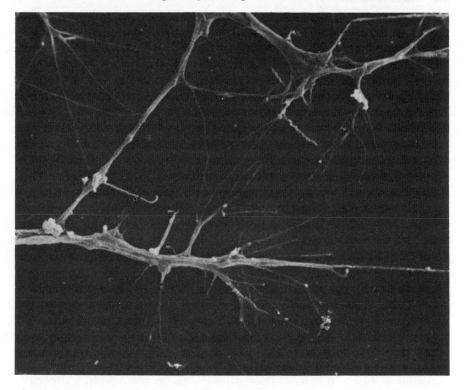

Figure 3.6.   Scanning electron micrograph of a growth cone from a dorsal root ganglion culture treated with NGF. Magnification, 2600×. Courtesy of Dr. F. J. Roisen.

growth cone to support neurite outgrowth (Bray, 1970), and the growth cone is also a site of active endocytosis (M. Bunge, 1977). Finally, the growth cone must lead the axon to its proper synaptic site during development, or to the previous location of the synapse during reinnervation. The growth cone must therefore be able to respond properly to external cues. The structure, synthesis, and function of neurites and growth cones are outlined in the following sections.

3.5.1.   *Neurite and Growth Cone Structure.* The unique architecture of nerve cells requires intracellular structures for maintaining the stability of long axons and dendrites as well as a force-generating system to pull the growth cone and neurites away from the cell body. The cytoskeletal components of nerve cells are clearly involved in both processes. Their functions are probably similar to the ones they have in fibroblasts. Indirect evidence for the roles of tubulin in maintaining neurite stability and actin in growth cone mobility comes from studies using reagents which dissociate microtubules and microfilaments, respectively. More direct evidence comes from ultrastructural studies. Colchicine, which depolymerizes microtubules, causes neurite retraction without altering the structure or activity of growth cones, suggesting that microtubules are required for the stabilization of neurites (Yamada et al., 1970). In addition, when neurite retraction is induced by colchicine, long portions of the retracting processes become actively motile, suggesting that the microtubule system may be responsible for restricting the motile activity of the neurite to its distal end and that the entire length of the neurite is capable of forming growth cones (Bray et al., 1978). Cytochalasin B, which alters the structure of microfilaments, has no effect on the neurite *per se*, but inhibits growth cone mobility, causes retraction of microspikes, and makes the growth cone become rounded (Yamada et al., 1970). This observation suggests that microfilaments play an active role in growth cone functioning. These inferences have been confirmed more generally by the electron-microscopic examination of cultured cells and by immunofluorescence studies using antisera against cytoskeletal proteins.

Time lapse photography has shown that nerve cells placed under culture conditions in which they can produce neurites extend filapodia from the cell margin and that these frequently make contact with the substratum. At some contact points the adhesive interaction is stronger, and the tension at these adhesions is sufficiently great to pull the nascent growth cone forward. If the adhesion to the substratum is weak, then the force generated by the growth cone pulls the growth cone off the substratum rather than forward. The growing neurite would then follow the more adhesive substratum—a conclusion supported by substantial experimental evidence.

The area where the neurite emerges from the body of cultured sympathetic neurons contains smooth endoplasmic reticulum, mitochondria, polysomes, and the major filamentous structures of nerve processes, microtubules and neurofilaments (M. Bunge, 1973). Microtubules originate at initiation sites in the cytoplasm and continue to elongate as the neurite grows (Heidemann et al., 1981). They frequently extend for the entire length of the neurite (Letourneau, 1982). The main shaft of the neurite contains numerous microtubules and neurofilaments oriented parallel to the long axis of the neurite and embedded in a fiber network similar to the microtrabecular lattice of fibroblasts (Ellisman and Porter, 1980; Letourneau, 1982). The membrane margin of the neurites sometimes contains parallel arrays of microfilaments, but most of the microfilaments form a three-dimensional network between the longer filaments and the membrane. The ratio of microtubules to neurofilaments in the processes of cultured cells varies, and in some cases the neurites appear to be completely filled with the filaments. Long, thin mitochondria, smooth endoplasmic reticulum, and a variety of vesicles are found scattered throughout the nerve processes (Figure 3.7). At the distal end of the neurite the density of the microtubules and neurofilaments decreases, and only a few filaments extend into the central portion of the growth cone; apparently none reach the filopodia.

The expanded portion or "palm" of the growth cone usually contains vesicles of various sizes and densities, lysosomes, smooth endoplasmic reticulum, vacuoles, mitochondria, and occasional polysomes (M. Bunge, 1973). Some of the numerous vesicular elements found in the growth cone may be precursors of the new membrane which is inserted into the growing neurite. These organelles are generally not found in filopodia and lamellipodia. In cultured sympathetic neurons the growth cone palm possesses dense-core (40–60 nm in diameter) vesicles containing the neurotransmitter norepinephrine. These vesicles are rarely seen in filopodia. Two forms of actin filaments are seen in growth cones. Fine meshworks of filaments are associated with the filopodia and lamellipodia of growth cones, in much the same way that they are associated with the leading edges of motile fibroblasts. In addition, parallel arrays of actin filaments extend into the filopodia and are frequently associated with areas of linear adhesion with the substratum (Letourneau, 1979, 1982). The polarity of these filaments, as defined by decoration with heavy meromyosin, is the same in the filopodia of both fibroblasts and growth cones (Isenberg and Small, 1978). A simplified composite model of the growth cone of a cultured cell is shown in Figure 3.8.

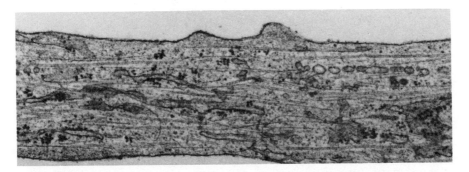

Figure 3.7. Ultrastructure of a neurite. Section of a neurite from a cultured CNS nerve cell line. The thicker filaments are microtubules, 25 nm in diameter. Courtesy of F. George Klier.

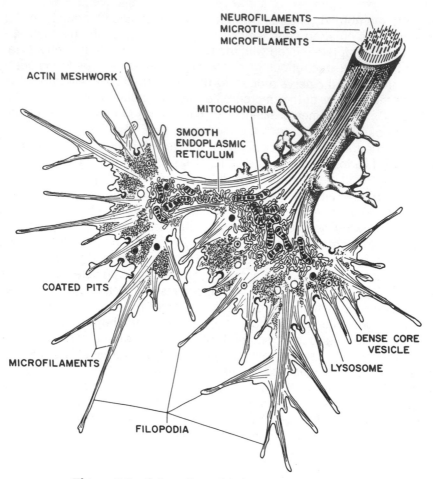

Figure 3.8.   Schematic model of a typical growth cone.

The actin meshwork concentrated in the peripheral margins of the growth cone and the linear arrays of microfilaments in the filopodia probably constitute the force-generating system responsible for neurite elongation. Myosin is colocalized with actin in the growth cone, and by analogy with the experiments demonstrating the contractile nature of fibroblast microfilaments, it is quite likely that the contractile properties of the actomyosin system pull the neurite away from the nerve cell body. Because actin filaments are linked to the plasma membrane with the same polarity as in their attachment to the Z lines of muscle sarcomeres (Begg et al., 1979), the microfilament system cannot cause the protrusion of the filopodia. Their attachment and polarity are in the wrong direction to push the filopodium forward. Since actin filaments are associated with the majority of the growth cone adhesion sites, the work of elongation is done against these sites. The cell—substratum adhesion plaques are localized primarily to the filopodia, but continue beneath the palm of the growth cone and lie also at the edges of the growth cone margin (Letourneau, 1981). The palm of the growth cone can also make "spotty" contacts with the substratum, but these are rare.

As a classical actomyosin contractile system is employed to generate movement, activation of the system probably involves the release of calcium within the filopodia. Reagents which elicit neurite outgrowth and stimulate cell—substratum adhesion, such as NGF and cAMP, cause a release of bound calcium ions from cells (Schubert et al., 1978). It is therefore likely that the release of calcium ions at the growth cone by trophic factors such as NGF mediates the directed outgrowth of neurites by both increasing growth cone adhesiveness and eliciting the contractile movements required to pull the neurite forward. Since growth cone movement does not demand the breaking of posterior adhesions, the rate of neurite extension frequently increases with increasing substratum adhesiveness. The translocation of fibroblasts *decreases* with increased adhesion, for they must release their posterior adhesion sites to move forward, and increasing the adhesiveness of the substratum makes this process more difficult.

The morphological differentiation of nerve cells is accompanied by a very large increase in the number of microfilaments and microtubules in both the cell body and the newly formed neurites. It was therefore thought possible that there is an accompanying increase in the synthesis of these cytoskeletal proteins. An examination of the accumulation of actin and tubulin showed that this was not the case. During the differentiation of C1300 mouse neuroblastoma cells the amount of actin, which comprises 7% of the total protein, remains constant. The ratio of β- to γ-actin also remains constant at 3 : 2 (Rein et al., 1980). An examination of tubulin synthesis during neurite outgrowth from C1300 neuroblastoma cells showed that there was no difference in the amount of tubulin present in undifferentiated round neuroblasts and in cells with long neurites (Morgan and Seeds, 1975). An analysis of protein synthesis during the NGF-induced differentiation of a clonal sympathetic nervelike cell line also showed a constancy of actin and tubulin synthesis (Garrels and Schubert, 1980). There does, however, appear to be unique isoform of β-tubulin which is found only in cells with neurites (Edde et al., 1981). On the basis of differences in isoelectric focusing characteristics and peptide mapping, there are three forms of α-tubulin and five forms of β-

tubulin. β2-tubulin in C1300 cells was uniquely associated with morphologically differentiated C1300 cells.

3.5.2. Microtubule Organization and Neurite Outgrowth.   A major sign of nerve cell differentiation is a change in cell shape. During embryogenesis cells are transformed from spherical neuroblasts to highly asymmetric neurons. Since the cytoskeleton is thought to play a major role in cell morphology, it is likely that there are large changes in its intracellular arrangement during nerve differentiation. One of the major structural elements in differentiated nerve cells is the microtubule. Neurites contain large numbers of microtubules, and if nerve cells are treated with microtubule-dissociating agents such as colchicine neurite outgrowth is inhibited and extending neurites are retracted (Yamada et al., 1970). In interphase fibroblastlike cells, microtubules radiate out from the perinuclear region toward the plasma membrane. In nerve cells a similar pattern is observed, with the exception that the microtubules continue down the neurites toward the growth cones.

What is the origin of microtubules and how does their growth relate to neurite formation? There are two arrangements of microtubules in cultured cells—the delicate network observed in interphase cells and the well-structured mitotic spindle apparatus in metaphase cells. At the electron-microscopic level, microtubules appear to be organized around discrete foci such as centrioles. These foci have been called microtubule-organizing centers (MTOCs). MTOCs may be located and counted by immunofluorescence techniques using anti-tubulin antisera following treatment of cells with microtubule-disrupting agents. When cells are exposed to colcemid, vinblastin, or low temperature and then placed in normal medium to allow for partial recovery before fixation and immunostaining, microtubules grow toward the cell periphery from a limited number of sites near the nucleus (Frankel, 1976). In addition, if cells are treated with colcemide and extracted with Triton X-100, and purified tubulin protein is then added to the cells, initiation and assembly of microtubules occurs at the cell's normal organizing centers (Brinkley et al., 1981). In fibroblastlike cells there is usually only one MTOC; it corresponds to the centriole region (Watt and Harris, 1980; Brinkley et al., 1981). However, larger numbers of initiation sites (five to 10) have been reported in some fibroblast clones (Spiegelman et al., 1979a).

When C1300 mouse neuroblastoma cells were examined for MTOCs by the technique of microtubule regrowth following colcemid treatment, the undifferentiated round neuroblastoma cells had approximately 12 initiation sites, while cells induced to morphologically differentiate by serum deprivation had only a single MTOC (Spiegelman et al., 1979*b*). Comparison of the time course of neurite formation with the number of MTOCs per cell led to the conclusions that (1) the multiple initiation sites in the round cells aggregated to form the single MTOC in differentiated cells, (2) the aggregation step occurred before the initiation of neurite outgrowth, and (3) cells with multiple neurites still had only one MTOC. A summary of the data relating neurite outgrowth and MTOCs is shown in Figure 3.9. Steps 1 and 2 represent the migration of the multiple MTOCs around the nucleus. Migration is followed by the directed (oriented) outgrowth of microtubules toward the potential neurite outgrowth site (step 3). Neurite elongation then continues in association with microtubule growth (steps 4 and 5). The mobility of the MTOCs during nerve differentiation may serve to facilitate the rearrangement of neurite function in differentiating cells. In contrast to the work of Spiegelman, Sharp et al. (1981), using both serial section electron microscopy and immunofluorescence to visualize centrioles, showed that both differentiated and undifferentiated C1300 neuroblastoma cells have four to five centrioles per cell. Although the electron microscopy verified that the colcemid method for assaying MTOCs was correct, the reason for the experimental discrepancy is not clear. The most likely explanation is that the different neuroblastoma clones used had different phenotypes.

Once neurite outgrowth is initiated, the growth of the microtubules in the neurite keeps pace with growth cone extension. Since microtubules

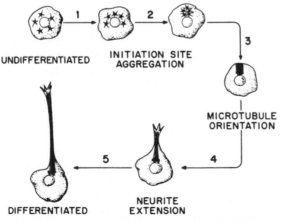

Figure 3.9.   A model for neurite outgrowth. A clone of the C1300 mouse neuroblastoma was induced to differentiate by the removal of serum from the culture medium. Steps 1 and 2 take 24–48 hr, 3 and 4 are complete 24 hr later, and neurite outgrowth (step 5) continues for many days. From Spiegelman et al. (1979*b*).

grow only by the addition of new subunits at their ends, and since the tubulin subunits are transported via the slowest axoplasmic transport system, it follows that cytoskeletal growth may occur from the cell body. Alternatively, microtubules may contain breaks along the neurite and subunits may be added throughout the neurite.

Finally, elevated cAMP can increase the length of microtubules severalfold in transformed fibroblasts (Tash et al.; 1980). Cyclic AMP also mediates neurite outgrowth in all cultured nerve cell populations.

The axons of neurons in different animals of the same species frequently have very similar morphologies and branching patterns. There are two hypotheses which could explain the shape-relatedness of different cells and there is evidence in support of each. One is that the morphology of the cell is determined by its immediate environment—neighboring cell surfaces, extracellular matrix, or other exogenous factors. In cultured cells adhesiveness to the substratum is largely responsible for determining the overall morphology and spreading pattern of cells. For example, neurites tend to follow scratches in the surface of the culture dish, presumably because the scratches have rougher and thus more adhesive surfaces than the culture substratum (P. Weiss, 1941). The other hypothesis is that cells are intrinsically programmed to assume a given morphology. An example of this was discovered by Albrecht-Buehler (1977), who showed that daughter fibroblastlike cells have similar cell shapes and similar geometries of microfilament bundles.

Do similar morphological relationships exist in daughter nerve cells? Low-density cultures of C1300 mouse neuroblastoma cells were induced to differentiate by lowering the serum concentration in the medium. The nerve cells flattened and elaborated between one and nine processes (Figure 3.10). When mitotic sister cells were examined and scored for neurite number, thickness, length, and branching patterns, about 60% were found to be closely related, either as identical twins or as mirror images of each other (Solomon, 1981). Cells which were not mitotically related were not morphologically related. This inherited relatedness lasts for two divisions, generating four morphologically similar granddaughter cells (Solomon, 1979).

The ability of cultured nerve cells to recapitulate neurite morphologies could be due to an ability to "copy" or redistribute their cytoskeleton in a similar way between daughter cells. Solomon (1980) ruled out this possibility by reversibly disrupting the cytoskeleton of differentiated neuroblastoma cells, allowing the cells to regrow their neurites, and observing whether the morphologies of the regenerated neurites were similar to the originals. When cells were treated with the microtubule-depolymerizing drug Nocodazole® the neurites retracted to generate

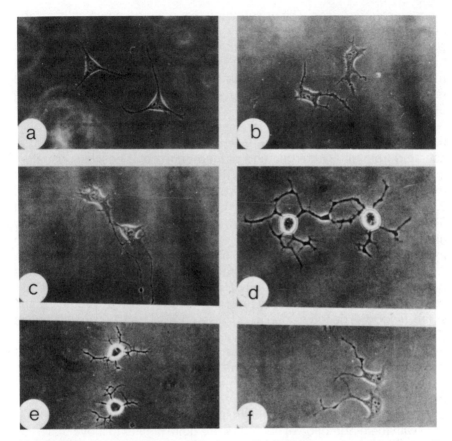

Figure 3.10.   Neurite morphology of sister cells. C1300 mouse neuroblastoma cells were plated at a low cell density, allowed to divide once, and induced to differentiate by lowering of the serum concentration in the culture medium to 0.1%. (a)–(f) show pairs of cells photographed 8 hr later. From Solomon (1981).

round cells. When the drug was removed and the neurites were allowed to regrow, 60% of the cells recapitulated their original morphologies. The reformation of similar branching patterns was not due to regenerating neurites following clues left by retracted neurites on the substratum, because the same frequency of recapitulation was observed in cells which moved across the substratum during the incubation with Nocodazole. These results show that the heritable morphological information of neurite morphologies is not contained within the complete cytoskeletal network. The information may, however, be derived from MTOCs, which survive Nocodazole treatment, or from unidentified determinants within the plasma membrane. The observation that cultured nerve cells exhibit heritable similarities in morphology is compatible with the rapidly expanding list of related nerve cell morphologies in genetically identical organisms. These results do not, however, rule out a role for the environment in directing neurite outgrowth during embryogenesis and nerve regeneration. Both intrinsic and extrinsic factors are responsible for the ultimate morphologies of neurons.

3.5.3. *The Kinetics of Neurite Outgrowth.*   When nerve cells are dissociated from ciliary ganglia and placed upon an adhesive substratum in culture, the round cells attach to the substratum, spread, and extend many active filopodia from the entire circumference of the cell (Figure 3.11). Time lapse cinematography has shown that while hundreds of filopodia spontaneously extend and retract from the cell surface, a few attach to the substratum and fail to retract. Broad lamellae are then formed on the distal ends of these filopodia, giving rise to growth cones. As the growth cones migrate away from the cell body, frequently at rates between 50 and 100 μm/hr, the activity of the filopodia around the circumference of the cell between growing neurites is greatly reduced; the small filopodia become rare in all areas except at the growth cones (Collins, 1978*a*). Because the sequence of neurite initiation shown in Figure 3.11 was observed with dissociated ciliary ganglia neurons, it could be argued that it is representative of nerve regeneration, for the cells previously possessed neurites *in vivo*. However, when neurite formation is induced in clonal nerve cell lines which had not previously been morphologically differentiated, the morphological changes associated with the elaboration of neurites were quite similar to those observed in the dissociated ciliary ganglion cultures. Figure 3.12 shows neurite formation induced in the C1300 mouse neuroblastoma cell line by the removal of the serum from the culture medium. The serum depletion increases cell–substratum adhesion, causing cell flattening and neurite outgrowth (Seeds et al., 1970; Schubert et al., 1971).

The growth of neurites from dissociated ganglion cells appears to be random (Figure 3.11), whereas neurite outgrowth from organ cultures of whole ganglia is radial (Figure 1.1). Since growth cones have a great deal of morphological and ultrastructural similarity to the leading edge of motile fibroblasts, growth cones may exhibit the type of contact inhibition of movement observed in fibroblasts. In fibroblast cultures cell–cell contact leads to a cessation of forward movement and a subsequent reinitiation of movement in another direction. This type of reaction leads to centrifugal cell migration, which becomes more pronounced with increasing numbers of cell–cell contacts. Mutual contact inhibition of nerve

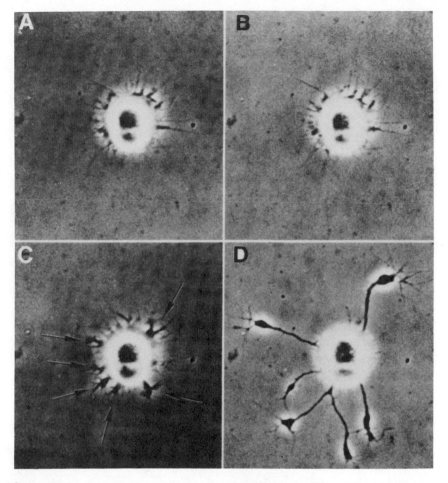

Figure 3.11. Neurite outgrowth from ciliary ganglion cells. Chick ciliary ganglion neurons were dissociated and plated into heart cell growth-conditioned medium, which promotes neurite outgrowth of this cell type. (A) and (B) were taken 21 and 25 min after exposure to the conditioned medium, and (C) and (D) at 32 and 56 min, respectively. From Collins (1978a).

fibers has been demonstrated in organ cultures of explanted chick sensory ganglia, and was used to explain the radial outgrowth of nerve fibers from these ganglia (Dunn, 1971). The simplest explanation that is consistent with the data for the radial outgrowth of neurites in this system is that the outward direction of growth is the one in which the frequency of contacts with neighboring nerve fibers is the lowest.

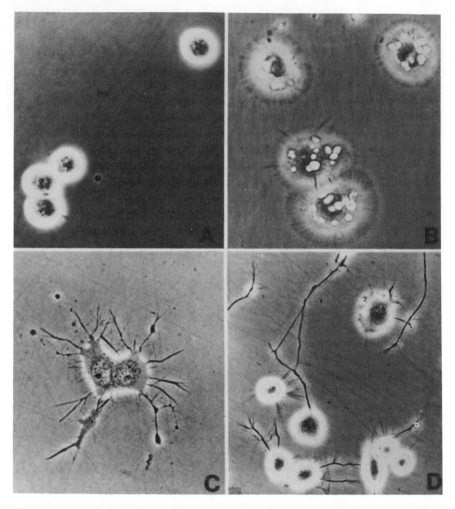

Figure 3.12. Neurite outgrowth from clonal C1300 mouse neuroblastoma cells. Exponentially dividing cells were dissociated and plated into serum-free medium, which promotes neurite formation with these cells. Phase photomicrographs were taken 5 min (A), 15 min (B), 5 hr (C), and 18 hr (D) after the cells were plated. From Schubert et al. (1971).

Although the above data suggest that mutual contact inhibition of growth cone movement may be involved in pattern formation in some organ cultures, it is clear that most neuronal cells do not exhibit contact inhibition of movement of the type demonstrated by cultured fibroblasts. Nerve and glia must make contact with other cells as they migrate during development. Cell migration is typified by the migration of Schwann cells along the axons of peripheral nerve and by the migration of nerve cells along the processes of Bergman glial cells. An example of growth cone migration is the outgrowth of neurites along the axon of the first nerve to reach the peripheral field—the "pioneer" axon. This interaction results in the formation of axon bundles, or fascicles, which are common both *in vivo* and in cultured cells.

A detailed description of the interactions between nerve growth cones and various cell types was made possible by time lapse photography of low-density dissociated ciliary ganglion cultures (Wessells et al., 1980). Four types of interactions between moving cells were observed in these heterogeneous cultures: (1) Between non-neuronal cells (glia or fibro-blasts), there was a paralysis of movement indistinguishable from that described for fibroblasts, with a cessation of membrane movement in the area of contact. (2) Between growth cone and axon, no contact inhibition was observed, and the growth cone frequently adhered tightly to and migrated down the axon. (3) Between growth cone and growth cone, there was no contact inhibition of movement; the growth cones usually simply crawled past each other. (4) When a moving non-neuronal cell encountered a neurite or nerve cell body there was no inhibition of move-ment of the non-neuronal cell. Since an interaction between a non-neuro-nal pair results in contact inhibition, this observation suggests that the surfaces of nerve cells are distinct from those of other cell types in that they do not inhibit the forward motion of fibroblasts. The chemistry re-sponsible for this possibly unique characteristic of nerve cells is un-known.

Another feature of the growth cone is its ability to release at least one proteolytic enzyme into its immediate environment. Plasminogen activa-tor is an enzyme which cleaves plasminogen to produce plasmin, a proteolytic enzyme which in turn hydrolyzes fibrinogen. Krystosek and Seeds (1981) used a simple visual assay to determine the hydrolysis of fibrinogen around morphologically differentiated nerve cells. Although plasminogen activator activity was found around the body and the growth cone, the proteolytic activity was frequently much higher around the growth cone than around the cell body. The activity was only rarely observed along the length of the neurite. It is likely that the release of proteolytic activity facilitates the advance of the growth cone through extracellular matrix material.

A common observation and two experiments have shown that neurites grow by the deposition of new membrane at the growth cone. An early observation by many neurobiologists was that the point of bifurcation of a neurite does not move away from the cell body, whereas the neurite distal to the bifurcation continues to elongate. This rules out the possibility that new membrane is inserted between the cell body and bifurcation of the neurite and suggests that growth must take place at the distal end of the neurite. Two experiments support this conclusion. The first was done by marking a small area along the neurite with finely ground pieces of glass wool and following the movement of the particles and cell with time lapse photography (Bray, 1970). Like the neurite bifurcations, the particles did not move away from the cell body while the neurite continued growing, even if the marker was deposited very close to the growth cone. This datum showed that new membrane was inserted into the neurite at or near the growth cone. Feldman et al. (1981) came to the same conclusion from experiments using a different method of marking old neurite surface membrane. Growing neurites were first treated with the lectin concanavalin A, which binds to cell-surface carbohydrate, and then with a rabbit antibody which bound specifically to the concanavalin A. This procedure marked all existing membrane. The excess reagents were washed away, and the cells were allowed to continue extending neurites for another 24 hr. At this time, fluorescent-labeled antiserum against the rabbit anti-concanavalin A was added to stain all of the membrane previously labeled with concanavalin A. Assuming that the concanavalin A was not freely mobile in the membrane following its reaction with the first antibody, three different results were possible. (1) If the neurites were labeled throughout their length, then new membrane must be inserted all along the process. (2) If the neurites were labeled toward their distal end and there was a loss of label near the soma, then new membrane was added at the cell body. (3) Finally, if the label appeared only in the proximal membrane of the neurite and not in the area behind the growth cone, new membrane must be added at the growth cone. The third alternative was the only one compatible with the data, again showing that the growth cone was the site of new membrane insertion. Since the new axon is produced behind the advancing growth cone the proximal neurite can remain in stable association with other neurites when neurite bundles or fascicles are formed during neurogenesis.

In 1926 Levi observed that if a nerve process is cut the severed portion of the neurite reattaches to the substratum and continues to grow and move about, in much the same way that enucleated fragments of fibroblasts (minicells) continue to exhibit movements and contact inhibition similar to those of whole cells. Shaw and Bray (1977) repeated the experiments of Levi with dissociated cell cultures of chick sensory ganglia. When a single neurite was cut, there was an initial collapse of the distal piece, followed by the formation of new growth cones at both ends (Figure 3.13). The reformed neurite fragment regrew to approximately the

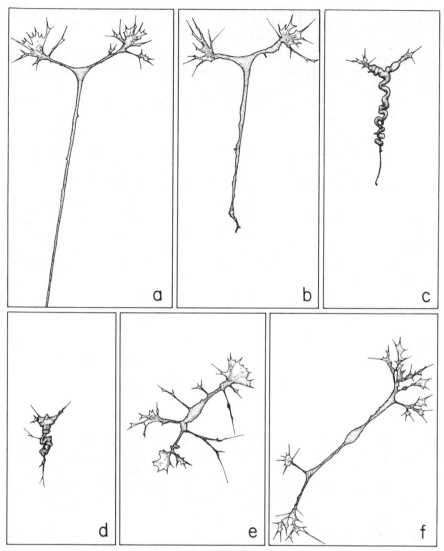

Figure 3.13. Regrowth of cut neurite. A neurite of a cultured chick sensory ganglion cell was cut away from the cell body and photographed at various times after the lesion. Tracings were made from the film. (a) Just after lesion. (b) 1 min. (c) 6 min. (d) 11 min. (e) 24 min. (f) 41 min. Redrawn from Shaw and Bray (1977).

length of the original cut segment and continued to migrate over the substratum for several hours. Additional evidence shows that neurites have the potential to form growth cones along their entire length, for chick dorsal root ganglion cells treated with colchicine generate filopodia similar to those of growth cones along the length of the neurite (Bray et al., 1978).

Neurites generally grow in straight lines and at a constant rate. If they branch, the branching occurs at the growth cone; only rarely are collateral branches derived directly from the neurites of cultured cells. Since cultured neurons are generally anchored only at the cell body and at the growth cones, the neurites are under tension or mechanical stress (Bray, 1979). In neurons with extensive branching, the network of neurites appears to be in mechanical equilibrium. If neurites are experimentally displaced, a predictable compensatory rearrangement of the remaining network brings about a new state of equilibrium. This drive to maintain a mechanical equilibrium and the tension exerted by the growth cone can determine the direction of neurite outgrowth in cultured cells. Although this phenomenon is not as obvious *in vivo,* where there are a large number of lateral associations between neurites and neighboring cells, it is possible that the requirement for the growth cone to maintain tension on the neurite could force the growth cone to select the pathway of highest adhesiveness, where it is able to maintain the best grip on the substratum and can exert the maximum pull. A wide variety of evidence from clonal and dissociated nerve cell cultures has shown that both the rate of neurite outgrowth and its guidance are regulated by the adhesiveness of the substratum.

3.5.4. *The Role of Cell–Substratum Adhesion in Neurite Outgrowth and Axon Guidance.*

Three mechanisms have been proposed to explain the directed outgrowth of axons toward their appropriate targets *in vivo*—extracellular channels, selective adhesion, and gradients of diffusible substances. Extracellular channels (holes) may be present through which the initial, or "pioneer," axon can migrate. Extracellular channels exist before the outgrowth of optic nerve fibers in chicks (Krayanek and Goldberg, 1981), and Singer and his colleagues have demonstrated extracellular guidance channels in frog spinal cord (Singer et al., 1979). These channels must be the path of least resistance, have a unique surface, or have both properties. The first neurites to follow the path adhere tightly to the neuroepithelial cell processes which line the spaces. Once the initial pioneer axons have traversed the channels, the following fibers adhere preferentially to these axons or to the glial cells which are tightly associated with them (Goldberg, 1977).

Nerve cells may be guided directly by contact with neighboring cells or extracellular matrix material. Since the growth cone must adhere to a substratum to move forward, all forms of neurite outgrowth require an adhesive surface. During the past 50 years it has frequently been observed that guidance cues for axon outgrowth appear to be located in the substrate on which axons grow (see, Katz et al., 1980, for a review). Glial cell surfaces may also guide migrating cells. For example, Bergman glia appear to direct the migration of neuroblasts in the cerebellum (Schmechel and Rakic, 1979). Numerous other experiments have demonstrated that outgrowing embryonic, transplanted, and regenerating axons behave as if they are following defined pathways in the substrate. It has been argued that the precursor substrate pathways are established in the neural plate and that these precursor patterns are continually modified during development to generate the ultimate pattern of nerve-fiber tracks (Katz et al., 1980).

In addition to guiding axons, substrate pathways can guide migrating cells. For example, there are defined pathways for neural crest cell migration (LeDouarin and Teillet, 1974). It has been suggested on the basis of microscopic studies that extracellular matrix material initiates crest cell migration and serves as a guidance mechanism during the early stages of crest migration (Lofberg et al., 1980).

Finally, guidance may be mediated by a diffusible substance through the establishment of a gradient. The best example is nerve growth factor (NGF). NGF is a protein which is responsible for the maintenance and differentiation of sympathetic nerve cells. Since the target tissues for sympathetic innervation probably produce low levels of NGF, it has been proposed that the released NGF molecules serve as chemoattractants and guide incoming fibers to their targets via a diffusion gradient. Direct evidence for this possibility was obtained by the intracerebral injection of NGF into neonatal rodents (Chen et al., 1978). The animals were examined 2 weeks later and it was shown that noradrenergic fibers of the sympathetic ganglia had entered the spinal cord and grown toward the site of NGF injection. Normally these fibers remain outside the cerebrospinal axis. The data strongly suggest that NGF can direct the growth of axons, presumably through the establishment of a diffusion gradient.

We have presented evidence for each of the three mechanisms— channels, specific adhesion, and gradients. It is likely that all are employed at some point during the development of the nervous system. Due to the availability of cultured nerve cells, perhaps the best understood aspect of the three mechanisms is the role of cell–substratum adhesion in neurite guidance.

As indicated in the previous section, neurite outgrowth and growth cone mobility share many characteristics with the locomotion of fibroblastlike cells. Cell–substratum adhesion affects both processes by stabilizing the various cellular protrusions and by anchoring the cell or growth cone margin during the actomyosin-driven translocation process. In both cellular locomotion and neurite outgrowth adhesive interactions regulate the rate and direction of movement. In addition, cell–substratum adhesion mediates the induction of neurite outgrowth in some types of nerve cells. The study of the role of cell–substratum adhesion in neurite outgrowth and guidance has been pursued in a parallel and complementary fashion in two culture systems—clonal cell lines and primary cultures of embryonic chick spinal ganglia. The results of these studies suggest that substrate adhesiveness plays a critical role in the regulation of neurite outgrowth and in the guidance of growing axons to their appropriate target.

Experiments with clonal nerve cells and primary cultures have shown that conditions which increase the intensity of cell–substratum adhesion increase both the frequency and the rate of neurite elaboration. There are two distinguishable mechanisms involved in neurite outgrowth; both are observable in clonal cells. In the case of the PC12 sympatheticlike clone, growth on an adhesive surface is necessary but apparently not sufficient for neurite extension. PC12 cells plated on very adhesive substrata, such as polylysine, flatten, but do not extend neurites. They do, however, send out processes when NGF is added to the cultures, and the rate of neurite extension is proportional to the adhesiveness of the substratum (Figure 3.14). PC12 cells adhere more rapidly to polylysine-coated dishes than to plastic tissue culture dishes, and the rate of

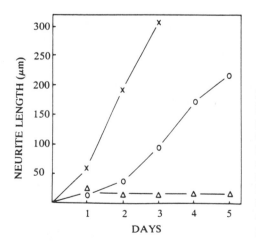

Figure 3.14. Effect of substratum adhesiveness on the rate of neurite outgrowth. Exponentially dividing PC12 cells were plated onto plastic tissue culture dishes and poly-L-lysine coated tissue culture dishes. NGF was added to some cultures, and the average neurite length was determined as a function of time. △, poly-L-lysine; ○, standard tissue culture dish plus NGF; ×, poly-L-lysine plus NGF. From Schubert (1979).

neurite extension is proportionally greater on that surface (Schubert, 1979). In contrast to the PC12 cell line, which requires exogenous NGF for neurite outgrowth, increased cell–substratum adhesiveness is both necessary *and* sufficient for the induction of neurite outgrowth in the C1300 mouse neuroblastoma and five rat CNS nerve cell lines. It was initially shown that C1300 neuroblastoma cells extend neurites on tissue culture plastic to which they adhere well, but not on the less adhesive (more hydrophobic) Petri dish (Schubert et al., 1969). It was then established that removing part or all of the serum from the culture medium further enhances neurite outgrowth (Seeds et al., 1970). Decreasing the amount of serum in the culture medium probably strengthens the interactions between the cell surfaces and the substratum through the removal of proteins which block this interaction. Adsorbed proteins alter the net surface charge of many substrata, and numerous serum and nonserum proteins can block the enhanced rate of neurite extension caused by serum deprivation (Schubert et al., 1971). Reagents which cause neurite extension in C1300 cells, such as cAMP and 5-BUdR, also cause cell flattening and increase cell–substratum adhesion in essentially all cell types.

The effect of substratum adhesiveness on the neurite outgrowth of embryonic chick dorsal root ganglia (DRG) cultures is virtually identical to its effect on clonal cell lines. To determine the relationship between adhesion and neurite outgrowth, dissociated DRG neurons were cultured in the presence of NGF (required for neurite formation in older DRG cultures) on substrate of tissue culture plastic, polylysine (PLYS), polyornithine (PORN), or collagen. When the adhesion of growth cones to these substrata was assayed, it was found that they adhered best to PORN and PLYS, less well to collagen, and least well to tissue culture dishes (Letourneau, 1975*a,b*). When neurite formation was followed on these surfaces, the following observations were made: (1) The rate of neurite growth is directly proportional to the adhesiveness of the substratum. (2) The whole neurite adheres to PORN- and PLYS-coated substrata. Therefore, the shape of the neurite recapitulates the path of the extending growth cone, producing crooked neurites. On tissue culture dishes axons are straight and adhere to the substratum only at the growth cone and cell body. (3) The fraction of cells extending neurites is greater on the more adhesive surfaces. (4) There is much more extensive branching on PORN and PLYS than on the tissue culture dish surface. *In toto,* these results again suggest that there is a high correlation between cell–substratum adhesion and the dynamics of nerve differentiation.

An alternative way to examine the role of adhesion in neurite outgrowth is to place cells on a surface consisting of a grid of two different adhesive surfaces and observe which surface best supports neurite growth. When chick sensory ganglion neurons were plated on culture dishes consisting of 80-μm square palladium grids overlying surfaces of tissue culture plastic (sulfonated, hydrophilic plastic), Petri dish plastic (unsulfonated, hydrophobic plastic), PORN, or collagen, neurite extension was preferentially localized to one surface (Letourneau, 1975b). For example, when plated on a grid consisting of palladium and PORN, neurites preferentially grew on the PORN. On palladium and Petri dish plastic, the neurites preferred the palladium substratum. By examining neurite outgrowth on various binary combinations of surfaces, it was determined that the order of neurite preference was: PORN > collagen > tissue culture plastic = palladium > Petri dish plastic. This same order was demonstrated for the relative adhesive strengths of the growth cones to the various substrata. It follows that neurites given a choice between two substrata grow upon the one to which their growth cones adhere more strongly. Analogous experiments done with fibroblasts many years before had yielded similar results (Carter, 1967). When fibroblasts were plated on a grid of combinations of material with different adhesiveness, the fibroblasts always migrated to and remained upon the surface to which they adhered best. This phenomenon, termed haptotaxis, is discussed in detail in Chapter 6. The orders of preference of fibroblasts and growth cones for the various artificial surfaces are essentially the same.

It should be pointed out that although fibroblast translational movement and axon elongation are similar in many respects, they also exhibit differences. The major difference is that cell locomotion is inhibited by high cell–substratum adhesiveness, but the rate of neurite elongation is increased. This is probably because fibroblastlike cells, unlike growth cones, must break their posterior contacts before locomotion can occur. Second, lanthanum (La^{3+}) ions differ in their effects on cell locomotion and neurite outgrowth (Letourneau and Wessells, 1974). La^{3+} binds to the outer surface of cells and displaces other cations, such as Ca^{2+}. It also competes with calcium for various conductance channels in the membrane. Although La^{3+} increases the adhesion of both neuronal and glial cells to the substratum, it inhibits the locomotion of the glial cells but does not alter the movement of growth cones. The mechanism of this difference has not been studied, but these results suggest that the processes involved in cell movement and growth cone movement are not identical. Nerve cells also grow neurites on an agar surface which does not support the locomotion of fibroblastlike or glial cells.

The term "contact guidance" was coined by Paul Weiss in 1961 for the phenomenon in which fibroblastlike cells moving on an oriented substratum such as aligned collagen fibrils assume the orientation of the substratum. This experiment was later reproduced using growing axons. Spinal ganglia were explanted to a surface containing aligned collagen fibrils, and NGF was added at two concentrations to produce low and high densities of neurite outgrowth (Ebendal, 1976). At a low neurite density on aligned collagen there was a pronounced orientation of fiber outgrowth parallel to the collagen matrix. On nonoriented collagen the outgrowth was radial, presumably because of contact inhibition between growth cones. Scanning electron microscopy showed that neurites in the low-density cultures were frequently associated with the surfaces of the aligned collagen fibers. In contrast, explants with a high density of neurite outgrowth were largely radial on both aligned and nonaligned substrata. On the basis of these data, it was argued that at high neurite density contact inhibition was the dominant mechanism in neurite outgrowth, while contact guidance was the major guiding component at low neurite densities (Ebendal, 1976). These results could also be explained in terms of fasciculation, because at high neurite density the growth cones may adhere preferentially to the neurites of other cells.

Fasciculation is intimately associated with cell–substratum adhesion and contact guidance. Rarely in the vertebrate nervous system do axons extend for any distance as individual neurites; rather, tracts of nerves are made up of bundles or fascicles of nerve processes. The fibers within one bundle are all derived from the same type of nerve in terms of ontogeny and function. In addition, during the development of the nervous system the growth cones of homologous axons grow out along the axons of the first pioneer axons to leave the area of the nerve cell bodies. These observations suggest a selective adhesion mechanism in which the growth cones of like axons preferentially grow on the same cell type, and show that the side-to-side adhesion of the axons may also be cell-type specific.

Experiments with explants of rat neural retina and sympathetic ganglia support the arguments for selective fasciculation (Bray et al., 1980). The two types of neurites can be distinguished on the basis of catecholamine content. When small explants of sympathetic ganglia were placed next to each other in culture, the outgrowing neurites from the pieces of tissue grew into each other and formed common fascicles. Similar results were obtained with paired retina explants. However, when neural retina and sympathetic ganglia were placed adjacent to each other, the two types of axons did not interact, but rather avoided one another. There appeared to be a form of contact inhibition operating, because growth cones partially retracted and changed their direction of movement when they encountered axons of the opposite cell type. Finally, when retinal and ganglion cells were dissociated and allowed to reaggregate, the fiber outgrowth from the mixed aggregate formed separate bundles of retina and sympathetic axons. These data show that mechanisms for selective fasciculation exist, again suggesting that growth cones have the ability to recognize homologous axons and use this information to determine their path of growth.

In summary, complex interactions of growth cones are extremely difficult to study *in vivo,* but the mechanisms of neurite guidance may be amenable to study in various cell culture systems. It is clear that any detailed understanding of the molecular interactions involved in contact guidance, fasciculation, and contact inhibition of movement will have to be gleaned from studies on cultured cells. It is also likely that the relative contribution of each mechanism to axon elongation is dependent on the surfaces in the immediate environment of the outgrowing neurite. Given the choice of several substrata, neurites will probably grow on the most adhesive. If the strength of adhesion is great, then interference from other processes such as contact inhibition may be proportionally less. Contact inhibition and other cell–cell interactions become more important when the cell–substratum interactions of the growth cone are relatively weak.

3.6   Regulation of Nerve Differentiation by Exogenous Factors

It is believed that during the development of the nervous system a series of molecules are released from cells which alter the growth, migration, differentiation, and synaptic interactions of neurons. These molecules may be synthesized and released by the appropriate target for innervation to form a gradient to guide the incoming nerves and affect their differentiation, or they may have a more widespread distribution to affect the survival and growth of certain classes of nerve cells. A good example of the latter class of compounds is NGF. NGF is a protein which affects in a highly specific manner the survival and differentiation of a few nerve cell populations. It is to date the best-characterized molecule of this type in the nervous system. Its purification was dependent on a tissue culture assay for nerve differentiation (see Chapter 1 and later in this chapter). Since cultured cells were so successfully used for the characterization of NGF, it was believed that similar assays could be used to isolate the large number of "factors" similar to NGF which are likely to be present in the nervous system and its target organs. Although this experimental approach is still a promising area of research, substantial progress has been made with only one other factor—material which induces neurite outgrowth in primary cultures of parasympathetic ganglion neurons.

3.6.1. *Neurite Outgrowth in Primary Cultures.* As outlined above, the rate of neurite outgrowth can be regulated in a quantitative manner by varying cell–substratum adhesion. For a number of nerve cell types, however, an adhesive substratum is necessary for neurite extension, but not sufficient. For example, sympathetic neurons can be plated upon very adhesive substrata such as PLYS or PORN, yet they will not extend neurites unless NGF is present. Since NGF induces the differentiation of only a limited number of nerve cell types, investigators have undertaken a search for molecules which regulate the differentiation of other groups of cells. These experiments have been limited by the kinds of nerve cell populations that can be maintained in culture. To date sympathetic, parasympathetic, and motor neurons and heterogeneous cultures of brain neurons have been examined. The major source of materials that regulate the differentiation and survival of these nerve cells has been growth-conditioned medium from cultures of the cells that they innervate.

Cells release a large number of molecules into their culture medium, including various growth factors, cell-adhesion molecules, enzymes, and cell-surface components (see Chapter 6 for a complete discussion). Some of these extracellular macromolecules can cause the morphological differentiation of some nerve cells. The following paragraphs outline what is known about the material in heart muscle growth-conditioned medium that causes neurite extension in chick parasympathetic neurons.

When parasympathetic ciliary ganglion cells of 8-day chick embryos are dissociated with trypsin and placed in PORN-coated tissue culture dishes, the cells attach to the substratum within 15 min and extend filopodia around the circumference of the cell (Figure 3.11) (Collins, 1978a). Time lapse microcinematography reveals that the filopodia are in constant motion, extending and then retracting into the cell body with half-lives of less than a minute. If the cells remain in the initial plating medium containing 10% fetal calf serum, they die within 24 hr. However, when cells are exposed to medium in which embryonic chick heart cells have been grown for 48 hr, the nerve cells survive and neurite formation is initiated within 20 min (Helfand et al., 1976, 1978; Collins, 1978a). This effect is equally pronounced whether serum-free or serum-containing growth-conditioned medium is employed. By 20 min after the addition of heart-conditioned medium (HCM), some of the filopodia attach to the substratum and become relatively immobile. By 30 min new filopodia and membrane ruffles are localized to small areas where growth cones are forming, and by 1 hr the majority of the cells have extended neurites (Figure 3.11). Neurite extension and growth cone activity continues while movement around the rest of the cell body ceases. The rate of movement of the growth cone away from the cell body is ~50 μm/hr. The material in the conditioned medium which affects nerve differentiation is thought not to be NGF for the following reasons (Helfand et al., 1978): (1) Ciliary ganglia do not respond to NGF. (2) An antiserum against NGF does not block the effect of HCM. (3) Conditions which inactivate NGF (e.g. pH 1.5) do not reduce HCM activity.

The response of ciliary neurons to HCM is dependent on two developmental parameters—the age of the heart muscle from which the neurite-promoting material is isolated and the developmental stage of the ciliary neurons. The biological activity of chick heart cell extracts in promoting embryonic ciliary ganglion survival and neurite production increases approximately 10-fold from embryonic day 8 until hatching (day 21), and it remains at the high level for at least one more week (Hill et al., 1981). When ciliary ganglia from Stage 33 embryos (approximately 12 days of incubation) are placed in culture in HCM, over 80% of the neurons initiate neurite outgrowth. Under identical conditions, less than 20% of the cells from Stage 40 neurons (14 days of incubation) begin to grow neurites within the first 72 hr of culturing (Collins and Lee, 1982). There is thus a decrease between embryonic days 10 and 14 in the ability of ciliary ganglia neurons to initiate outgrowth. These times correlate well with the period in which ciliary neurons are forming synapses with their peripheral targets. Chick DRG cells also lose their ability to extend neurites in culture with increasing embryonic age, but do so with a slightly different time course from that of ciliary ganglia cells (Greene, 1977). There are two alternatives to explain the loss of responsiveness; neither has been ruled out. Either there is a loss of capacity of the older neurons to initiate rapid neurite outgrowth under any conditions, or the proper inducing material is missing from the culture medium (Collins and Lee, 1982).

In addition to acting on sympathetic, parasympathetic, and sensory ganglia cells, HCM apparently induces neurite outgrowth from spinal cord, retina, and optic tectum neurons (Obata and Tanaka, 1980). These results indicate that essentially all types of nerve cells respond morphologically to molecules in HCM. Furthermore, a wide variety of cell types release material into their culture media that causes neurite outgrowth. Conditioned media from rat skeletal muscle, fibroblasts, and lung cells all increased neurite formation in organ cultures of rat spinal cord (Dribin and Barrett, 1980). In the absence of conditioned medium, the short neurites extending from the pieces of spinal cord were dense and tangled. In the presence of conditioned medium, the neurites were twice as long and extended radially from the piece of tissue. Skeletal-muscle-conditioned medium also increased the activity of CAT, the enzyme which synthesizes ACh, in spinal neuron cultures approximately 10-fold relative to normal medium (Giller et al., 1977). Finally, an examination of conditioned media from nearly 100 different cell cultures for their ability to induce neurite outgrowth from embryonic chick ciliary ganglia showed that this activity was widely distributed among both neuronal and non-neuronal cells (Adler et al., 1981). The neurite-promoting activity was found in many rat and human cell lines, including fibroblastlike cells, hepatomas, and kidney cells, but was not found in any of the 10 lymphoid cell lines tested. Since the lymphoid lines grow in suspension and cells which make neurite-promoting factors all grow attached to the substratum, there is a correlation between the adhesive properties of cells and their ability to make neurite-promoting material. Neurite-promoting material has also not been found in conditioned media from chick CNS, sympathetic, and parasympathetic nerve cell cultures, all of which are only weakly adherent to the culture substrata. These results suggest that there is a very limited cell-type specificity with respect to the synthesis and release of neurite-promoting materials.

What is the nature of the material in growth-conditioned medium and how does it promote neurite extension? Three effects of growth-conditioned media on the differentiation and survival of ciliary neurons have been observed: (1) initiation of neurite outgrowth (initiation factor); (2) nerve cell survival (survival factor), and (3) an increase in the rate of neurite extension over that produced by material which initiates neurite extension (rate factor).

When embryonic ciliary ganglion cells are dissociated with trypsin, allowed to recover their surface molecules which were removed by trypsin, and plated on PORN-coated tissues culture dishes in the presence of HCM, they initiate neurite outgrowth within 20 min. The neurite-inducing material in the conditioned medium must attach to the substratum of the culture dish to function (Collins, 1978b). This was demonstrated by exposing PORN-coated culture dishes to HCM, washing the dishes extensively, and then plating dissociated ciliary ganglion cells. The cells were also plated into dishes to which HCM had been added recently. Figure 3.15 shows that neurite outgrowth was initiated much more rapidly when the substratum was pre-coated with HCM than when complete HCM was used. The material bound to the substratum that was essential for neurite outgrowth could be removed from HCM by sequential passage over a number of PORN-coated culture dishes. The material in HCM responsible for neurite extension does not bind to normal tissue culture dishes or to collagen-coated dishes. The neurite-initiating factor in HCM can also direct neurite outgrowth. When the material is painted

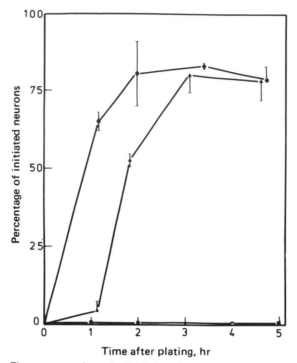

Figure 3.15. Time course of neurite initiation. Ciliary ganglion cells were plated on PORN-coated culture dishes in whole HCM, on dishes pretreated with HCM, or on PORN-coated dishes alone. The percentage of all cells that had initiated neurites was plotted as a function of time after plating. ▲, HCM on an untreated, PORN-coated dish; ●, PORN-coated dish previously exposed to HCM, washed, and plated in the absence of HCM; ■, control: PORN-coated dish alone. From Collins (1978b).

on a PORN-coated substratum near ciliary ganglion cells, the growing neurites follow the painted pathway exactly (Figure 3.16) (Collins and Garrett, 1980). The chemical nature of the material in HCM which adheres to the substratum and initiates ciliary ganglion cell neurite growth has not been defined. Preliminary evidence suggests, however, that it may be a macromolecular complex containing proteoglycans (F. Collins, personal communication; Lander et al., 1982). Large glycoprotein complexes containing proteoglycans and proteins have been isolated from growth-conditioned medium of muscle cells and termed adherons; they are able to induce neurite extension in a variety of clonal cell lines (LeBeau et al., 1984).

The fact that dissociated ciliary ganglion neurons plated on PORN-coated substrata in normal serum-containing media survive for less than

Figure 3.16. Directed outgrowth of ciliary neurons. An 8-day chick embryo ciliary ganglion was placed at the bottom of a long narrow area of the tissue culture dish which was coated with partially purified substratum-conditioning factor from HCM. The growth of the neurite was largely confined to the area covered with the factor (see Collins, 1978*b*). This photograph is from unpublished material prepared by Frank Collins.

24 hr has led to the development of an excellent assay system for the detection and isolation of factors which promote neuron survival beyond 24 hr. Applying growth-conditioned media from heart muscle cultures to ciliary ganglion cultures enables the cells to survive for several weeks; complete HCM also stimulates neurotransmitter synthesis (Helfand et al., 1976; Collins, 1978*b;* Nishi and Berg, 1979). The survival factor and the neurite-initiating factor are distinct, for the two activities can be separated from each other by quantitative adsorption of the neurite-initiating material from growth-conditioned medium to PORN-coated substrata, which leaves all of the survival factor in the supernatant (Collins, 1978*b;* Adler and Varon, 1980). Medium which has been depleted of the neurite-initiating material is able to support the long-term viability of the neurons but unable to induce neurite outgrowth. The molecular nature of the survival-promoting material in HCM is not known.

An apparent third activity in heart-conditioned medium increases the rate of neurite extension by ciliary neurons approximately threefold within the first hour of neurite outgrowth (Collins and Dawson, 1982). When neurite elongation is monitored by time lapse photography, neurite outgrowth is seen to be much more extensive when complete growth-conditioned medium is present than when only the substrate-attached neurite-initiation factor is used in conjunction with fresh culture medium. The rate-promoting factor is not removed when all of the neurite-initiating material is adsorbed by passage over a number of PORN-coated culture dishes, and like the survival factor, the rate-promoting factor does not itself induce neurite extension. The chemistry of this material which increases the rate of neurite elongation has remained elusive. That the survival and rate promoting activities are identical has not been ruled out.

3.6.2. *The Role of Cyclic AMP in Nerve Differentiation.* Cyclic nucleotides appear to play a central role in the regulation of many biological processes, including cellular division, morphological differentiation, and enzyme activity (see Friedman, 1976, and Willingham, 1976, for reviews). In particular, cyclic adenosine 3',5'-monophosphate (cyclic AMP) is thought to be an intracellular "second messenger" for many hormonally regulated changes in cellular biochemistry (Robison et al., 1968). Some hormones interact with their membrane receptors to activate adenylate cyclase, the enzyme which synthesizes cyclic AMP from ATP. The resulting elevation of intracellular cyclic AMP probably affects the activities of cyclic-AMP-dependent protein kinases, which, in turn, regulate many aspects of cellular metabolism. For example, glycogen stimulates cyclic AMP synthesis in the liver, and the cyclic AMP activates phosphorylase (the first and rate-limiting step in glycogen break-down) and inactivates glycogen synthetase, resulting in an increased degradation of glycogen to glucose. Through the use of cultured cells, the widespread involvement of cyclic AMP in processes other than en-zyme regulation has been elucidated.

The initial observations on cultured cells were made simultaneously by two groups who showed that an active derivative of cyclic AMP, N^6, O^2 dibutyryl cAMP (dibutyryl cyclic AMP) altered the shape and growth pattern of transformed fibroblasts from a polygonal morphology and a random cell distribution on the culture dish to the elongated cells grow-ing in parallel arrays characteristic of normal fibroblasts (Hsie and Puck, 1971; G. Johnson et al., 1971). These experiments stimulated a great deal of interest in the role of cyclic AMP in cell division, morphology, and transformation (see Friedman, 1976, for a review). The massive amount of data generated through the mid-1970's showed that in most (but not all) cases, cyclic AMP levels were (1) high in confluent, contact-inhibited cells, (2) low in exponentially dividing cells, and (3) lower in transformed cells than in their "normal" counterparts. If the intracellular levels of cyclic AMP were experimentally raised or lowered, there was inhibition or stimulation, respectively, of cell division. Although the above relation-ships between cell growth and cyclic AMP levels are generally true for fibroblastlike cells, completely opposite correlations have been re-ported in other cell types. For example, cyclic AMP stimulates division of some glial cells (Raff et al., 1978a).

Since the dibutyryl derivative of cyclic AMP is frequently used to change the intracellular cyclic AMP content of cells artificially, some-thing should be mentioned about its mode of action and possible arti-facts. Because the stability and membrane permeability of cyclic AMP in culture medium may be limited, more stable and more lipid-soluble derivatives, such as dibutyryl cyclic AMP, are frequently employed. Di-butyryl cyclic AMP is effective in culture medium for several days. It is hydrolyzed in the cell through both deacylation and phosphodiester bond cleavage (O'Neill et al., 1975). Both the N^6 and O^2 monobutyryl derivatives are produced, but the O^2 compound is rapidly hydrolyzed by phosphodiesterase. The N^6 derivative is not as good a substrate for phosphodiesterase, and is the major breakdown product found in cells. It probably exerts its effects on cells by mimicking cyclic AMP and by elevating intracellular cyclic AMP by acting as a phosphodiesterase inhibitor. Butyric acid, a breakdown product of dibutyryl cyclic AMP, can produce some of the effects of cyclic AMP in cultured cells. For example, it can cause reverse transformation in fibroblasts (Storrie et al., 1978) and induce hemoglobin synthesis in erythroleukemic cells (Reeves and

Cserjesi, 1979). Although this hydrophobic fatty acid may have a variety of effects on membrane structure, many of the similarities between responses to cyclic AMP and those to butyric acid are due to butyric acid's ability to increase intracellular cyclic AMP (Storrie et al., 1978).

In addition to regulating cell division, cyclic AMP may have a critical role in the determination of cell shape and adhesion. In transformed fibroblastlike cells, exogenous dibutyryl cyclic AMP converts cells that are round, poorly spread on the substratum, and covered with surface blebs to flattened, elongated, blebless cells (Bloom and Lockwood, 1980). The morphological change of transformed cells to a more normal phenotype, termed reverse transformation, is accompanied by an appearance of stress fibers within a few hours after the addition of dibutyryl cyclic AMP. There are a variety of experiments that show that cyclic AMP directly affects the change in fibroblast morphology and that the phenotypic change is not due simply to the indirect effects of the reagents used to elevate intracellular cyclic AMP (Willingham, 1976).

Exogenous cyclic AMP usually increases the adhesiveness and inhibits the translational mobility of fibroblasts (G. Johnson et al., 1972). However, if cells are exposed to a gradient of cyclic AMP, differences in the local cyclic AMP concentration may occur across the cell, resulting in directed migration, as opposed to the complete inhibition of movement observed with a global increase.

Finally, dibutyryl cyclic AMP promotes the extension of existing cell processes in fibroblastlike cells, in part through the inhibition of their spontaneous retraction rate (Willingham and Pastan, 1975). These morphological changes do not require new RNA or protein synthesis, and occur equally well in enucleated cells (Schroder and Hsie, 1973).

All of the above morphological and ultrastructural changes induced by cyclic AMP and its analogues in fibroblastlike cells have also been observed in cultured nerve cells. Cyclic AMP induces neurite outgrowth in essentially all cultured nerve cells which have been examined, and the overall response of nerve cells to cyclic AMP is very similar to that of fibroblastlike cells (Table 3.2). Cyclic AMP, its active derivatives, and agents which elevate intracellular cyclic AMP all inhibit cell division and stimulate neurite formation. In nerve cells the production of long processes or neurites has been equated with cellular differentiation, the morphological change from a round, neuroblastlike cell to a morphologically complex cell with one or more neurites. The observation that dibutyryl cyclic AMP and other agents which increase intracellular cyclic AMP also promote neurite outgrowth in clonal nerve cells was made in 1971 (Furmanski et al., 1971; Prasad and Hsie, 1971). These experiments were done with the C1300 mouse neuroblastoma, but have since

Table 3.2. Similarities in the Responses of Fibroblastlike Cells and Nerve Cells to Dibutyryl Cyclic AMP

Response to Dibutyryl Cyclic AMP	Cell Type					References
	Fibroblast	C1300 Neuroblastoma	Primary CNS Nerve	PC12	Human Neuroblastoma	
Process or neurite outgrowth	+	+	+	+	+	Prasad and Hsie (1971); Willingham (1976)
Increased cell–substratum adhesion	+	+	?	+	?	Willingham (1976); Schubert et al. (1971, 1978)
Inhibition of cell division	+	+	NA[a]	+	+	Willingham (1976); Schubert et al (1978); Bottenstein (1981).
Microtubule and microfilament organization	+	+	+	+	+	Willingham (1976); Luckenbill-Edds et al. (1979); Schubert, unpublished
Inhibition of net cell translocation	+	+	?	+	?	Willingham (1976); Schubert, unpublished
Induction of ornithine decarboxylase	+	+	?	+	?	Bachrach (1975); LaCorbiere and Schubert (1981); Rumsby and Puck (1982)

[a] NA, not applicable; these cells normally do not divide. PC12 is a sympathetic nervelike cell line. The responses of five rat neuroblastoma cell lines are like those of the mouse C1300 neuroblastoma.

been extended to essentially all cultured cells of neuronal origin. These include human neuroblastomas (Prasad and Mandal, 1972), five clonal nerve cell lines from the rat (Schubert, 1974), and primary cultures of CNS nerve cells (Shapiro, 1973; Obata, 1981) and dorsal root ganglia (Roisen et al., 1972). In addition to exogenous cyclic AMP, such reagents as phosphodiesterase inhibitors and prostaglandins E_1 and E_2 also stimulate neurite outgrowth. Serum deprivation stimulates neurite outgrowth in C1300 mouse neuroblastoma cells (Seeds et al., 1970); this response is probably due to the elevation of cyclic AMP caused by low-serum medium (Sheppard and Prasad, 1973). As in fibroblasts, the initial cyclic-AMP-induced morphological changes in nerve cells do not require RNA or protein synthesis and occur in enucleated cells (Miller and Ruddle, 1974).

The ubiquitous effect of cyclic AMP in promoting nerve differentiation suggests that it may be acting as a classical "second message" for exogenous factors which regulate nerve differentiation *in vivo*. To date only one molecule has been purified that promotes the differentiation of a subset of nerve cells; that is NGF. During the past few years a compelling amount of experimental evidence has been obtained that suggests that cyclic AMP is a functional intermediate in the NGF response. In addition, the "common denominator" of a wide variety of reagents that induce the differentiation of neuroblastoma cells also appears to be cyclic AMP.

3.6.3. The Differentiation of C1300 Neuroblastoma. If the differentiation of nerve cells is defined as an increase in morphological and ultrastructural complexity, then a wide variety of reagents and growth conditions are able to induce the differentiation of clonal nerve cell lines such as the C1300 mouse neuroblastoma (Table 3.3). The variety of "inducers" makes it unlikely that they all directly affect gene expression to cause the morphological changes in the cells. The most likely explanation is that they individually stimulate a common pathway leading to the changes in phenotype. For example, all of these experimental manipulations may decrease the translational mobility of proteins in the membrane and therefore increase the stability of neurites once they are formed. Dimethyl sulfoxide and butyric acid both induce erythropoiesis in a clonal cell line of Friend-virus-transformed erythroleukemia cells; they also induce neurite outgrowth in the C1300 neuroblastoma cells. Both compounds increase the phase transition temperature of the plasma membranes of Friend cells, indicating that there is a decrease in fluidity and an increase in stability (Lyman et al., 1976). Similar mechanisms may be involved in nerve differentiation. Another alternative,

Table 3.3.   Induced Differentiation of Cells

Inducer	Cell Division	cAMP Content	References
Adhesive substratum	No change	?	Schubert et al. (1969)
Serum deprivation	Decrease	Increase	Seeds et al. (1970); Slepp and Prasad (1973)
Growth-conditioned medium	No change	?	Schubert et al. (1971)
Cyclic AMP	Inhibits	Increase	Prasad and Hsie (1971) Furmanski et al. (1971)
Prostaglandin E_1	Inhibits	Increase	Gilman and Nirenberg (1971)
Cholera toxin	Inhibits	Increase	Dawson (1979)
Dimethylsulfoxide	Inhibits	?	Kimhi et al. (1976)
Sodium butyrate	Inhibits	Increase	Prasad and Sinha (1976)
Hexamethylene bisacetamide	Inhibits	?	Storrie et al. (1978)
Bromodeoxyuridine (BUdR)	Inhibits	No change	Prasad et al. (1973); Schubert and Jacob (1970); Giotta et al. (1980a)
X-Irradiation	Inhibits	Increase	Prasad et al. (1973)
Substance P	Inhibits	Increase	Narumi and Maki (1978)

which is not incompatible with the first, is that all of the conditions favoring differentiation increase intracellular cyclic AMP. As indicated in Table 3.3, most of the conditions that induce the morphological differentiation of C1300 clones also increase levels of intracellular cyclic AMP. These include serum deprivation and the addition of prostaglandin E_1, sodium butyrate, or substance P. With one exception, the other conditions which promote neurite outgrowth have not been examined with respect to cyclic AMP metabolism at times immediately after the alteration of the growth medium. The exception is exposure to BUdR, after which no change in intracellular cyclic AMP has been detected. It is possible, however, that a rise in intracellular cyclic AMP is caused by BUdR but is too small or too rapid to detect. The less likely alternative is that BUdR bypasses the cyclic AMP second message and directly affects nerve differentiation. Exogenous dibutyryl cyclic AMP and serum deprivation also cause the morphological differentiation of five clonal nerve cell lines from the rat CNS; other chemical inducers have not been examined. Finally, virtually all of the conditions listed in Table 3.3 cause neurite outgrowth in human neuroblastomas (see Bottenstein, 1981, for a review).

Changes in two additional parameters have been associated with the morphological differentiation of C1300 cells: an alteration in the chemistry of the plasma membrane and the induction of certain enzymes. Brown (1971) was the first to demonstrate that a new glycopeptide appeared on the surface of neuroblastoma cells after they were exposed to BUdR, and since then numerous reports have shown that the chemical composition of the plasma membrane differs between the round neuroblastoma cells and the differentiated cells with elaborate neurites (see, for example, Truding et al., 1974; Zisapel and Littauer, 1979). Although these experiments were done using different techniques, making it impossible to reach a consensus about which proteins in the membrane change and by how much, there is no doubt that the morphological differentiation of mouse neuroblastoma cultures is accompanied by changes in the plasma membrane. The morphological differentiation of C1300 cells is also accompanied by increases in the specific activities of several enzymes. It is, however, necessary to distinguish between alterations in enzyme activity that are directly coupled to nerve differentiation and those that are caused by the reagent used to induce differentiation but are not necessarily related to the developmental changes.

The C1300 mouse neuroblastoma clones contain large amounts of acetylcholine esterase (AChE), the enzyme which hydrolyses ACh to acetate and choline. The specific activity of this enzyme invariably increases when C1300 cells are subjected to conditions which induce the morphological differentiation of the cells (see Bottenstein, 1981, for a review), and it has been assumed that the induction of AChE is a valid marker for nerve differentiation. However, since conditions which induce nerve differentiation also inhibit cell division, two possibilities exist concerning the increase in AChE activity: (1) It is a result of and is necessarily coupled with morphological nerve differentiation. (2) It is associated only with the cessation of cell division.

Neurite formation can be induced in neuroblastoma cells by plating the cells in tissue culture dishes, either in media containing 10% serum or in low-serum media. In contrast, cells plated in Petri dishes or in agar-coated tissue culture dishes neither attach nor extend neurites. The rate of cell division is identical in all of these conditions. If the cessation of cell division is solely responsible for the increase in AChE activity, cells maintained in identical growth conditions should show similar increases in enzyme activity independent of the substrate on which they were plated. If, however, enzyme activity is a function of neurite formation (morphological differentiation), the enzyme activity in cells plated on agarless tissue culture dishes should be different from that of cells plated on agar-coated dishes.

To distinguish between these alternatives, exponentially growing cells in suspension were plated in 1% serum in tissue culture dishes and in agar-coated dishes. Cell number, percentage of cells with neurites longer than 40 μm, protein per cell, and the specific activity of AChE were then followed as a function of time. Figure 3.17c shows that there is a faster initial increase in the specific activity of AChE in cells extending neurites than in the round-cell population; cell number and viability were identical in both populations. Since there was an approximately fourfold increase in the amount of protein per cell in the population possessing neurites after 4 days (Figure 3.17d), there must be an approximately 20-fold difference in enzyme activity per cell between the two populations. This strongly supports the hypothesis that attachment to a surface, and not just cessation of cell division, exerts a strong influence in the control of protein synthesis in differentiating neuroblasts. These results do not, however, establish that AChE is controlled during neurite formation by a

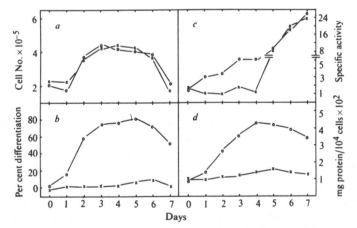

Figure 3.17.   Induction of AChE in neuroblastoma cells. Exponentially growing cells were dissociated and plated in adhesive tissue culture dishes (O) or in nonadhesive agar-coated dishes (×). Cell number (a), percentage of cells with neurites (b), the specific activity of AChE (c), and the protein per 10⁴ cells (d) were then determined as a function of number of days in the two culture conditions. From Schubert et al. (1971).

mechanism other than neurite growth *per se,* for if most of this enzyme is membrane-bound, then the observed increase in esterase accompanying neurite formation may result simply from the increased ratio of membrane (surface area) to cytoplasm.

The increase in AChE activity can be divided into two phases (Figure 3.17c). The first, relatively slow increase is correlated with neurite formation and the concomitant increase in protein per cell; the later, more rapid increase in specific activity is accompanied by a loss of viable cells and a general deterioration of the cells in culture. This latter rapid increase in esterase activity is not a result of neuroblast differentiation, for the rates of induction are equal in both attached cells with neurites and the round-cell suspension culture.

Thus, increased AChE activity in mouse neuroblastoma cells is correlated with two separable events. The first is the morphological differentiation of the cells, defined by the outgrowth of neurites, and the second is the aging of the culture and eventual cell death. An increase in AChE *per se* cannot be used as a marker for nerve differentiation. A similar conclusion was reached through the study of neural crest differentiation (Ziller et al., 1983).

3.7 A Nerve Growth Factor

The peripheral nervous system contains two groups of cells localized in ganglia along the length of the spinal cord. These are the sensory neurons, which transmit information from the sense organs to the brain, and the sympathetic neurons, which innervate the circulatory system, organs, glands, and smooth muscles. In 1948, Elmer Bueker was examining the effects of altering the peripheral fields of motor neurons and the spinal ganglia by ablation of chick embryo limb buds followed by replacement with pieces of tissue derived from mouse and avian tumors. An examination of the embryos showed that the sensory ganglia adjacent to a piece of mouse sarcoma 180 had grown extensively into the neoplastic tissue and that the sensory ganglia themselves were greatly enlarged. A repetition of this experiment a few years later showed that the effect of the tumor was much greater than originally reported (Levi-Montalcini and Hamburger, 1951). Both the sensory and sympathetic ganglia were enormously enlarged relative to the contralateral ganglia, and the viscera of embryos exposed to the tumor contained large numbers of sympathetic nerve fibers. These data suggested that the tumor cells were releasing some compound which greatly enhanced the growth of sympathetic and sensory nerve fibers. To isolate this hypothetical factor, however, a simple assay that did not involve the use of whole embryos was needed.

A cell culture assay was devised by placing sensory or sympathetic ganglia from 8-day chick embryos in semi-solid culture media (Levi-Montalcini et al., 1954). When the explants were cultured with pieces of sarcoma 180, there was extensive neurite outgrowth, but only minimal outgrowth occurred in controls (see Figure 1.1). Since this cell culture assay takes less than 24 hr to complete, it became possible to begin purification of the factor and to screen alternative sources for the activity. Much larger amounts of the activity were discovered in snake venoms and in an organ phylogenetically related to the poison gland of snakes, the submaxillary salivary glands of mice (S. Cohen and Levi-Montalcini, 1956; S. Cohen, 1958). Despite an extensive search for richer sources, the mouse submaxillary gland has remained the major source for the protein which is now known as nerve growth factor (NGF).

The most direct evidence that NGF plays a significant role in the development and maintenance of the sympathetic nervous system is the destruction of the sympathetic ganglia caused by the injection of anti-NGF antibody into neonatal rodents (Levi-Montalcini and Booker, 1960). The destructive effects of antibodies against NGF on immature sympathetic nerve cells could be due to the inactivation of endogenous NGF or to the antibody-mediated cytolysis of the target cells. It is likely that immune cytolysis is not involved, as (1) anti-NGF does not destroy adult neurons even though their receptor density is similar, (2) the effects of the antisera in young animals can be reversed by injecting excess NGF up to 2 days after introducing the antibody, and (3) the effects of the antibody were observed in complement-deficient mice.

The destruction of sympathetic cells by the antisera is age dependent, for the proportion of nerves that degenerate following the administration of the antiserum decreases with age, becoming negligible in adults. NGF is also required for the survival of sympathetic and sensory ganglion cells in culture, and this dependency also decreases with increasing age of the donor animals (Lazarus et al., 1976). The NGF requirement of sensory cells in culture may, however, be replaceable by other growth factors *in vivo,* for exposure of neonatal animals to anti-NGF sera does not cause the degeneration of sensory ganglia.

Positive effects of NGF have been demonstrated by injection of NGF into young animals. NGF increases the amount of tyrosine hydroxylase and dopamine β-hydroxylase in sympathetic ganglia, and also increases the size of the ganglia and the extent of innervation of their target tissues. Although NGF increases the size of sympathetic and sensory ganglia of chick embryos, it does not increase the mitotic rate of neuronal precursors. Rather, exogeneous NGF promotes the survival of neurons which would normally die, and it also increases the diameters of the nerve cell bodies.

3.7.1. *Molecular Structure, Synthesis, and Tissue Distribution.* The

NGF activity in mouse submaxillary glands is stored as a multimeric protein with a sedimentation value of 7S. It consists of three subunits with a stoichiometry of $\alpha_2\gamma_2\beta$, and it retains 1 or 2 atoms of zinc (Varon et al., 1967). The β-subunit, which has a sedimentation value of 2.5S, contains all of the neuroendocrine activity and is itself a dimer of two identical subunits of molecular weight 13,259 (Angeletti and Bradshaw, 1971). The two β-NGF monomers, each containing 118 amino acids and three intrachain disulfide bonds, are held together by noncovalent bonds. They have a net positive charge at neutral pH, making them nonspecifically adhesive to negatively charged substrata such as glass and plasma membranes. The β-NGF dimer is very stable, with a dissociation constant of 10^{-13} M (Bothwell and Shooter, 1978). An initial comparison of the amino acid sequence of β-NGF with the A and B chains of insulin suggested that the two proteins have a common ancestral gene. Although this contention is still controversial (Thomas and Bradshaw, 1980), insulin and NGF do have very similar effects on the physiology of the NGF-sensitive neuron (see Section 3.7.2).

Both the α- and γ-subunits of 7S NGF bind noncovalently to β-NGF but not to each other. The α-subunit has a molecular weight of about 26,000 and is composed of two chains of 17,000 and 9,000 daltons, and aside from a possible role in stabilizing the 7S complex, its function is unknown. In contrast, the γ-subunit has been shown to be an endopeptidase with a specificity for arginine (Greene et al., 1968). The function of this peptidase activity may be to convert a high-molecular-weight precursor of β-NGF to the 2.5S β-NGF dimer. Finally, in the presence of chelating agents the dissociation constant of the α- and γ-subunits from the 7S complex is about 10^{-7} M. However, in the presence of zinc ions at neutral pH, the dissociation constant of 7S NGF decreases to 10^{-12} M (Bothwell and Shooter, 1978). The 7S NGF complex is apparently a stable storage complex which protects the β-NGF from proteolytic degradation. The 7S complex dissociates once it has been secreted from the submaxillary gland, and the β-subunit becomes bound to α_2-macroglobulin in plasma (Ronne et al., 1979).

If submaxillary glands are labeled with [^{35}S]methionine for several hours, a 13,000-dalton protein can be detected after specific precipitation with anti-NGF antibody; it comigrates with authentic β-NGF on SDS–

acrylamide gels and contains NGF peptides. However, if the tissue is pulse-labeled with isotope for a few minutes, a 22,000-dalton protein is immune precipitated, the radioactivity of which can be chased into the 13,000-dalton β-NGF subunit. This high-molecular-weight pro-NGF contains all of the β-NGF peptides, which strongly suggests that it is a synthetic precursor to β-NGF (Berger and Shooter, 1977). It therefore appears that the β-subunit, and perhaps the others as well, is synthesized as a higher-molecular-weight precursor and then proteolytically cleaved to generate the final product.

A most significant question regarding the function and mode of action of NGF within the animal revolves around the site of its synthesis. For example, do the target tissues for sympathetic and sensory innervation synthesize NGF and use this NGF to attract incoming nerve fibers and maintain the nerve population once synapses are formed, or is NGF carried from its site of synthesis to the appropriate target via the circulatory system? There is evidence that supportive cells such as glia make NGF. All of the alternatives are controversial, for to date the only unambiguous site of NGF synthesis is a target tissue for sympathetic innervation, the mouse submaxillary gland.

There is voluminous but only circumstantial evidence that NGF is synthesized by other target tissues (for a discussion, see Thoenen and Barde, 1980). The difficulty with the experiments generating this evidence is that very small amounts of NGF-like activity are present, and the immunological techniques used to detect the NGF are frequently suspect. NGF activity was discovered in a mouse tumor, and it has subsequently been demonstrated that many rodent tumors and a variety of permanent clonal cell lines derived from neoplastic tissue synthesize NGF-like activities. However, a comparison of NGF production by the cells in culture and those *in vivo* indicates that the tissues in the animal from which the cultured cells are derived do not synthesize NGF. This suggests that NGF synthesis in culture may be a response to the culture conditions rather than a valid reflection of the normal phenotype. It follows that either the NGF levels in the target organs are too low to detect by current techniques or no NGF is present. If the latter is correct, then the significance of the apparent transport of NGF from the target organ to the nerve cell body is in question.

The transfer of information between the target organ and the innervating neuron takes place in both directions along the axon. For example, the electrical activity of motor neurons flows from the nerve cell body to the muscle, but the volume of the peripheral muscle tissue which is innervated determines the extent of development and the number of motor neurons which survive within the spinal cord. The retrograde transport of NGF from the tissue which it innervates to the nerve cell body is thought to play a role in the maintenance of sympathetic and sensory neurons. This conclusion is based on experiments conducted both *in vivo* and in tissue culture systems. When NGF isotopically labeled with ^{125}I is injected unilaterally into the anterior eye chamber of rodents, the majority of the isotope accumulates in the superior cervical ganglion which innervates the iris of the injected side (for a review, see Thoenen and Barde, 1980). The isotope which accumulates in the ganglion is associated mostly with NGF, according to molecular weight on acrylamide gels and reactivity with antibodies specific for NGF. The transport of the NGF is apparently specific, for it is dependent on NGF surface receptors and independent of nerve activity. Essentially all macromolecules can be endocytosed by active nerve terminals, for the exogenous molecules are taken up during the process of the retrieval of synaptic vesicle membrane at the terminal. In contrast, NGF is taken up by nerve terminal in a manner which is independent of synaptic activity. Electron-microscopic autoradiography has demonstrated that NGF is localized to small vesicles and other smooth membranous structures in the axon and nerve terminals. In the cell body the transported NGF is found in lysosomal vesicles and other vesicular structures; none is found free in the cytoplasm or associated with the nucleus. Finally, there is an increase in the levels of tyrosine hydroxylase in superior cervical ganglion neurons the nerve terminals of which have been exposed to NGF. It should be pointed out, however, that while the data on the retrograde transport of NGF are interesting, the experimental evidence is not compelling. For example, it has not been shown directly that the retrogradely transported NGF is biologically active, nor has the possible involvement of a second message, such as cyclic AMP, been ruled out.

The role of NGF in the maintenance of long sympathetic axons has also been studied in culture (Campenot, 1977). Dissociated sympathetic neurons from newborn rats were placed in the center chamber of a three-chamber culture system that allowed the outgrowth of neurites from cells in the center chamber to the side chambers through a fluid-impermeable barrier. When NGF was present in all three chambers, neurites grew into both side chambers. However, neurites never grew into chambers containing no NGF. To determine whether the continual presence of NGF was required for the outgrowth of neurites, nerve processes were allowed to grow into chambers containing NGF, and the NGF was then removed from one of the side chambers containing neurites but no cell bodies. On the removal of NGF the neurites stopped elongating, and either degenerated or retracted within 2 weeks after the withdrawal of NGF. This occurred even when NGF was present in the chamber which contained the cell bodies and the proximal portions of the neurites. In the converse experiment, NGF was withdrawn from the chamber containing the cell bodies and proximal neurites but left in the side chambers containing the distal neurites and their growth cones. In this case the cells survived the withdrawal of NGF, but only if their neurites had crossed into the chambers containing NGF. These results show that the continued presence of NGF is required for the maintenance of neurites, an observation consistent with the conjecture that target organs supply NGF to maintain their synaptic interactions. Although NGF is clearly required for the survival of these cultures, it is possible that the disappearance of distal neurites on the withdrawal of NGF was due to the requirement for NGF for cell–substratum adhesion of the growth cones (see Section 3.5.4).

3.7.2. *Mechanism of Action of NGF.*   Although NGF has been studied for many years, until recently very little information was available concerning the molecular mechanisms which lead to the morphological and biochemical responses to NGF of sympathetic and sensory neurons. This dearth of information was due at least partially to the lack of a suitable experimental system. Explanted ganglia are very heterogeneous in cell type, and dissociated sympathetic ganglia generate relatively few cells. Since ganglia isolated from young animals need NGF for survival and older ganglia grow neurites in the absence of NGF, it has been difficult to determine the immediate effects of NGF on a naive cell. This problem was partially surmounted by the establishment of a clonal cell line called PC12 (Greene and Tischler, 1976). These cells were adapted to culture from a neoplasm called a pheochromocytoma, a spontaneous tumor thought to be derived from the adrenal medulla chromaffin cell. Although the pathologists may have classified it as a chromaffin-cell tumor, the PC12 cell line derived from the tumor more closely resembles sympathetic nerve cells than it does chromaffin cells. It is unlikely that much could be learned about the NGF response without the services of PC12 or a clonal cell line similar to it. It should be stressed again that results with clonal cells must be confirmed, when possible, with data obtained from *in vivo* experiments and with primary cultures. However, if the physiological response of the cells in culture is the same as that *in vivo,* then the underlying chemistry is almost certainly the same. The following paragraphs outline what is known about the interaction of NGF with responsive cells that results in the cells' morphological and biochemical differentiation.

The initial interaction of most peptide hormones is with a cell-surface receptor specific for that hormone. This is also the case with NGF. Using

[^{125}I]NGF which had full biological activity (iodination frequently alters the functional properties of proteins), it was shown that NGF binds to chick sensory ganglia at two independent sites. Site I has a K_d of 2.3 × 10^{-11} M and Site II a K_d of 1.6 × 10^{-9} M (Sutter et al., 1979). NGF dissociates from Site I in 10 min and from Site II in a few seconds. The associations of NGF with the two sites are diffusion limited and require exogenous calcium ions. In contrast to chick sensory neurons, the binding of NGF to the clonal PC12 sympathetic nerve cell line is calcium independent, but there may be two distinguishable NGF-binding sites (Landreth and Shooter, 1980; Schechter and Bothwell, 1981). The 2.5S β-NGF subunit binds directly to the NGF receptor, for if the 7S NGF complex is maintained as 7S by using an excess of γ- and β-subunits, it does not bind; neither does covalently cross-linked 7S NGF. Since 7S NGF binds to cultured cells, it must dissociate spontaneously.

When [^{125}I]NGF is incubated with sensory or sympathetic nerve cells for more than a few minutes, the bound NGF loses its ability to dissociate from its receptor and becomes insensitive to exogenous trypsin (Olender and Stach, 1980). This tight binding of NGF has been termed sequestration and occurs primarily at the high-affinity sites. Sequestration must be dependent on energy, for it is inhibited by dinitrophenol and sodium fluoride. It is not, however, temperature dependent, because it can occur at 4°C. The sequestration of NGF could result from either a change in the NGF–receptor complex, such as covalent cross-linking, or the internalization of NGF. Internalization of NGF is temperature dependent (Levi et al., 1980). It is therefore likely that sequestration is a result of a physical rearrangement of the NGF–receptor complex within or near the plasma membrane. Since the high-affinity (Type I) site may be associated with the cytoskeleton (Schechter and Bothwell, 1981), sequestration may reflect the association of receptors with the cytoskeletal elements needed to cap and internalize hormone–receptor complexes. By using fluorescently labeled NGF it is possible to determine the surface distribution of the bound NGF and to follow its processing by the cell. Three steps have been observed which are very similar to those involved in the processing of other growth hormones such as epidermal growth factor: (1) NGF binds to receptors diffusely distributed over the cell surface, including neurites, the cell body, and the growth cones. (2) The receptor–NGF complexes cluster to form patches. (3) The patches are endocytosed in an energy-dependent manner and make their way to the lysosomes, where the NGF is hydrolyzed.

In addition to binding to NGF, sympathetic and sensory neurons bind and respond to both insulin and epidermal growth factor (EGF). The metabolic responses of nerve cells to all three hormones are very similar, except that only NGF induces neurite outgrowth. All three hormones generate an anabolic response that includes increased rates of macro-molecular synthesis, carbohydrate and amino acid transport, enzyme induction, and the stimulation of Na^+,K^+-dependent ATPase (see, for example, Frazier et al., 1972; Greene and Shooter, 1980; Schubert et al., 1980; Boonstra et al., 1981; LaCorbiere and Schubert, 1981; Yarden et al., 1982). In addition, EGF and insulin increase the extent of phosphoryl-ation of a subset of those proteins which are phosphorylated by NGF (discussed later in this section). These common sets of specificities should make it possible to distinguish experimentally the general ana-bolic effect of NGF on cells from those metabolic changes directly asso-ciated with neurite extension.

Once a hormone has interacted with its receptor on the external plasma membrane, there are several basic sequences of events which can lead to the phenotypic response of the cell. (1) The hormone may be internalized and the hormone–receptor complex transported to the nu-cleus, where it directly affects gene expression. For example, steroid hormone–receptor complexes are thought to interact directly with chro-mosomes to modify RNA synthesis. (2) The hormone may be internalized and the hormone, the hormone–receptor complex, or the modified re-ceptor liberated into the cytoplasm, where it alters protein synthesis. Diphtheria toxin binds to cell-surface receptors, is internalized, and is apparently released into the cytoplasm, where it effectively shuts down protein synthesis (Pappenheimer and Gill, 1973). (3) Finally, the hor-mone may bind to plasma-membrane receptors and by means of this interaction stimulate the synthesis of a "second message." This second message, which presumably is free in the cytoplasm and able to modify protein translation as well as migrate to the nucleus, could then stimulate the phenotypic changes associated with the hormonal response. There is an increasing number of experimental results which rule out the first two alternatives.

Autoradiographic experiments employing [^{125}I]NGF and cytochemical data obtained by coupling NGF to horseradish peroxidase showed that NGF taken up *in vivo* by nerve terminals and transported back to the cell body is localized neither to the cytoplasm nor to the nucleus, but to closed membranous vesicles (Thoenen et al., 1981). In an analogous study, which used dissociated cultures of rat sympathetic neurons, [^{125}I]NGF internalized by complete cells or taken up exclusively by nerve terminals and retrogradely transported to the cell body was localized primarily to lysosomes and multivesicular bodies (Claude et al., 1982). There was no accumulation of the isotopically labeled NGF by the nu-cleus or the nuclear membrane. Other studies with the clonal PC12 cell line also indicate that isotopically labeled NGF does not accumulate on the nucleus after its initial interaction with cell-surface receptors and subsequent internalization, and that "nuclear receptors" for NGF are nonexistent (Schechter and Bothwell, 1981; Rohrer et al., 1982). Finally, enucleated cells still respond to exogenous NGF by extending neurites (Nichols et al., 1981; Schubert and LaCorbiere, unpublished results).

To rule out the possibility that free NGF in the cytoplasm plays a role in the NGF response, either NGF or antibodies against NGF were injected into chick DRG cells by fusing the cells with erythrocyte ghosts loaded with these reagents (Heumann et al., 1981a). If free cytoplasmic NGF were functional, then fusion of DRG cells with ghosts containing NGF should elicit neurite outgrowth and promote survival. It did not. Conversely, cytoplasmic anti-NGF did not inhibit the ability of exogenous NGF to cause the morphological differentiation of the cells. Although the possibility of direct interaction of a very few molecules of NGF per cell with nuclear material has not been formally eliminated, the biochemical and ultrastructural experiments outlined above clearly rule out the possibility that a major fraction of the NGF within a cell is associated with the nucleus or is free in the cytoplasm.

It has not formally been ruled out that the internalization of the NGF–receptor complex may be required for some of the long-term effects of NGF on nerve differentiation, but the direct NGF–receptor interaction must mediate the more rapid responses to NGF, such as alterations of the cell surface, which occur well before any NGF is internalized. The internalization of NGF may, however, have no function other than the removal of the bound NGF and its receptor from the cell surface; essentially all hormone receptors are metabolized in this manner. It follows that at least the early effects of NGF are mediated by one or more "second messages" that result from the initial interaction of NGF with its cell-surface receptor. The best candidates for these intermediates are cyclic AMP and Ca^{2+} ions.

At low concentrations of NGF half-maximal binding to cell-surface receptors occurs within a few minutes and saturation by 12 min. If the initial binding of NGF to cells triggers an immediate cellular response via either a second message or a conformational change of the NGF receptor, then phenotypic changes should be seen within this period of time. The first detected changes are the appearance of increased membrane ruffling and the increased formation of coated pits (Connolly et al., 1979, 1981). As viewed by scanning electron microscopy, the surface of PC12 cells is smooth. Within 30 seconds after the addition of NGF, microvillus formation and ruffling occurs. It is maximal at 7 min, and the ruffles disappear within 15 min. An increase in the number of coated pits is also observed within 30 seconds. Both ruffling and the formation of additional coated pits have been associated with the internalization of ligand–receptor complexes on the cell surface. Because a similar sequence of cell-surface changes took place in primary cultures of sympathetic neurons exposed to NGF, it is likely that these events reflect very early responses to NGF.

Concomitantly with the cell-surface changes, and probably reflecting them, are increases in cell–substratum adhesion and lectin agglutinability (Schubert et al., 1978). To demonstrate that NGF causes a rapid increase in the cell–substratum adhesion of PC12 cells, Schubert and his collaborators labeled low-density PC12 cultures overnight with [^{3}H]leucine, washed them with medium containing 10% fetal calf serum, and plated them into tissue culture dishes containing the same medium and NGF. They then determined the number of attached cells as a function of time and plotted the data as the percentages of the input cells which adhered. Figure 3.18 shows that NGF enhanced the initial rate of

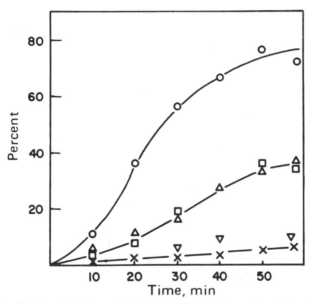

Figure 3.18. Effects of NGF, cyclic AMP, and 50 mM K$^+$ on cell–substratum adhesion. Exponentially growing cells were labeled overnight with [^{3}H]leucine in complete growth medium. The cells were washed and plated in tissue culture dishes containing HEPES-buffered modified Eagle's medium plus 10% fetal calf serum and the reagents indicated below. The initial rate of adhesion was determined at the indicated intervals. The data are plotted as the percentage of the input cells remaining attached to the surface as a function of time. ×, Control, no additive; □, 1 × 10^{-3} M dibutyryl cyclic AMP; O, 10 ng ml^{-1} β-NGF; △, 50 mM potassium substituted for sodium; ▽, 10 ng ml^{-1} β-NGF preincubated for 30 min with a 1:40 dilution of heat-inactivated anti-NGF serum. From Schubert and Whitlock (1977).

cellular adhesion to the tissue culture substratum by about 10-fold over the control cells.

Alterations of the cell surface were assayed also by lectin-induced agglutination. Plant lectins are multivalent proteins which bind to carbohydrate moieties on the surface of cells and in some cases induce the aggregation of the cells. Like cellular adhesiveness, their binding to the cell surface is an indirect indicator of the geometry and chemistry of the external plasma membrane. Therefore, an alteration in the ability of lectins to agglutinate cells reflects a change in cell-surface structure. When the lectin agglutinability of NGF-treated cells was compared with that of controls, the cells exposed to NGF agglutinated more rapidly. Because the number of lectin receptors remained constant, the conformation of the surface must have changed.

From the data on cell–substratum adhesion and lectin agglutinability, and from direct observations with scanning electron microscopy, it follows that one of the first phenotypic changes following the interaction with NGF is a change in cell-surface structure. This is followed by the initiation of neurite outgrowth.

Ultrastructural observations of PC12 cells following exposure to NGF show that the sequence of changes associated with neurite outgrowth are similar to those of neurite regeneration in cultures of dissociated sensory and sympathetic ganglia. The time course for PC12 on collagen is, however, much slower. On a collagen substratum, small neurites between 0.5 and 1 cell diameter in length appear on day 1 after NGF exposure. The cells are flattened due to increased cell–substratum adhesion, and they develop an extensive meshwork of microfilaments above the plasma membrane near the substratum (Luckenbill-Edds et al., 1979). Small neurites with microspikes and which contain dense-core vesicles are present. By day 3 the neurites are several cell diameters long, with a cytoplasm containing few ribosomes and many smooth-membrane vesicles, which are thought to be involved in plasma-membrane assembly. By 1 week, the neurites contain many microtubules, which may stabilize the long cellular protrusions. Since NGF does not cause the appearance of any new intracellular structures, it must cause an organizational or quantitative change in pre-existing organelles. Because the rate of neurite outgrowth is regulated by cell–substratum adhesion, the time course of the events outlined above can be sped up by a factor of approximately two by plating the cells on a more adhesive polylysine surface (Figure 3.14).

Besides changing the morphology of target cells, NGF also increases the activities of several enzymes and the rate of protein synthesis. When NGF is injected into young rats, there are increases in the specific activities and amount of TH and dopamine β-hydroxylase (DBH) in sympathetic ganglia (Otten et al., 1977; Thoenen et al., 1971). NGF applied to the terminals of superior cervical neurons also causes the levels of TH to increase (Paravicini et al., 1975). Since DBH and TH are key enzymes in norepinephrine synthesis, these data show that not only is NGF able to affect the survival and differentiation of sympathetic ganglion cells, but it also may modulate the level of neurotransmitter synthesis by the cells.

An extension of the *in vivo* studies to organ cultures of superior cervical ganglion (SCG) neurons shows that exogenous NGF is able to elicit a small increase (50–100%) in the amount of TH synthesized over the amount accumulated by control cultures after 2 days (Max et al., 1978). If the rat SCG neurons are dissociated before being placed in culture, they have a strict dependency on the presence of low levels of exogenous NGF. However, if NGF in the culture medium is increased above the level required for survival (0.5 μg 7S NGF/ml), there is a proportional increase in catecholamine synthesis which saturates at approximately 5 μg/ml (Chun and Patterson, 1977). The reason that extremely high levels of NGF are needed to increase catecholamine synthesis is not clear; the presence of an active contaminant in the NGF preparation has not been ruled out. The clonal PC12 cell line, which shares most of the phenotypic traits of cultured sympathetic ganglion cells, does not respond to NGF by induction of TH. There are, however, subclones of PC12, and additional clonal cell lines derived from the same tumor as the PC12 clone, which do respond to NGF with an increase in the specific activity of TH (Goodman and Herschman, 1978; Hatanaka, 1981).

The last NGF-inducible enzyme we will consider is the nearly ubiquitous ornithine decarboxylase (ODC). ODC is responsible for the conversion of ornithine by decarboxylation to the diamine putrescine, which is the rate-limiting reaction in the synthesis of the polyamines spermine and spermidine. Elevated levels of ODC appear to be associated with the rapid growth and multiplication of cells. ODC also has the shortest half-life (10 min) of any known eukaryotic enzyme. NGF produces an increase in the specific activity of ODC of rat SCG both *in vivo* and in cultured cells (MacDonnell et al., 1977). NGF also induces ODC in the clonal sympathetic nerve cell line PC12, in which the phenomenon can be studied more unambiguously than in heterogeneous primary cultures. The induction of ODC in PC12 cells apparently requires RNA synthesis (induction is blocked by actinomycin D), and maximum enzyme activity occurs within 4 hr of exposure to NGF (Guroff et al., 1981; LaCorbiere and Schubert, 1981). Due to the rapid induction of ODC by NGF and its activation by a cyclic-AMP-dependent protein kinase (Kudlow et al., 1980), ODC has been extensively studied with regard to the possible role of cyclic AMP as a second message for NGF. The data suggesting that cyclic AMP is an intermediate in ODC induction and in most of the phenotypic responses of sympathetic nerve cells are outlined below.

The experiments discussed in the preceding sections show that NGF does not exert its effects on cells by a direct interaction with chromatin, nor does it function as a free molecule in the cytoplasm. Since protein synthesis must be controlled at the level of either nuclear RNA synthesis (transcription), nuclear RNA processing, or protein synthesis (translation), the physiological effects of NGF must be mediated by one or more signals produced as a direct result of the binding of NGF to its receptor. We will therefore consider three aspects of the NGF response.

1. What is the nature of these signals?
2. How do they function to produce the phenotypic responses to NGF?
3. Can the signals for the morphological and enzymatic responses of cells to NGF be separated experimentally?

Exogenous cyclic AMP and reagents which elevate intracellular cyclic AMP cause the morphological differentiation of all clonal nerve cell lines and most, if not all, nerve cells in primary cultures. In addition, dibutyryl cyclic AMP induces morphological and enzymatic changes in PC12 cells that are similar to those induced by NGF (Schubert, 1979), and cyclic AMP analogues also cause the morphological differentiation of sympathetic and sensory neurons in primary cultures (Roisen and Murphy, 1973). Since cyclic GMP analogues are ineffective, and since cyclic AMP is a known regulator of cellular differentiation in many cell types (McMahon, 1974), cyclic AMP has been considered the most likely candidate for a second message in NGF-responsive cells.

There are four criteria which should be met if cyclic AMP is a second message for a hormone (Robison et al., 1971): (1) Exogenous hormone leads to an increase in intracellular cyclic AMP. (2) The action of the hormone is mimicked by cyclic AMP or dibutyryl cyclic AMP. (3) The action of the hormone is potentiated by inhibitors of cyclic nucleotide phosphodiesterase. (4) The hormone activates adenylate cyclase in broken-cell preparations. To date the first three criteria have been satisfied in sympathetic neurons and the clonal PC12 sympathetic-neuronlike cell line. The hormonal stimulation of cyclic AMP synthesis in broken cells is experimentally very difficult, and has not been achieved with several hormones, including NGF, whose actions are thought to be mediated by cyclic AMP. The following paragraphs outline some experiments showing that the first three criteria are met for NGF-induced neurite outgrowth, enzyme induction, and protein synthesis and phosphorylation. The possible role of cyclic AMP in the other aspect of the NGF response, nerve cell survival, has not been examined.

Perhaps the first data suggesting that cyclic AMP mediates the NGF response were obtained with chick DRG cultures (Roisen et al., 1972a,b). Dibutyryl cyclic AMP, cyclic AMP, and NGF all induce neurite outgrowth, and electron microscopy of the resultant neurites showed that the ultrastructures of cyclic-AMP-induced processes are very similar to those induced by NGF. In addition, both dibutyryl cyclic AMP and NGF reverse the inhibitory effect of colcemid on axon elongation, suggesting that NGF-induced cyclic AMP may mediate microtubule polymerization in growing neurites. Essentially identical results were obtained with the clonal sympathetic nerve cell line PC12 (Schubert and Whitlock, 1977). NGF and dibutyryl cyclic AMP both increase the rate of neurite outgrowth. In addition, NGF and cyclic AMP increased the cell—substratum adhesiveness of PC12 cells (Fig. 3.18), as well as their rate of aggregation. Since plating cells on adhesive substrata initiates neurite outgrowth from some nerve cells, the fact that NGF and cyclic AMP also increase the rate of cell adhesion in PC12 cells shows that adhesion can be enhanced by alteration of the surface properties of either the substratum or the cell.

PC12 cells appear to require NGF or cyclic AMP for neurite extension, as plating them on a very adhesive surface, such as polylysine, does not promote neurite outgrowth. There is, however, a claim that extracellular matrix from bovine corneal epithelium promotes neurite outgrowth in PC12 cells in the absence of NGF (Fuji et al., 1982), but it has not been ruled out that some component in the matrix stimulates cyclic AMP synthesis or calcium ion flux in the cells by a mechanism independent of NGF, resulting in neurite outgrowth. The less likely alternative is that the matrix substratum is much more adhesive than the other surfaces tested, and that this increased adhesiveness can enable the cells directly to bypass the requirement for NGF or cyclic AMP.

There is a slight difference in morphology between the neurites produced in both PC12 and primary sympathetic neurons by dibutyryl cyclic AMP and those produced by NGF. This is to be expected, since the presentation and concentrations of cyclic AMP and NGF are completely different and the two reagents have quantitatively different effects on cell—substratum adhesion (Figure 3.18). In addition, some of the active cyclic AMP analogues have effects not evident with the hormone, for most of the analogues are toxic over periods longer than a few days. For this reason little significance should be attributed to small quantitative differences between the phenotypes produced by exogenous cyclic AMP and those produced by the hormone.

The initial observation which led to the discovery of NGF was that pieces of tumor tissue grafted to chick embryos tended to attract the growth of sympathetic nerve fibers. A more formal demonstration of the chemotactic activity of NGF was made by injecting NGF into the brain of neonatal animals (Levi-Montalcini, 1976). Following the injection, nerve fibers from sympathetic ganglia entered the spinal cord and ascended the dorsal column to the site of NGF injection—a territory far removed from their normal targets. To determine whether cyclic AMP was involved in the chemotactic response to NGF, Gunderson and Barrett (1980) placed NGF or cyclic AMP in a micropipette located next to the growth cones of dissociated DRG neurons and assayed the growth response to the local gradient. The growth cones reoriented themselves and grew toward the sources of both NGF and cyclic AMP, showing that cyclic AMP could reproduce the NGF response. In addition, a similar chemotactic response was observed to calcium combined with a calcium ionophore, but not to calcium alone. These data suggest that both cyclic AMP and calcium may be intermediates in the orientation of outgrowing neurites toward NGF. A similar attraction for fiber outgrowth has been observed when sympathetic ganglia and explants of sympathetically innervated tissues are cocultured (Chamley, et al., 1973). The neurites extend more rapidly toward the normally innervated tissue than toward noninnervated material.

With respect to enzyme induction, both NGF and dibutyryl cyclic AMP stimulate TH activity in rat SCG cultures (Yu et al., 1975). There was a debate in the literature about whether or not NGF could induce TH in SCG organ cultures. As in many controversial experiments, the culture medium turned out to be the problem, for serum can mimic some of the effects of NGF on both energy metabolism and neurite outgrowth as well as inhibit NGF-induced neurite outgrowth (see Yu et al., 1975, for a discussion). However, in the absence of serum both NGF and dibutyryl cyclic AMP can induce TH in SCG cultures. Finally, the induction of CAT and ODC by NGF in PC12 cells is mimicked by cyclic AMP analogues and reagents such as cholera toxin that increase intracellular cyclic AMP (Schubert et al., 1977; LaCorbiere and Schubert, 1981). In all three examples (TH, CAT, and ODC), the induction of the enzyme by NGF is enhanced by phosphodiesterase inhibitors, showing that cyclic AMP is directly involved in the response and fulfilling the third requirement for a cyclic-AMP-mediated hormone response.

An additional criterion for a cyclic-nucleotide-mediated response is that the hormone causes an increase in intracellular cyclic nucleotide. NGF-induced increases in intracellular cyclic AMP have been observed in rat SCG organ cultures (Nikodijevic et al., 1975) and in organ cultures of chick DRG (Skaper et al., 1979). Similar increases in intracellular cyclic AMP have been observed in the clonal PC12 line (Schubert and Whitlock, 1977). In all three cases the cyclic AMP increase is relatively small (two- to fivefold) and transient, with the peak occurring between 5 and 12 min after the addition of NGF and rapidly returning to base line. The transient kinetics of cyclic AMP have been observed in several hormone responses, and the brief elevation may be sufficient to maintain protein kinases in an activated state for extended periods of time. There are several papers claiming that NGF does not increase intracellular cyclic AMP in responsive cells. However, the majority of the experiments did not examine cyclic AMP until 15 min or longer after the addition of the hormone, and in others the experimental errors in the cyclic AMP determinations were sufficiently great that a response could not be seen even if one were present.

Before the phenotypic changes caused by NGF can be understood, its effect on total macromolecular synthesis must be defined. To assay the effect of NGF on protein synthesis, the rate of synthesis was assayed by quantitative 2-D gel electrophoresis (Garrels and Schubert, 1979; also see Section 2.8 for methods). When PC12 cells were pulse-labeled for 2 hr with [^{35}S]methionine in the presence or absence of NGF and the proteins synthesized during this period separated by 2-D gel electrophoresis, no new proteins were seen to be synthesized after exposure to NGF, nor were any proteins qualitatively repressed by NGF. However, significant changes in the rates of synthesis of approximately 30% of the proteins were noted when the gels were analyzed by a computerized scanning system. If dibutyryl cyclic AMP produces changes in protein synthesis like those caused by NGF, it is likely that the effect of NGF on protein synthesis is mediated by cyclic AMP. The only alternative is that cyclic AMP fortuitously induces the "true" second message for NGF. A quantitative comparison of the NGF- and dibutyryl-cyclic-AMP-induced responses demonstrated that the two are highly correlated. No proteins were significantly altered by dibutyryl cyclic AMP (or 8-bromo-cyclic AMP) but not by NGF. These data were presented as scatter diagrams, in which the proteins are plotted according to their intensities in the first sample on the horizontal axis and according to their intensities in the second sample on the vertical axis. If the two samples were identical, each point would fall on the diagonal line. The scatter due to experimental error can be seen in Figure 3.19A, in which two controls are plotted against each other. In Figure 3.19B (control vs NGF) and 3.19C (control versus dibutyryl cAMP), a "halo" of points is seen scattered about the cluster of points on the diagonal. These represent the proteins that respond most to NGF and dibutyryl cAMP, respectively. In Figure 3.19D

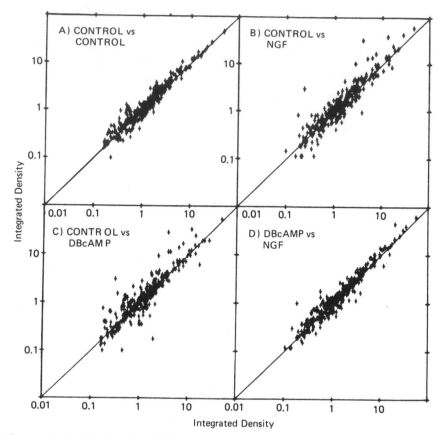

Figure 3.19. Scatter plots of NGF and cyclic AMP responses. PC12 cells were isotopi-cally labeled after 3 days in the presence of the reagents indicated and the amounts of isotope in about 600 proteins were determined by quantitative 2-D gel electrophoresis. Each cross represents one protein plotted on the horizontal axis according to its intensity in the first sample and on the vertical axis according to its intensity in the second sample. Proteins with equal intensity on each axis fall on the diagonal line. Each axis is in units of absorbance × mm². DBcAMP is dibutyryl cAMP. From Garrels and Schubert, 1979.

(dibutyryl cAMP versus NGF) the halo of points is gone and the points are again tightly clustered, indicating that the NGF-treated and dibutyryl cAMP-treated samples are nearly as similar to one another as are dupli-cate controls. These results demonstrate that NGF causes only quantita-tive modulations of protein synthesis in PC12 cells, and are perhaps the strongest evidence that the response of PC12 cells to NGF is mediated via cyclic AMP.

NGF stimulates cyclic AMP synthesis in sympathetic nerve cells, but how does cyclic AMP alter the metabolism of the cell to elicit the phenotypic response? There are two known mechanisms by which cyclic AMP can influence the cell; one has been studied extensively in bacteria (Adhya and Garges, 1982). (1) Cyclic AMP can interact with a receptor protein which binds to specific sites on DNA and modifies gene transcription. (2) Cyclic AMP can interact with protein kinases which phosphorylate proteins, a process which has been demonstrated only in eukaryotic cells. Although cyclic-AMP-binding molecules that interact with DNA have not been identified in mammalian cells, protein kinases activated by cyclic AMP can phosphorylate chromosomal proteins.

Since the only demonstrated actions of cyclic AMP in mammalian cells are mediated via cyclic-AMP-dependent protein kinases, the involvement of protein phosphorylation in the NGF response was investigated. If PC12 cells are labeled for 1.25 hr with radioactive [^{32}P]orthophosphate in the presence or absence of NGF and the isotopically labeled proteins are separated by acrylamide gel electrophoresis, incorporation of ^{32}P is seen to be altered in at least seven proteins (Halegoua and Patrick, 1980). Those whose labeling increases include TH, ribosomal protein S6, histones H1 and H3, and the nonhistone chromosomal high mobility group (HMG) 17 protein; the phosphorylation of histone H2A is reduced. The degree of stimulation is between 2.7 and nine times the control levels, depending on the protein. Maximum stimulation occurs within 15 min, but the changes are observed for up to 3 days in the presence of NGF. The changes in phosphorylation could have been due either to alterations in the rates of protein synthesis or to altered levels of phosphorylation of pre-existing proteins. To decide between these alternatives, Halegoua and Patrick blocked protein synthesis with puromycin or cycloheximide, and treated the cells with NGF as before. The inhibition of protein synthesis had no effect on protein phosphorylation. If these NGF-induced protein phosphorylations were mediated by cyclic AMP, then it would be expected that the changes in phosphorylation caused by NGF, cyclic AMP, and cholera toxin would be similar. The results showed that they were virtually identical. Since all three reagents modify protein phosphorylation in a similar way it is likely that the NGF-induced changes in phosphorylation were mediated by cyclic AMP.

As outlined above, insulin and EGF share with NGF some metabolic effects on PC12 cells. They also cause the phosphorylation of a subset of the same proteins as NGF. The phosphorylation of HMG 17 is stimulated only by NGF, cyclic AMP, and cholera toxin, whereas insulin and EGF stimulate the phosphorylation of ribosomal protein S6 and histone H1.

Insulin and EGF have different effects on the phosphorylation of TH, H2A, and H3 (Table 3.4). That EGF and insulin enhance the phosphorylation of a subset of the proteins phosphorylated by NGF suggests that the mechanisms are different. The changes caused by NGF, cyclic AMP, and cholera toxin are the same, suggesting a common mechanism. If the mechanisms are the same, then stimulation of phosphorylation by saturating amounts of NGF and cyclic AMP should not be additive, whereas saturating amounts of NGF should be additive with EGF or insulin. The data show that the NGF and cyclic AMP combination is not additive, whereas NGF plus EGF or insulin is. These results again support the idea that the actions of NGF are mediated by cyclic AMP.

Although cultured perinatal sympathetic neurons require NGF for survival and differentiation, cultured embryonic mouse SCG sympathetic neurons do not require NGF for survival and neurite outgrowth. This allows the study of intrinsic ganglionic development. Under NGF-free conditions, neurite formation by sympathetic neurons is not inhibited by actinomycin D, puromycin, or cycloheximide, suggesting that neither *de novo* protein synthesis nor RNA synthesis is required (Bloom and Black, 1979). In contrast, the spontaneous increase in TH is inhibited by all three reagents. Similarly, neurite formation by clonal nerve cell lines does not require protein or RNA synthesis (Schubert, 1974; Schubert et al., 1971), which suggests that neurites are derived from the assembly of preformed components. Similar experiments with embryonic chick sympathetic ganglia also showed that RNA synthesis is not required for NGF-induced neurite outgrowth (Partlow and Larrabee, 1971), consistent

Table 3.4. Phosphorylation of Specific Proteins by NGF, Cyclic AMP, EGF, and Insulin[a]

	Protein					
Condition	TH	S6	H1	H3	HMG17	H2A
NGF	+	+	+	+	+	−
Dibutyryl cyclic AMP	+	+	+	+	+	−
Cyclic AMP + theophylline	+	+	+	+	+	−
Cholera toxin	+	+	+	+	+	−
EGF	+	+	+	NC	NC	NC
Insulin	NC	+	+	+	NC	−

[a] PC12 cells were labeled for 1.25 hr with ^{32}P with or without the indicated reagent, and the extent of phosphorylation of each identified protein was quantitated. + indicates increased phosphorylation relative to untreated controls; − indicates decreased phosphorylation. NC indicates no change in phosphorylation. From Halegoua and Patrick (1980).

with the observation that enucleated cells still respond to NGF with neurite outgrowth. However, NGF-induced neurite outgrowth does require new protein synthesis in this system, for the inhibition of protein synthesis by cycloheximide is closely paralleled by a reduction in fiber outgrowth (Partlow and Larrabee, 1971). The NGF inductions of ODC and TH in SCG are both transcription and translation dependent (Mac-Donnell et al., 1977; Otten et al., 1977). These results suggest that two distinguishable processes may be involved in the NGF response: (1) neurite outgrowth, which does not require RNA synthesis in its initial stages; and (2) the RNA-dependent induction of some enzymes. These results do not rule out a requirement for new RNA synthesis for long-term neurite extension and maintenance, which is in fact likely but is difficult to test because of the toxicity of the RNA-synthesis inhibitors.

In most cell types in which cyclic AMP has been implicated in a hormone response, calcium ions play an equally important role in that response (for an excellent review, see Rasmussen and Goodman, 1977). For example, when glucagon interacts with liver cells, there is an increase in intracellular cyclic AMP, but there are also changes in membrane potential, fluxes of Na^+, K^+, and Ca^{2+}, and intracellular pH (Rasmussen and Goodman, 1977). A rise in intracellular free Ca^{2+} is caused by exogenous cyclic AMP, which, like glucagon, also stimulates gluconeogenesis and produces ionic changes. Variations on the theme of associated cyclic AMP and ion movements have been documented in a variety of hormone-sensitive cells. Ca^{2+} is also involved in nerve differentiation, for it is required for neurite outgrowth and activation of the contractile apparatus of growth cones during neurite formation. Ca^{2+} fluxes are also involved in the response of cells to NGF. Elucidation of the role of Ca^{2+} in NGF-induced neurite outgrowth resulted from three observations (Table 3.5) (Schubert et al., 1978): (1) All experimental conditions which lead to increased PC12 cell–substratum adhesion also stimulate Ca^{2+} fluxes. (2) NGF and cyclic AMP each stimulate the release of cellular calcium into the culture medium, suggesting that cyclic AMP, rather than the ligation of NGF with its receptor, is stimulating Ca^{2+} release. (3) Finally, elevated exogenous K^+ ions are able to stimulate neurite outgrowth in PC12 cells without detectably increasing intracellular cyclic AMP. K^+ depolarizes the cells, resulting in the flux of Ca^{2+} through voltage-sensitive Ca^{2+} channels. Since neurite outgrowth is reflected by an increase in cell–substratum adhesion, if Ca^{2+} is involved in neurite extension by promoting adhesion, then the inhibition of Ca^{2+} movement should block adhesion. Cobalt is a known blocker of potential sensitive Ca^{2+} channels, and Co^{2+} inhibits K^+-induced cell–substratum adhesion of PC12 cells. These results show that neurite extension per se

Table 3.5.   Intracellular Cyclic AMP, Neurite Formation, and Enzyme Induction

	Response				
Condition	Neurite Outgrowth	Ca^{2+} Flux	Cyclic AMP	CAT	ODC
NGF	+	↑	↑	↑	↑
Dibutyryl cyclic AMP	+	↑	↑	↑	↑
50 mM K$^+$	+	↑	—	—	—

NGF, dibutyryl cyclic AMP, and elevated exogenous K$^+$ all induce neurite outgrowth and Ca^{2+} flux, but K$^+$ does not increase cyclic AMP or induce ODC or choline acetyltransferase in PC12 cells (Schubert et al., 1978 and unpublished observations).

can take place in the absence of increased intracellular cyclic AMP, and suggest that the apparent trigger for increased cell–substratum adhesion and neurite outgrowth is the mobilization of Ca^{2+} by the cells.

As outlined in Section 3.5, the movement of growth cones is probably achieved by an actomyosin-driven system similar to that of muscle. It is also known that Ca^{2+} regulates cell motility in all cell types, from mammalian muscle to protozoa (see, for example, Porter, 1976). It is therefore likely that Ca^{2+} mediates growth cone motility and neurite outgrowth, including that induced by NGF. There is ample experimental evidence, complementary to the data obtained with PC12, to support this conclusion.

An examination of the ionic basis of action potentials in outgrowing neurites shows that the action-potential current is carried by Ca^{2+}, whereas that of the axon and cell body is Na$^+$ dependent (Meiri et al., 1981). This Ca^{2+} action potential may be needed to maintain the high intracellular free Ca^{2+} pool required for generating growth cone mobility. Voltage-sensitive Ca^{2+} channels are, in fact, the first channels to be detected during embryogenesis (Spitzer, 1979).

In a series of elegant experiments with a clone of a C1300 mouse neuroblastoma cell it was again shown, by both optical and electrophysiological techniques, that voltage-achieved Ca^{2+} channels are more abundant in growth cones than in the rest of the cell (Grinvald and Farber, 1981; Anglister et al., 1982). To determine if Ca^{2+} entry into the growth cone plays a direct role in neurite extension, Ca^{2+} entry was induced by four methods: (1) depolarization with K$^+$; (2) stimulation of the soma with an intracellular microelectrode; (3) the application of a Ca^{2+} ionophore plus Ca^{2+}; and (4) application of a depolarizing solution plus Ca^{2+} directly to the growth cone. The morphological responses of the growth cones were followed by time-lapse video recordings. In all cases there was a significant increase in the area of the growth cones and, as in PC12 cells under similar conditions, there was a pronounced flattening of the membrane and an associated increase in neurite elongation. These results support the hypothesis that Ca^{2+} entry is the trigger for increased cell–substratum adhesion and neurite extension.

NGF also stimulates Na^+,K^+-dependent ATPase (Brown and Scholfield, 1974) and the methylation of phospholipids in the distal portions (growth cones) of growing SCG neurites but not in the cell bodies (Pfenninger and Johnson, 1981). Maximum phospholipid methylation occurs within 10 seconds after the addition of NGF, and is followed by a decline in the rate of methionine incorporation into lipids. Since phospholipid methylation may stimulate Ca^{2+} entry into cells (Hirata and Axelrod, 1980), it is possible that the methylation step precedes the other intracellular responses to NGF. Finally, Gunderson and Barrett (1980) showed that growing chick DRG neurites in culture reorient themselves toward local high concentrations of NGF released from a micropipette. The neurites also grow toward sources of cyclic AMP and high Ca^{2+} plus a calcium ionophore (but not to Ca^{2+} alone), showing that both cyclic AMP and Ca^{2+} flux across the growth cone membrane can mimic the effect of NGF.

Alterations in adhesive properties certainly play key roles during embryogenesis, and the above data suggest that NGF-induced modification of membrane structure can trigger specific changes in cellular adhesion. Peptides that stimulate the chemotactic mobility of carcinosarcoma cells also enhance cell–substratum adhesiveness (Varani et al., 1981), and it is possible that the ability of such material to increase the adherence of cells may generate the chemotactic response. Since NGF reversibly increases cell–substratum adhesion, growing axons could be "guided" by the NGF-induced increase in the adhesiveness of the growth cone to the underlying cells, rather than the growth cone following a substratum-associated gradient. Growth cones elongate preferentially on surfaces to which they adhere more firmly, and it should make no difference whether the growth cone or the substratum is responsible for initiating the adhesive response. This type of NGF effect on cell–substratum adhesiveness could also explain the requirement for local exposure to NGF for continued neurite outgrowth in cultured sympathetic ganglion cells.

Given the above information about the ability of NGF to induce the morphological and biochemical differentiation of cells, is it possible to construct a workable, testable model for NGF action which is compatible with the data? Before constructing a model, it is necessary to consider the sequence of the events which are known to follow the interaction of cells with saturating amounts of NGF. Figure 3.20 shows that the first event detected after the addition of exogenous NGF is an increase in phospholipid methylation (Pfenninger and Johnson, 1981). These data were, however, obtained from rat SCG cells (which require NGF for survival) by deprivation of the cells for NGF for 1.5 hr and then addition of NGF back to the cultures. This type of experimental paradigm makes it impossible to determine whether methylation is a response associated with the NGF-induced survival effects (analogous to adding serum back to serum-starved cultures) or is an initial event in the NGF-induced differentiation of the cells. The first response unambiguously associated with NGF-induced differentiation is membrane ruffling, for this occurs in PC12 cells, which do not require NGF for survival in serum-containing medium. Between 1 and 10 min after the addition of NGF a whole series of biochemical and morphological changes has been detected, and it is difficult to clearly define a temporal relationship among them. However, by artificially changing one parameter and observing the effect this has on the others, one can deduce a probable sequence of events. For

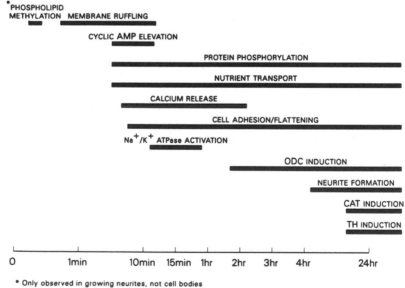

TIME COURSES OF NGF RESPONSES

*Only observed in growing neurites, not cell bodies

Figure 3.20. Time course of NGF responses.

example, increasing intracellular cyclic AMP increases Ca^{2+} release, but increasing intracellular Ca^{2+} by depolarizing the cell does not increase intracellular cyclic AMP. Therefore, the cyclic AMP increase probably precedes that of Ca^{2+}.

The events which occur less than 1 hr after the addition of NGF have been termed "early" functions. The "late" functions, which appear after several hours, include enzyme inductions and neurite formation. On the basis of these observations and a limited number of extrapolations from better-studied hormonal responses, it is possible to construct a tentative outline of what transpires in NGF-responsive cells (Figure 3.21).

1. NGF interacts with the high affinity (Type 1) receptor (R) on the cell surface.

2. As the direct result of this interaction there may be an alteration of phospholipid methylation or some other immediate response (?) that reflects a conformational change in the membrane receptor. This may cause membrane ruffling.

3. The NGF–receptor complex then moves in the membrane and interacts with adenylate cyclase (AC). The NGF–receptor–AC complex probably becomes associated with the cytoskeleton (CYTO), as this association is required for internalization and NGF receptors have

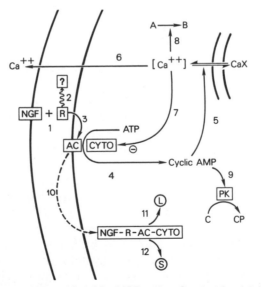

Figure 3.21. Model for NGF action. See text for details.

been detected in cytoskeletal preparations. This interaction may also render the bound NGF inaccessible to exogenous trypsin, thus accounting for sequestration.

4. Cyclic AMP is generated.

5. The elevated intracellular cyclic AMP causes the release of Ca^{2+} from intracellular (or possibly plasma-membrane) stores (CaX).

6. This results in the efflux of Ca^{2+} and a reorganization of cell-surface molecules, yielding increased lectin agglutinability and increased cell–substratum adhesion.

7. The rise in Ca^{2+} inhibits the activity of AC—a negative feedback relationship documented in several systems and explaining the transient increase in intracellular cyclic AMP.

8. The elevated intracellular Ca^{2+} acts as a cofactor in numerous biochemical reactions (A → B) such as actomyosin contraction and phospholipase activation.

9. Cyclic AMP stimulates protein kinases (pk), which phosphorylate a variety of proteins (C → CP), changing their activities and/or specificities so that they affect a number of biochemical parameters within the cell, including transcription.

10. The NGF–receptor–AC complex is then internalized, where it either

11. is degraded in lysosomes (L) or

12. remains in a relatively stable state (S) by which cyclic AMP may be produced locally, although necessarily for limited periods of time, because the effects of NGF are readily reversible.

At least one more intracellular "signal" in addition to cyclic AMP and Ca^{2+} may be needed to produce all of the long-term effects of NGF, such as the maintenance of long neurites. This conclusion is based on the negative experimental evidence that the manipulation of Ca^{2+} and cyclic AMP alone has not been able to produce cultures which have the same long-term viability and morphological stability as those exposed to NGF.

It is, however, possible to distinguish experimentally those phenotypic changes which require only elevated Ca^{2+} from those requiring elevated cyclic AMP. Conditions which increase intracellular Ca^{2+} stimulate cell–substratum adhesion, cell aggregation, and neurite extension. All of the surface-membrane properties which are altered by NGF are also altered in the same manner by artificially induced Ca^{2+} fluxes. These ionic changes do *not* increase cyclic AMP, cause protein phosphorylation, or induce any enzyme. The latter three parameters of the NGF response require the elevation of intracellular cyclic AMP.

The above outline for the intermediate steps in NGF-induced nerve differentiation is a plausible scenario for what is going on in the responsive cell. It should be viewed as a "working model," and it is possible to design experiments which would rule out any steps not in fact involved in the NGF response. This might require, for example, the selection of AC or protein kinase mutants and the use of cell-free systems to examine various aspects of the response. It is, however, quite certain that any progress in the understanding of NGF action will require the use of clonal nerve cell lines.

3.8   Neurotransmitter Synthesis in Cultured Cells

A neurotransmitter classically has been defined as a compound which is synthesized by a nerve, is released by electrical activity in the nerve, and interacts with specific receptors in the postsynaptic cell to generate a change in the cell's electrical activity. There has been a popular notion in neurobiology that a neuron is able to synthesize and secrete only one neurotransmitter. This idea has been ascribed to Sir Henry Dale, and is known as Dale's Principle. It is clear from the writings of Dale, however, that he proposed the idea of neurotransmitter unity within a single cell as a summary of the data existing at that time and as a working hypothesis rather than a scientific doctrine (Dale, 1935). Eccles (1957) extended Dale's very conservative views to form the doctrine that "a cell cannot make one kind of transmitter substance for some of its terminals and another kind for others of its terminals." The suggestion that one cell can make and secrete only one neurotransmitter is, however, not known to be valid. Data derived both from immunohistological data *in vivo* and from cultured nerve cells have shown that nerve cells can synthesize multiple neurotransmitters.

The *in vivo* evidence for the coexistence of transmitters is derived largely from studies showing that cells which stain histochemically for catecholamines or with antibodies to enzymes involved in catecholamine biosynthesis costain with antibodies against one of several putative peptide neurotransmitters, such as somatostatin (see Hokfelt et al., 1980, for a review). Evidence for the coexistence of ACh and vasoactive intestinal peptide (VIP) in parasympathetic neurons has also been presented (Lundberg, 1981). Finally, there is consistent but not irrefutable evidence that many invertebrate neurons release multiple transmitters (Osborne, 1979).

The *in vivo* data rely upon indirect immunological techniques for the identification of neurotransmitters in individual cells, and therefore cannot be used to rule out the possibility that cells concentrate the peptide hormones but do not synthesize them. In contrast, data with cultured cells demonstrate quite unambiguously that some nerve cells do synthesize simultaneously more than one neurotransmitter. This information came initially from studies with clonal cell lines and more recently from studies using dissociated primary cultures of rat sympathetic ganglion cells.

The first clonal cell lines established in culture were derived from the C1300 mouse neuroblastoma. Cell lines derived from mouse tumors generally have very unstable karyotypes. In addition, this tumor had been passaged in animals for over 30 years before it was grown in culture and cloned. As a result of these two factors the hundreds of clones isolated from the original tumor had widely different characteristics. An analysis of neurotransmitter synthesis in these clones revealed that the clones fell into four subsets: (1) cholinergic; (2) adrenergic; (3) those which synthesized both catecholamines and ACh; and (4) those which synthesized neither. All four groups of cells synthesize and secrete into the culture medium large amounts of AChE. In addition, a limited number of clones of the C1300 tumor are able to synthesize serotonin (Breakefield and Nirenberg, 1974) and VIP (Said and Rosenberg, 1976). The syntheses of catecholamines and ACh both tend to increase with the morphological differentiation of the cells, although no instance has been reported in which an exogenous agent that causes the morphological differentiation of the cells qualitatively induces or suppresses the synthesis of a neurotransmitter. Neurotransmitter synthesis in C1300 clones is limited largely to ACh and/or catecholamines, and is similar to those of dissociated primary cultures of sympathetic ganglion cells and the clonal PC12 sympathetic-nervelike cell line. This is to be expected, since neuroblastomas are thought to be derived from cells of the neural crest. Cell lines derived from human neuroblastomas also synthesize predominantly ACh and/or catecholamines (Bottenstein, 1981).

In contrast to C1300 mouse neuroblastoma clones, clonal cell lines derived from the rat CNS make primarily the inhibitory neurotransmitter γ-aminobutyric acid (GABA). This is not surprising, since GABA is the transmitter with the highest molar concentration in the brain. Of the five CNS nerve cell lines studied, all make appreciable amounts of GABA, up to 20% of the total free amino acid pool, and some make smaller amounts of ACh and catecholamines (Schubert et al., 1974b, 1975). To determine if exponentially dividing cells were able to synthesize more than one neurotransmitter, Kimes et al. (1974) recloned the initial clones and assayed the subclones for transmitter synthesis. All of the subclones were able to synthesize two or more neurotransmitters, showing that individual CNS cells, like those from the neural crest, are able to make multiple neurotransmitters. The CNS nerve lines and most CNS glial clones are also able to concentrate exogenous GABA (Schubert, 1975).

An examination of the free amino acid pools of cell lines derived from several embryonic primordia showed that all neuroectodermally derived cells, including glial clones and melanocytes, were able to make small amounts of GABA; no GABA was detectable in mesodermally derived cells (Schubert et al., 1975). The finding of a low level of a highly "specialized" molecule such as GABA in all derivatives of an embryonic tissue may be an indication that a block of genes which may be needed later in development is turned on at a low level at an early stage of development, with their activity later being enhanced in those parts of the lineage that require the specialized product.

Another group of cultured cells able to synthesize multiple neurotransmitters is the sympathetic ganglia. As outlined above (Section 3.2), dissociated SCG cells from newborn rats are able to synthesize both ACh and catecholamines. Evidence that a single cell within this heterogeneous population can synthesize both neurotransmitters has been gleaned from four experiments.

1. *Electrophysiology.* If single SCG neurons are plated on a monolayer of heart cells, the electrical response of the heart cells following nerve stimulation can serve as an assay for ACh and norepinephrine (NE) release. If ACh is released there is a slow hyperpolarization of the muscle which is blocked by atropine. If NE is released the heart cells are depolarized, there is an increase in number of spontaneous action potentials, and the overall response is blocked by propranolol. When this experiment was done with single SCG cells, many released both NE and ACh (Furshpan et al., 1981).

2. *Ultrastructure.* Individual SCG neurons contain separate populations of the synaptic vesicles thought to contain ACh (clear vesicles) and those thought to contain NE (dense-core vesicles) (Johnson et al., 1980a).

3. *Biochemistry.* Single-cell rat SCG cultures have been described that synthesize both ACh and NE from isotopically labeled precursors (Reichardt and Patterson, 1977).

4. *Immunology.* Over 90% of SCG cells that form cholinergic synapses with other SCG neurons contain immunocytochemically defined TH within their cytoplasm (Higgins et al., 1981).

On the basis of these results it is clear that sympathetic nerve cells from neonatal rats placed in cell culture have the capacity to synthesize and release both NE and ACh. These data, in conjunction with those obtained with clonal cell lines, suggest that sympathetic neurons have the ability to make multiple neurotransmitters.

Like the C1300 mouse neuroblastoma and dissociated primary cultures of neonatal rat SCG, the clonal PC12 cell line has the ability to synthesize both ACh and NE (Greene and Rein, 1977; Schubert et al., 1977; see also Section 3.2). In addition, an examination of ACh and NE storage in this cell line solved an old problem related to the early stages of the formation of cholinergic synapses—the observation that although cholinergic miniature endplate potentials (MEPPs) are observed at the young synapse, clear vesicles typical of those thought to store and release ACh are frequently not present. Only dense-core vesicles, which are more frequently associated with catecholamine storage, are detected. Dense-core vesicles of variable size and unknown function are found in numerous nonadrenergic nerves, such as sensory fibers (Yamada et al., 1971), motor nerve terminals (Pappas et al., 1971), and autonomic preganglionic fibers (Birks, 1974). In addition, dense-core vesicles are numerous in growing or regenerating peripheral nerve fibers (Lentz, 1967; Yamada et al., 1971), and are invariably associated with newly formed cholinergic nerve–muscle synapses in cell culture (Shimada and Fischman, 1975; Kidokoro et al., 1975).

ACh and the catecholamines could be either present in vesicles or free in the cytoplasm of cultured nerve cells. To determine whether these compounds are associated with a particulate subcellular fraction, PC12 cells were labeled for 18 hr with [^{3}H]tyrosine and [^{14}C]choline (Schubert and Klier, 1977). They were then lysed and centrifuged into linear gradients from 0.6 to 1.8 M sucrose, and the distribution of radioactivity in ACh, choline, tyrosine, NE, and dopamine (DA) was determined. More than 80% of the catecholamines and ACh banded with particles of similar but not identical densities in the gradient (Figure 3.22). The equilibrium density for catecholamines was approximately 1.2 and that of ACh was 1.0. The detergent Triton X-100 released the particulate neurotransmitters into the soluble fraction.

Although the differences in buoyant density make it unlikely, it is possible that both ACh and catecholamines are stored in the same vesicle. For example, some vesicles could contain both ACh and catecholamines and others could contain only catecholamines. Alternatively, the two classes of transmitter could be stored in different vesicles. To distinguish between these possibilities, cells were labeled with [^{3}H]choline and [^{3}H]tyrosine in the presence or absence of 2 μM reserpine, a reagent known to deplete catecholamine-storage vesicles. If the same vesicles contain both ACh and catecholamines, then reserpine should

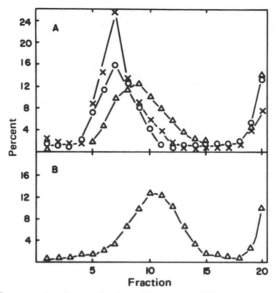

Figure 3.22. Sucrose density gradient fractionation of ACh and catecholamine storage vesicles. Cells were labeled for 18 hr with [³H]tyrosine or [¹⁴C]choline. One set of cultures was labeled the same way but also included 2 μM reserpine. The cells were washed and homogenized, and the lysate was centrifuged into a 0.6–1.8 M linear sucrose gradient. The data are plotted as the percentage of the radioactivity recovered for each molecular species vs the fraction number. (A) Normal distribution: ×, NE; ○, DA; △, ACh. (B) Sedimentation of ACh after reserpine treatment. Intracellular DA and NE levels were decreased by reserpine to 3.1% and 2.4%, respectively, of the control levels, as defined by isotopic labeling; these data are not plotted. From Schubert and Klier, 1977.

deplete both transmitters. Reserpine blocked the accumulation of intracellular DA and NE by 97%, but ACh synthesis and storage was decreased by less than 15%. When the reserpine-treated cells were homogenized and run on a sucrose gradient, the catecholamine-containing particulate fraction was decreased by 97% and the ACh storage particles remained (Figure 3.22).

If there are two classes of vesicles within PC12 cells, it should be possible to detect these differences morphologically. When PC12 cells grown in the absence of NGF were examined by electron microscopy, a large number of dense-core vesicles were observed (Schubert et al., 1980). When vesicle diameter was plotted against frequency of appearance, there was a wide distribution of particle sizes, with a peak frequency at a diameter of 150 nm (Figure 3.2). Because reserpine depletes the catecholamine content of PC12 cells, it follows that if one size class of vesicles was eliminated by this compound, it represents the catecholamine-containing particles. Figure 3.2 shows that the larger-diameter vesicles were greatly decreased in frequency after reserpine treatment and that vesicles averaging 75 nm in diameter become the predominant class. If the smaller vesicles contain ACh, then they should comigrate on a sucrose gradient with radioactive ACh after reserpine treatment. Reserpine-treated cells were homogenized and the vesicles sedimented on a sucrose gradient. Each gradient fraction was then fixed with glutaraldehyde and repelleted. The fractions corresponding to the sedimentation position of ACh contained the 75-nm class of dense-core vesicles found in reserpine-treated cells, along with free ribosomes and very few other membrane structures.

In summary, it is clear that some individual cultured cells from both the central and peripheral nervous systems are able to synthesize and store two or more neurotransmitters. It is therefore likely that at one stage in their development, probably an early one, nerve cells are able to synthesize multiple neurotransmitters and that the synthesis of signifi-cant amounts of all but one is suppressed by environmental influences or by the nature of the neuron's synaptic field. It is quite likely that low levels of several putative neurotransmitters are synthesized throughout the life of most neuroectodermally derived cells. The significance of this low-level synthesis is not known. Finally, ACh can be stored in dense-core vesicles similar to those which contain catecholamines.

3.9 Electrical Excitability

The ability of nerve cells to generate an action potential is a second differentiated membrane property, in addition to neurotransmitter syn-thesis, which distinguishes this cell type from most others. Electrical excitability and morphology are, in fact, the criteria used to distinguish between neuronal and glial cells in the nervous system. Once the cells are placed in culture the morphological criterion is no longer functional, and the nerve cells are usually defined by their ability to propagate an action potential. Since the nervous system apparently functions by virtue of its ability to use the electrical properties of cells to convey and pro-cess sensory information as well as to control the motor output, it is mandatory that the differentiation of the excitable membrane be understood. A cell may become more differentiated or specialized either through the loss of properties which it already possesses or through the acquisition of new traits. With respect to membrane properties there are examples of both. For example, the membranes of the initial cell—the fertilized egg—are electrically excitable in tunicates, coelenterates, echinoderms, annelids and some vertebrates. The eggs of chordates and some arthropods, however, are not excitable (Hagiwara and Jaffe, 1979). In all excitable egg cells the action potential current is carried by Ca^{2+} ions or by a combination of Ca^{2+} and Na^+ ions. No pure Na^+ action potentials have been observed in eggs. It follows that the development of nonexcitable tissues during embryogenesis must result in the loss of the egg action-potential mechanism. Little information has been obtained about this process.

In contrast, the acquisition of electrical excitability has been examined in some detail, both in nerve cells (see Spitzer, 1979, for a review) and in skeletal muscle (see Chapter 2). Of the nerve cell lineages examined, perhaps the best example is the Rohon-Beard neurons of amphibian embryos. The differentiation of this cell has also been examined in cell culture, allowing a direct comparison to be made between the *in vitro* and *in vivo* developmental sequences, and answering questions about the role of the environment on the differentiation of the nerve cell membrane.

3.9.1. *The Differentiation of Rohon-Beard Neurons.* Rohon-Beard cells are primary sensory neurons in *Xenopus,* the African clawed toad, which undergo their last mitosis during the gastrula stage (Spitzer, 1979). The *Xenopus* egg is not electrically excitable, and the Rohon-Beard cells are not electrically excitable at the time of closure of the neural tube. Within a few hours after neurulation, pure Ca^{2+}-dependent action potentials appear. The action potential disappears if external Ca^{2+} is removed, and it is blocked by Ca^{2+}-channel blockers such as Co^{2+}. After 1 day the same cells exhibit both Ca^{2+} and Na^+ components. The mixed action potential lasts for another 2 or 3 days of development (Figure 3.23). The Ca^{2+} component is then lost and the action potential current carried exclusively by Na^+ (it is eliminated by the removal of external Na^+ and by tetrodotoxin, a specific blocker of Na^+ channels; Co^{2+} has no effect). Since this sequence of action-potential mechanisms takes place in the absence of cell division, it follows that it reflects either a restructuring of preexisting membrane or the insertion of new membrane components.

The next question asked about these cells was whether the above sequence of membrane differentiation requires external cues for it to occur or results from a preprogrammed set of instructions that are implemented independently of the environment. Spitzer and Lamborghini (1976) tested these alternatives by placing cells from the neural plate into tissue culture at a low cell density, so that the cells were not in

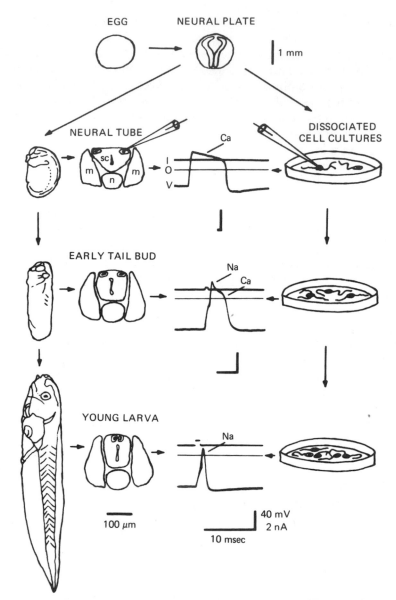

Figure 3.23. Developmental sequence of Rohon-Beard neurons. Cultured neurons prepared from the neural plate were compared with neurons in intact embryos of the same stage. The intracellular recordings illustrate the changes in ionic dependence of the action potential from Ca^{2+} to Na^+. sc, Spinal cord; m, myotome; n, notochord. From Spitzer (1981).

contact with each other, and following the development of the action potential. The neurons were identified in the cultures on the basis of morphology. Electrophysiological measurements from single cells indicated that the sequence of ionic dependence of the action potential was $Ca^{2+} \rightarrow Ca^{2+} + Na^+ \rightarrow Na^+$. This sequence occurred in cell culture with a time course essentially the same as that *in vivo* (Figure 3.23). The data on the Rohon-Beard cells show that a normal developmental sequence which occurs *in vivo* can be reproduced in a very simple cell culture environment, for the cells were cultured in a balanced salt solution plus 0.1% bovine serum albumin.

3.9.2.   *Clonal Cell Lines.*   The electrophysiological properties of many types of cultured cells have recently been summarized (Nelson and Lieberman, 1981) and will not be dealt with in detail here. There is an extensive literature on primary cultures from the mammalian CNS, but the cellular heterogeneity of this population limits its experimental usefulness to pharmacological and biophysical studies. Cells in dissociated CNS cultures do make synapses, and the electrical properties of the cells change with time in culture. It has not, however, become clear how information gleaned from this preparation can lead to a better understanding of the development of the nervous system. In contrast to heterogeneous primary cultures, clonal nerve cell lines offer the electrophysiologist the advantages of (1) a stable phenotype over long periods of time, (2) the ability to manipulate at least the morphological differentiation of the cells by such exogenous reagents as cyclic AMP, (3) the potential for genetic manipulation, and (4) in most cases, a relatively larger cell diameter than that of cells *in vivo,* facilitating electrical recording from the cells. To date, however, the usefulness of the clonal cell lines with respect to understanding electrical differentiation of nerve cells has been more limited than was hoped when the lines were initially established. The primary limitations are the generally unknown relationship between the clonal cell lines and their *in vivo* counterparts and the lack of reproducible evidence that any clonal cell line can develop in culture from a cell with completely passive membrane properties to a cell with a well-differentiated nerve cell membrane. Although there is some disagreement among laboratories, a conservative interpretation of the data suggests that all of the clonal nerve cell lines are able to generate action potentials at their most undifferentiated stages, and that any electrical differentiation of the cell is limited to quantitative changes in excitability, accompanied in some cases by changes in ion selectivity. Similar observations have been made with clonal myoblast cultures (Kidokoro, 1975a). These results are not surprising, as the criterion used to define or classify a cell line as neuronal has been its ability to generate action potentials; nonexcitable cells would have been classified as glia. In addition, it is likely that cells obtain most of their nervelike properties at a stage very early in development, preceding the transformation event that leads to the formation of the tumors from which the clonal cell lines are derived.

Initial studies on clones of the C1300 mouse neuroblastoma suggested that during the process of differentiation from a round cell to a morphologically complex cell with neurites there was a change in membrane properties from passive to "very excitable." This hypothesis was formally ruled out by the demonstration that C1300 cells growing for 20 generations in suspension culture, where they were unable to differentiate, were capable of generating action potentials (Schubert et al., 1974a). Even recordings from pairs of cells in the process of mitosis showed a functional action-potential mechanism.

That exponentially dividing C1300 clones generate action potentials has been confirmed, and it has been shown that the action potentials of the round cells are Ca^{2+} dependent (Miyake, 1978). On morphological differentiation cells of the clone used by Miyake produced mixed Na^+- and Ca^{2+}-dependent action potentials, and selection of the most morphologically differentiated cells indicated that they usually produced only Na^+-dependent action potentials. These results show that in at least one clone of C1300 there may be a differentiation of the neuronal membrane, but it may not develop from a passive cell. The sequence of ion-channel use for the action-potential mechanism is the same as in Rohon-Beard cells.

Similarly, it has been claimed that the PC12 clone is electrically inexcitable in the absence of NGF and acquires excitability after exposure to NGF (Dichter et al., 1977). Experiments in other laboratories, however, have shown that PC12 cells which have never been exposed to NGF are excitable by electrophysiological (A. Ritchie, personal communication) and ion-flux methods (Stallcup, 1979). The inability to record action potentials was probably, as with the C1300 clones, due to technical difficulties associated with penetrating the small non-NGF-treated cells compared with the relative ease with which the very large cells found after exposure to NGF are penetrated. There may, however, be quantitative differences in the degree of electrical excitability before and after exposure to NGF, as there is about a sixfold increase in the density of sodium channels after exposure of PC12 cells to NGF (Rudy et al., 1982).

The inability to detect electrically excitable cells in culture can be caused by a number of circumstances. (1) Culture conditions and the general state of health of the culture can affect the responses. It has repeatedly been observed in our laboratory that slight changes in culture conditions, such as pH shifts or varying serum concentrations, drastically effect the ability to record from cells, as evidenced by lower resting potentials and difficulty in obtaining stable penetrations. The fact that some electrophysiological work has been carried out under conditions of serum starvation may explain the diversity of observed responses. (2) The electrophysiological technique that is employed can lead to artifacts. Most of the published work on C1300 neuroblastoma cells has been done by penetration of the cell with a single electrode and use of a bridge circuit to both pass current and record voltage changes. The extreme fragility of nerve cells makes it necessary to use high-resistance electrodes. The long time constant of the potential change in these electrodes can mask the membrane response, with the result that an electrically active membrane can look like a passive membrane and will be classified as such. (3) Another problem is that many times the cell is damaged during penetration. The resting potential might be lowered, the action-potential mechanism could be inactivated, and again a normally excitable cell would be recorded as having a passive membrane. (4) In an attempt to get around the problem of low and variable resting potentials, a number of workers routinely pass a steady current into the cell to hyperpolarize the cell artificially to a standard value of -60 to -80 mV. They then pass pulses of current through the electrode to test for excitability. These experiments have all been done with a single electrode and a bridge circuit, and the electrode may not be measuring the true membrane potential, resulting in false negative assays for membrane excitability. The only valid control against this possibility is to use two electrodes and show that the current-passing electrode records the same voltage changes as the recording electrode.

Although the clonal lines may have been of limited interest for novel electrophysiological experiments, they could be of tremendous use in neurotransmitter and peptide pharmacology. Cell-surface receptors for virtually all known transmitters have been found on clonal cell lines, and receptors for neurally active peptides are likely to be found. Since the physiological responses of clonal cells can be measured by means other than classical electrophysiological methods, such as ion-flux measurements, they could be used both for the routine characterization of pharmacologically active reagents and for examination of the mechanisms by which these reagents work. Surprisingly little enthusiasm has been generated for the use of clonal lines for these types of studies.

—4—

Glia

4.1 *In Vivo* Origin of Glial Cells

Like neurons, glia are derived from the germinal cells near the surface of embryonic cerebral ventricles. Nerve and glia are generated in the ventricular germinal zone and migrate outward toward their ultimate positions in the nervous system. There are two alternative models for the origin of glial cells: (1) Schaper suggested in 1897 that the same germinal cells give rise to both nerve and glia. (2) In 1889 His proposed that there are separate classes of germinal cells producing neurons and glial cells. Since the time when these alternatives were clearly delineated, a number of experiments have been done to decide between them. To date, both models have accumulated supportive evidence. The more recent *in vivo* data strengthen the hypothesis that the proliferative zone contains two cell lineages, one giving rise to nerve and the other to glia. Using electron microscopy and an antibody to a glia-specific protein (GFAP) to identify cells, Levitt et al. (1981) showed that there are two distinct populations of mitotic cells in the ventricular zone. One is GFAP-positive and the other is GFAP-negative. These observations are compatible with His's idea that nerve and glia cells arise from separate precursor populations. They do not, however, eliminate the possibility that the GFAP-negative cells comprise several types of cells, some of which may differentiate into nerve or glia, nor do they address the identity of the common precursor to the two cell populations in the ventricular zone. In contrast to the *in vivo* evidence of Levitt et al., clonal cell lines have been isolated which are able to give rise to both nerve and glia, and several clonal cell lines have properties of both nerve and glia (see Section 3.2). Since nerve and glia must have a common precursor at some stage in development, it will be important to define its properties and determine the time of its appearance in the developing nervous system.

In any given area of the nervous system neurons are invariably present before glial cells. Glia then migrate into the region and divide. The presence of nerve is required for glial proliferation, for glia do not divide if the nerve cells with which they are normally associated are eliminated (Cowan et al., 1968). Finally, some glial cells divide in normal adult animals, while neuronal cells normally do not undergo mitosis. The function of this continuing gliogenesis is not known.

4.2 Types of Glial Cells

There are three classes of glial cells in the mammalian CNS—astrocytes, oligodendrocytes, and microglia. The astrocytes were the first group to be described. Brain sections stained by Cajal's gold-sublimate method contain a population of cells that are star-shaped and whose processes are closely associated with blood vessels and nerve cell bodies. These are astrocytes, and they are divided into two classes. Fibrous astrocytes contain large numbers of histochemically demonstrable cytoplasmic fibrils (IFs) and are found primarily in white matter. The astrocytes in gray matter contain fewer fibrils than fibrous astrocytes and are called protoplasmic astrocytes.

Astrocytes have several known functions in the brain, and it is likely that more will be discovered as these cells are studied in greater detail. The initial function assigned to astrocytes was providing for the physical integrity of the CNS. Although astrocytes can be viewed as the structural support for nerve cells, their role is not passive. They are capable of taking up extracellular neurotransmitters and influencing the local environment of nerve cells by limiting the free diffusion of neurotransmitters and altering the redistribution of ions around nerve cells.

It has been argued that a major role of glial cells in the CNS is the regulation of potassium ions in the extracellular space surrounding neurons (Kuffler and Nicholls, 1966). Electrically active nerve cells release K^+ into their immediate environment. If these ions were not removed, they would depolarize the nerve cell and cause an increase in the excitability of the nerve. In some invertebrates and amphibians, the glial membrane allows the passive movement of K^+ through the cell in response to a local increase in K^+ concentration. The membrane behaves as an ideal potassium electrode, which allows its membrane potential to be used directly to measure external K^+ concentration. This is also true of mammalian astrocytes, but the membrane characteristics of these cells are not as ideal as those of phylogenetically "lower" animals. Thus glia may serve as buffers which limit the build-up of external K^+ around nerve cells. Although the movement of K^+ across the glial cell membrane is passive (it does not require metabolic energy), the membrane potential must be maintained by an energy-dependent Na^+, K^+-dependent ATPase pump. This pump is much more active in glial cells than in nerve, and its activity is also more sensitive to K^+ ion concentration than is that of nerve cells (Henn et al., 1972). Alternatively, astrocytes may act as spatial barriers, keeping potassium released by firing nerve cells in the immediate area, where it can influence the electrical activity of neighboring nerve cells.

Astrocytes also serve as substrata to guide neurons to their proper destinations during development (Jacobson, 1978), and they divide to fill vacant space following the destruction of CNS tissue. Data obtained *in vivo* and from cultured cells suggest that astrocytes may also produce growth, maintenance, or differentiation "factors" that directly affect glial development. These factors will be discussed later.

Oligodendrocytes were first described in gold- and silver-stained preparations as spherical or polygonal cells with fewer processes than astrocytes. Oligodendrocytes are most abundant in white matter, where they are responsible for the myelination of axons. They also myelinate axons passing through gray matter and occur as satellite cells near nerve cell bodies. Myelinating glia in the peripheral nervous system are termed Schwann cells. Finally, there is an extensive and somewhat controversial literature suggesting that oligodendrocytes provide nutritional support for neurons and that there is a transfer of macromolecules from glia to nerve (Peters et al., 1976; Lasek and Tytell, 1981).

The remaining class of glial cells in the CNS is the microglia. Unlike oligodendrocytes and astrocytes, microglia are apparently not derived from neuroectoderm. They are probably of mesodermal origin and invade the nervous system early in development, probably from the vascular system. They remain dormant in the normal brain, but are activated by trauma or infection. At that time they are transformed into macrophages and digest the debris generated by the injury (Imamoto and Leblond, 1978). Since this class of glia is not of ectodermal origin and has not been described in culture, it will not be discussed further.

4.3   Glial Cultures

Glia were the first cell type from the nervous system to be adapted to continuous cell culture, primarily because of the ease with which glial brain tumors can be induced by chemical carcinogens. As outlined in Chapter 1, once a cell is transformed and capable of continuous neoplastic growth in the animal, it is usually easy to grow in culture. One of the first glial cell lines to be adapted to culture was the C6 clone of a chemically induced rat glioma (Benda et al., 1968). This cell line has been studied extensively with respect to its metabolism of glial-specific enzymes and its ability to synthesize a nervous-system-specific protein, S100. The characterization of the C6 glioma was followed by the establishment in clonal culture of a large number of glial cell lines from rats, mice, hamsters, and humans. The vast majority of these lines have phenotypes which more closely resemble that of astrocytes than that of oligodendrocytes. None of the putative oligodendrocyte or Schwann cell lines synthesize myelin, making their classification tenuous. Clonal glial lines have also been used to study the metabolic regulation of specific proteins and neurotransmitter uptake and to develop glial-specific antisera for developmental studies. Although most of the work on the glial cell lines has been descriptive in nature, it has led to an increase in our knowledge of glial functions. The clonal glial cell lines will certainly be used extensively in the future, for glial cells, which were once thought to be mere structural support for neurons, are just beginning to be studied in detail.

In contrast to clonal glial lines, which by definition have lost their ability to normally regulate division, primary cultures of some glia exhibit a more normal pattern of growth control and differentiation. Thus, primary cultures of glia have been employed successfully in the study of the regulation of cell division and myelination. Some of these studies were made possible by the development of cellular markers for glial cells in heterogeneous primary cultures. A number of procedures have been used; most involve staining cells with an antibody or toxin which has been coupled to a fluorescent dye. For example, antisera have been prepared against glia and the antisera adsorbed with nerve and fibroblasts so that they react exclusively with a specific type of glial cell (Schachner, 1974; Fields et al., 1975; Stallcup and Cohn, 1976). Other glial cell markers include GFAP, which is an IF protein found in astrocytes (Bignami et al., 1972; Raff et al., 1978b; see also Section 4.6) and myelin-specific antigens, such as myelin basic protein and galactocerebroside, which are localized to oligodendrocytes and Schwann cells. Most nerve cells, but very few glia, bind tetanus toxin (Dimpfel et al., 1977). Rat fibroblasts contain none of these markers, but do contain the surface antigen Thy 1. Thus, by employing antisera conjugated to different fluorochromes, one can identify the majority of the cell types in cultures containing nerve, two classes of glia, and fibroblasts.

Fibronectin is a high-molecular-weight protein that was first discovered on the surface of mesodermally derived cells. It is involved in the adhesion of cells to collagen substrata. Although there has been some confusion in the literature about the ability of glial cells to synthesize fibronectin, this appears to be due to species-specific differences. Cultured rat astrocytes apparently do not contain fibronectin (Stieg et al., 1980), but astrocytes from chicks do (Kavinsky and Garber, 1979). Some of the other cell types in CNS primary cultures of either species, such as fibroblasts, endothelial cells, and cells derived from the meninges, all stain positively for fibronectin. A clonal rat Schwann cell line also synthesizes fibronectin (Kurkinen and Alitalo, 1979), as do many glial cells derived from tumors of the rat CNS. It follows that the presence of fibronectin on the surface of cultured cells does not exclude the possibility that they are glia.

4.4   Glial Cell Proliferation

During normal development peripheral nerve axons contact and innervate their target tissues in the relative absence of glial cells. Schwann cells originating from the neural crest then migrate down the naked axons, become dispersed over the axon surfaces, and begin to proliferate (Webster, 1975). Once the fibers in the axon bundles are covered with Schwann cells, cell division ceases and the Schwann cells begin to myelinate the axons. There are at least three distinct processes during this developmental sequence—Schwann cell migration, the initiation of cell proliferation, and myelination. Very little is known about the migratory step, but considerable information has been obtained about the regulation of Schwann cell division and myelination. A large amount of this data has been collected from the study of cultured cells.

It has been known for many years that nerve cells are a source of mitogens. The best documented example is limb regeneration in amphibians (Singer, 1974). If a limb is amputated, it regenerates via the rapid proliferation of mesenchymal cells in a blastema arising at the point of amputation. If, however, the limb is denervated at the time of amputation, this mass of mitotic cells does not appear and the limb does not regenerate. Regeneration is limited in mammals, but glial cells are stimulated to divide in the immediate area of a crushed mammalian nerve. It is possible that the release of growth factors from the nerve initiates glial cell division.

To study the mitogenic activity of nerve in culture, two sources of Schwann cells have been exploited—cells from the rat DRG and cells from the rat sciatic nerve. The DRG is a sensory ganglion containing nerve cell bodies, glia, and supporting cells. The sciatic nerve is a major nerve in the leg and has a large number of myelinated axons. Salzer and Bunge (1980) isolated Schwann cells from DRG by explanting the complete mass of ganglion cells from 19-day embryos and placing it on a culture dish which had been coated with collagen to promote cell–substratum adhesion. They then treated the cultures for a few days with an antimitotic to kill the more rapidly dividing fibroblasts. Nerve fibers grew out of the explant, and, after the antimitotic agent was removed, there was an outward migration and proliferation of Schwann cells (Figure 4.1). Once the culture dish and fibers were covered with Schwann cells, nuclear DNA synthesis and cell division diminished and myelination was initiated. Wood and Bunge (1975) obtained a population of quiescent Schwann cells by removing the ganglion cell mass and allowing the nerve fibers to degenerate. When they added nerve cells back to these cultures, the Schwann cells started dividing. Autoradiographic analysis of [^{3}H]thymidine incorporation into DNA showed that most of the DNA synthesis occurred in cells closely associated with neurites. Similar results have been obtained with CNS nerve–glia cocultures (Hanson et al., 1982b). Mitogenic stimulation by neurites requires physical contact between neurite membrane and the Schwann cell, since the effect does not occur when the two are separated by a permeable membrane (Salzer et al., 1980a).

The direct role of neurites in the induction of sensory ganglion Schwann cell proliferation was tested by addition of sensory nerve membrane fractions to quiescent glial cultures (Salzer et al., 1980a; Cassel et al., 1982). Neurites from homogeneous cultures of DRG neurons were homogenized and the neurite membrane fraction added to subconfluent glial cultures. When low-density cultures were used, the percentage of the cells in the dish that incorporated [^{3}H]thymidine during a 24-hr labeling (the labeling index) increased about 20-fold. The extent of induced

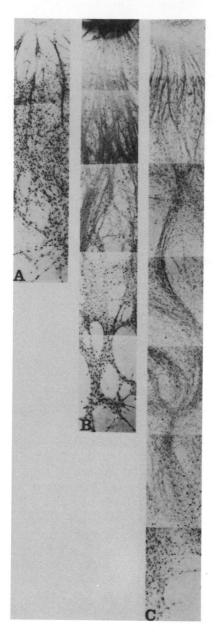

Figure 4.1. Glial outgrowth from rat DRG. Three stages in the development of the outgrowth from an explanted DRG are presented. Explants maintained in culture for 2 weeks (A), for 3.5 weeks (B), and for 5 weeks (C) were incubated with tritiated thymidine for 24 hr and processed for autoradiography. (A) At 2 weeks most cells throughout the outgrowth incorporate label; (B) at 3.5 weeks labeled nuclei are fewer and are concentrated at the periphery of the outgrowth; and (C) at 5 weeks labeled nuclei are rare and confined entirely to the periphery. Labeled nuclei appear as small dark spots at this low power (Toluidine blue stain, ×22). From Salzer and Bunge (1980).

DNA synthesis was dependent on the amount of neurite membrane added (Figure 4.2) and the density of the Schwann cells (Figure 4.3). There was a linear relationship between the amount of membrane added [expressed in units of membrane-associated enzyme, alkaline phosphodiesterase (PDE)] and the labeling index. The fact that confluent glial cultures were unresponsive to added neurites suggests either that growth inhibition by contact inhibition overrides neurite-induced mitogenesis or that the confluent cells have lost the hypothetical surface receptors for neurites. Cell-surface receptors are lost in fibroblasts; for example, the number of receptors for the mitogen EGF decreases 10-fold when the cells reach confluency (Holley et al., 1977). Membranes from fibroblasts and C1300 neuroblastoma cells were not mitogenic, so the mitogenic signal for the DRG Schwann cell preparation may be specific

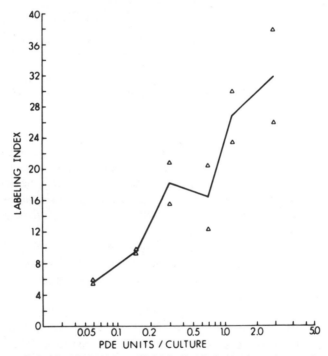

Figure 4.2. Relationship between glial labeling index and neurite membrane concentration. Increasing concentrations of neurite membranes were incubated with DRG Schwann cells. The amount of membrane used at each point is indicated on the x axis in terms of alkaline phosphodiesterase (PDE), a membrane-associated enzyme. The cells were pulse-labeled with [³H]thymidine and the labeling index determined. Duplicate points from a single experiment are shown at each concentration. From Salzer et al. (1980a).

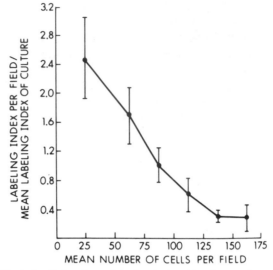

Figure 4.3. Effect of cell density on labeling index. Neurite membranes (1.2 PDE units/well) were added to four primary Schwann cell cultures. The labeling index decreased in regions of high cell density (as defined by the number of cells in the microscope field). The labeling index at each range of cell density was normalized to the mean labeling index per culture; the results for all cultures were then averaged. From Salzer et al. (1980b).

to neurites. The mitogenic potential of DRG nerve cell bodies has not, however, been ruled out. Finally, the mitogenic membrane fragments bound to the surface of the Schwann cells and were phagocytosed. Phagocytosis *per se* was not sufficient for the induction of Schwann cell proliferation, for the nonmitogenic membranes of fibroblasts and other nerve cells were also internalized.

In contrast to the DRG cultures described above, cultures of Schwann cells from sciatic nerve are obtained by treatment of the dissected nerve with trypsin and collagenase to dissociate the glial cells and plating of the resultant material in culture dishes. The cells are then exposed to the antimitotic drug cytosine arabinoside (Ara C) which selectively kills the rapidly dividing fibroblasts. Since fibroblasts but not Schwann cells have the Thy 1.1 surface antigen, the Ara C exposure is followed by a treatment with anti-Thy 1.1 serum plus complement, which together lyse the remaining fibroblasts. The resultant cultures are more than 99% Schwann cells, as determined by the presence of the Ran 1 glia-specific surface antigen on the remaining cell population (Brockes et al., 1979).

When grown in normal modified Eagle's medium containing 10% fetal calf serum, the doubling time of the sciatic nerve Schwann cells is a very slow 8 days. The division rate and incorporation of [^{3}H]thymidine into DNA are increased by the addition of two substances, cyclic AMP and a protein isolated from extracts of brain and pituitary. In most fibroblastlike cells the elevation of intracellular cyclic AMP inhibits cell proliferation and the incorporation of [^{3}H]thymidine into DNA (Pastan et al., 1975). However, reagents which raised intracellular cyclic AMP in sciatic nerve Schwann cell primary cultures stimulated cell division (Raff et al., 1978a). Cholera toxin [an activator of adenyl cyclase (AC)], a 3′,5′-cyclic nucleotide phosphodiesterase inhibitor, and the active dibutyryl derivative of cyclic AMP all stimulate cell proliferation and the incorporation of [^{3}H]thymidine into DNA. Since the amount of cyclic AMP may increase in areas adjacent to damaged peripheral nerve following a nerve crush (Appenzeller and Palmer, 1972), it is possible that this cyclic nucleotide serves as a mitogenic signal for the Schwann cell proliferation that normally follows nerve injury.

In addition to cyclic AMP, a protein isolated from the pituitary and brain also stimulates cell division in rat sciatic nerve primary cultures (Brockes et al., 1980). It was initially noted that a soluble extract of homogenized pituitary glands greatly reduced the 8-day division cycle of glial cultures derived from sciatic nerve. This mitogenic activity was not found in other tissues, nor did known pituitary trophic hormones or growth factors stimulate cell proliferation. The pituitary activity which stimulates sciatic nerve Schwann cell proliferation resides in a 60,000-MW protein that is a dimer of 30,000-MW subunits. This protein also stimulates cell division of CNS glial cells of the astrocyte class. The pituitary factor, termed glial growth factor (GGF), does not stimulate the incorporation of [^{3}H]thymidine into CNS microglia and oligodendrocytes. Like most factors which promote mitosis, GGF requires serum in the culture medium. However, it apparently does not increase intracellular cyclic AMP in the responsive Schwann cells. Since cyclic AMP directly stimulates mitosis, either there are two independent pathways for mitotic stimulation or the increase in intracellular cyclic AMP caused by GGF is too small to detect. Negative results mean very little when intracellular cyclic AMP is measured because of the inherent variability of the assay and the possibility of compartmentalization of cyclic AMP within the responsive cell. The physiological significance of the pituitary mitogen for glia is unknown, but it may be important to the normal development of the nervous system as well as serving as a signal for glial proliferation following injury.

4.5   Myelination and NGF Synthesis

Once Schwann cells are in their final location on the axon, they stop dividing and begin to elaborate myelin around the nerve. The individual Schwann cells lie in a row along the nerve, each separated from its neighbor by a small gap. The Schwann cell membrane then wraps around the axon repeatedly, leaving a flattened sheet of two cell-surface membranes at each circumvolution. The space between individual myelinating Schwann cells lying at intervals along the nerve axon is called a node of Ranvier. The number of layers in each myelin sheath is quite variable, but in general the larger-diameter axons have the thicker myelin. In the peripheral nervous system (PNS) each Schwann cell contributes one segment between the nodes of Ranvier to the myelin sheath. Since not all axons in the PNS are myelinated, it has been argued that the fibers that do become myelinated possess a unique quality that induces the Schwann cells to myelinate them (Spencer and Weinberg, 1978). In contrast to the situation in the PNS, individual oligodendrocytes of the CNS myelinate an average of 40 separate nerve fibers.

The function of the myelin is twofold. It increases the velocity of signal conduction along the nerve and reduces the energy requirement for conducting the impulse. The conduction speed is increased because the current does not have to travel continuously down the axon, but rather is able to jump between the nodes of Ranvier. Since the whole nerve is not involved in impulse propagation, the amount of energy needed to pump ions back into the axon to repolarize the membrane is greatly reduced.

Because myelin consists of multiple layers of cell-surface membrane, about 70% of its dry weight is lipid and the remainder is protein. Some of the lipids and proteins are unique to myelin and serve as useful biochemical markers for myelination in cultured cells. Antisera prepared against galactocerebroside, a myelin-specific glycolipid, and against proteins specific to CNS myelin (myelin basic protein) and PNS myelin (peripheral myelin glycoprotein, P_o) can be used to detect the synthesis of these proteins both *in vivo* and in cultured cells. *In vivo,* Schwann cells only make galactocerebroside and P_o once they have migrated down the nerves, have ceased dividing, and are developmentally committed to myelination. In contrast, myelin basic protein and galactocerebroside are detectable in the CNS before the onset of myelination. Myelin markers are also detected in glial cultures obtained from brain tissue several days before myelin is detected by electron microscopy. Galactocerebroside and myelin basic protein are synthesized by oligodendrocytes from neonatal rat brain for up to 10 weeks in culture in the absence of nerve, suggesting that their synthesis is constitutive in oligodendrocytes. In contrast, when Schwann cells were dissociated from neonatal sciatic nerve and placed in culture in the absence of nerve, they initially contained galactocerebroside and P_o, but the number of cells which contained these molecules then declined, with a half-life between 2 and 3 days (Mirsky et al., 1980). The number of Schwann cells in the cultures increased during this time, so it is not likely that the loss of P_o and galactocerebroside was due to cell death. It was probably due to a progressive loss of the antigens. It follows that the continual presence of nerve was required for Schwann cells to synthesize myelin constituents. Additional evidence for this conclusion was obtained from the study of rat DRG cultures.

As indicated earlier, Schwann cells migrate down nerve processes from their site of origin in the neural crest. The migrating glial cells lack a basal lamina. The basal lamina is an 80-nm-thick structure that lies immediately external to the plasma membrane. It is made up of collagen, glycosaminoglycans and glycoproteins, and it surrounds many tissues, creating spatial constraints on the tissue. When Schwann cells myelinate part of an axon, a basal lamina is formed immediately exterior to the myelinated axon. The whole bundle of axons is surrounded by another membrane called the perineurium. The space between the myelinated axons (the endoneurium) is filled with extracellular matrix material, collagen fibrils, and a few fibroblasts.

The origins of the Schwann cell basal lamina, the perineurium, and the endoneurium were discovered from the study of cultured rat DRG. (M. Bunge et al., 1980; R. Bunge and M. Bunge, 1981). This preparation has the advantage that cultures of nerve alone, Schwann cells alone, fibroblasts alone, or mixed cultures of defined cell types can be prepared. When these cultures were examined by electron microscopy, the pure neuronal cultures contained fascicles of neurites, but no basement membrane. Schwann cell cultures also lacked basement membrane, although the Schwann cells were able to synthesize collagen proteins. However, when nerve and Schwann cells were grown together, both basal lamina and collagen fibrils were associated with the Schwann cells, and myelinated fibers were observed. Since these cultures did not contain fibroblasts, it was concluded that the Schwann cells synthesize their own basement membrane, and that at least the assembly of this structure requires the presence of the nerve cells in the cultures. The cultures containing only nerve and glia did not, however, make the perineurium. This structure was observed in cocultures of nerve, glia, and fibroblasts, but not in cultures of nerve and fibroblasts. These experiments show that the fibroblasts or the glia make the perineurium and that the presence of both cell types is required for its synthesis. Finally, cultured Schwann cells need a collagen substratum to myelinate nerve fibers; Schwann cells located on axons that are not associated with the collagen-coated culture dish did not myelinate (R. Bunge and M. Bunge, 1981).

Occasionally DRG cultures contain fascicles of neurites suspended in the culture medium above the substratum of the culture dish. Whereas the attached neurites are myelinated, the Schwann cells associated with the suspended neurites appear to be blocked at a developmental stage prior to ensheathment and myelination. However, if slips of plastic covered with a collagen matrix are placed upon these arboreal fascicles, the block in Schwann cell development is removed and they myelinate normally (R. Bunge and M. Bunge, 1981). Since the area of normal development coincides only with that area in contact with the collagen, these investigators concluded that Schwann cells require contact with extracellular matrix material to develop normally. An alternative, that the nerve fiber requires contact with the matrix material to permit myelination, has not been ruled out.

A final requirement for Schwann cell myelination in culture appears to be the presence of the proteolytic enzyme plasmin in the growth medium (Kalderon, 1979). Plasmin has a trypsinlike enzyme activity which is normally associated with the liquefaction of blood clots by the hydrolysis of the insoluble clotting protein fibrin. Plasmin, however, acts on a wide range of protein substrates. Plasmin is derived from an inactive zymogen, plasminogen, by the action of another enzyme called plasminogen activator. In cultures of spinal cord neurons and Schwann cells, Schwann cell migration and myelination take place only if the culture system is able to produce active plasmin. If plasmin activity is prevented by protease inhibitors or by the removal of plasminogen from the culture medium, the movement of Schwann cells and their ability to myelinate nerve fibers are inhibited. Since plasminogen is required in the culture medium, one or more of the cell types within the cultures must make plasminogen activator. Although the site of action of plasminogen activator is unknown, it may promote cell motility by altering the cell–substratum adhesiveness of cells. The role of plasminogen activator and similar proteases has been demonstrated in many other biological processes, including normal cell migration, tumor cell invasiveness, blastocyst implantation, and basement membrane metabolism.

In addition to supporting the electrical function of nerve cells by myelination of axons, glial cells probably facilitate the differentiation of nerves by providing specific growth, differentiation, and maintenance "factors." Cultured glial cells should be ideal sources for such molecules, but to date only NGF has been isolated reproducibly (see Chapter

3 for extensive discussion of NGF). Evidence for the glial synthesis of NGF initially came from a clonal glioma cell line (Longo and Penhoet, 1974). The NGF derived from this cell line had the same isoelectric point as mouse NGF, reacted with antiserum to mouse NGF, and gave responses similar to true NGF in two bioassay systems. Other clonal glial cell lines also synthesize and secrete a biological activity similar to NGF (Murphy et al., 1977). Some peripheral nerves are dependent on NGF, and the regeneration of central noradrenergic neurons is enhanced by exogenously applied NGF (Stenevi et al., 1974); it follows that the surrounding glia may supply the NGF necessary for proper nerve growth and function. It should be pointed out, however, that glia are generally not present during the early phases of axon elongation. Another source of NGF may be required during the initial stages of nerve differentiation.

4.6 S100 and Glial Fibrillary Acidic Protein

In addition to myelin-specific proteins and cerebrosides, two other proteins appear to be synthesized largely by glial cells. These are the S100 protein and glial fibrillary acidic protein (GFAP) (for reviews see Eng, 1980, and Zomzely-Neurath and Walker, 1980). GFAP was originally isolated from plaques of brain tissue from multiple sclerosis (MS) patients. MS plaques arise from the division of fibrous astrocytes (a process termed gliosis) in an area of brain damaged by the disease. These astrocytes contain a high density of 10-nm glial filaments in their cytoplasm. When extracts of gliosed tissue were run on acrylamide gels, a major protein band of approximately 50,000 MW was found. This protein, GFAP, was purified and antiserum prepared against it. The serum reacted with astrocytic glial filaments but not with IFs of other cell types (Bignami et al., 1972). Through the use of these antisera and immunohistochemical techniques, it was established that GFAP is primarily localized to astrocytes. Antisera to GFAP are currently widely used to identify astrocytes both *in vivo* and in culture. Some human and rodent astrocytomas also contain GFAP, allowing for the unambiguous identification of these neoplasms.

In contrast to GFAP, tetanus toxin, and galactocerebroside, the nervous-system-specific protein S100 has been of little value as a marker for specific cell types (for a review, see Zomzely-Neurath and Walker, 1980). During their search for proteins that are localized to the nervous system, Moore and McGregor (1965) detected a protein which was soluble in saturated ammonium sulfate and was thus termed S100. It was found in both the CNS and the PNS, but at least 10,000-fold less was present in other tissues. It has a molecular weight of approximately

20,000, is highly acidic, and binds calcium ions (Starostina et al., 1981). To date no function has been ascribed to S100, although it may mediate protein phosphorylation (Patel and Marangos, 1982). In the bovine brain, S100 is a mixture of two proteins, S100-a and S100-b. Both are made up of 10,500-MW subunits. S100-b is a dimer of two identical subunits, while S100-a is a dimer consisting of one S100-b subunit and one unique subunit which is closely related in amino acid sequence to S100-b (Isobe et al., 1981).

In vivo, most of the S100 is found in glial cells of both the CNS and the PNS, but it occurs to a lesser extent in neurons as well. Developmental studies showed that the amount of S100 in rat brain increased more than 20-fold between birth and 25 days after birth, at which time it reached its adult level (Herschman et al., 1971). This postnatal increase in S100 protein occurred after the cessation of neuroblast proliferation and during the period of extensive glial cell division. The clonal C6 glial cell line also synthesizes S100 protein, but the synthesis of S100 is low during exponential growth. After the cells reach confluency there is an approximately sixfold increase in the rate of synthesis of S100 protein (Pfeiffer et al., 1970; Labourdette and Marks, 1975). Experiments showed that both the cessation of division and increased cell contact were responsible for the differences in S100 metabolism between dividing and stationary-phase glial cultures.

Studies with clonal cell lines have rigorously defined the types of cells that synthesize S100. The immunohistochemical methods used for localizing S100 *in vivo* only detect the presence of the antigen in a particular cell type. It is possible that the protein is synthesized by another cell and concentrated by the positively staining cell. To clarify the site of S100 synthesis, Schubert and colleagues (1974*b*) assayed S100 in nerve and glial cell lines derived from both the CNS and the PNS. In a large collection of rat clonal cell lines derived from chemically induced brain tumors, S100 was found in approximately 80% of the glial lines and 80% of the nerve cell lines. The concentrations of S100 in the glial cells were, however, generally higher than in the nerve cell lines. S100 protein was also detected in continuous cell lines of human melanomas (Gaynor et al., 1980). S100 protein is thus widely distributed among the various cell types derived from neuroectoderm.

4.7   Neurotransmitter Metabolism

For many years it was believed that glial cells played no role in neurotransmitter metabolism. This function was attributed uniquely to the electrically excitable nerve cell. However, with the advent of isotopic tracer techniques and homogeneous glial cell cultures, it has become clear that glial cells actively synthesize, concentrate, and release GABA, acetylcholine, and other putative neurotransmitters. However, they apparently do not accumulate or synthesize biogenic amines, such as NE and DA. We will discuss neurotransmitter uptake and then describe evidence that glial cells also synthesize and release GABA and ACh.

4.7.1. *Neurotransmitter Uptake.*   There are two classes of GABA-uptake systems in the nervous system, which are distinguished by their Michaelis-Menten constants. The high-affinity transport systems have K_m's around 10^{-6} M, whereas the low-affinity K_m's approach 10^{-3} M. These low- and high-affinity transport systems were initially characterized in several preparations, including brain slices, synaptosomes, and fractionated cell populations. The cellular heterogeneity of these preparations made it difficult to identify the types of cells involved in GABA uptake. However, by using low concentrations of exogenous [^{3}H]GABA and examining the accumulation of isotope in nervous tissue by autoradiographic techniques, Iversen and Kelly (1975) showed that GABA was taken up by high-affinity, saturable membrane systems of glia in both the PNS and the CNS. CNS GABA-releasing neurons and synaptosome preparations also concentrated exogenous GABA. Since the sensory and sympathetic ganglia of the PNS do not contain nerve terminals that concentrate GABA but do have glial cells that transport GABA, it was possible to study the glial GABA-uptake system independently of neuronal uptake. Like uptake into nerve, glial uptake of GABA was sodium and temperature dependent, and was blocked by a variety of GABA analogues. The specificity of the analogue inhibition, however, was different, for β-alanine was a potent inhibitor of glial uptake but was not effective in nerve terminal preparations. Conversely, 2,4-diaminobutyric acid was a much more potent inhibitor of GABA uptake in nerve cells than in glial cells. These results with pieces of tissue are in general agreement with those from dissociated cell cultures and clonal cell lines.

When a large number of clonal cell lines from the rat nervous system were assayed for their ability to take up exogenous GABA, both nerve and glial cells were found to concentrate GABA up to sevenfold (Hutchinson et al., 1974; Schrier and Thompson, 1974; Schubert, 1975). The ratio of GABA concentration in cells to that in medium was appreciably lower in cell lines than in explants of CNS tissue. Cell/medium uptake ratios were frequently greater than 100 in explant cultures. The uptake mechanism in the cell lines was, however, saturable, dependent on exogenous sodium, and temperature dependent. If calcium was deleted from the uptake buffer, the cells were able to equilibrate with exogeneous GABA but were unable to concentrate GABA. This suggests that the concentration process has a greater calcium ion dependence than the equilibration mechanism. The specificity of the GABA uptake processes in the clonal nerve and glial lines was, however, different from that observed *in vivo*. β-Alanine, which inhibits glia-mediated GABA uptake but not uptake by nerves *in vivo,* inhibited GABA uptake in both neuronal and glial cell lines. 2,4-Diaminobutyric acid, which blocks *in vivo* nerve uptake of GABA, had no effect on GABA uptake by cultured nerve cells. Clonal cell lines of other differentiated cell types, such as liver, plasmacytes, and pituitary cells, were impermeable to GABA or equilibrated with exogenous GABA but did not concentrate it (Schubert, 1975). These results with clonal cell lines support the *in vivo* data, which show that glial cells play a role in the removal of GABA from the extracellular space.

The exact functions of the glial GABA-uptake systems are unknown. One possibility is that they rapidly remove newly released GABA from synapses to reduce the signal-to-noise ratio at the synapse. Desarmenien and coworkers (1980) tested this hypothesis directly by blocking glial uptake of GABA in the rat DRG and then recording the electrophysiological responses to GABA released from micropipettes onto the surface of nerve cells. Under normal conditions GABA elicits a depolarizing response in the DRG. When GABA uptake by glia was blocked by 1 mM *l*-alanine, there was no effect on nerve cell membrane potential, input resistance, or the amplitude and time course of the GABA depolarizing response. These results show that the inhibition of glial GABA uptake does not alter the electrophysiological response of DRG neurons to iontophoretically applied GABA and, by extension, suggest that synaptic transmission between GABA DRG neurons is not directly altered by glial uptake.

As outlined above, CNS tissues have both high- and low-affinity GABA uptake systems. Experiments with a rat glioma cell line which has a high-affinity ($K_m = 30–70 \ \mu M$) uptake system for *l*-alanine and taurine and a low-affinity ($K_m = 700 \ \mu M$) uptake system for GABA show that the low-affinity of the uptake system for GABA may be a reflection of GABA being transported through the high-affinity uptake sites for the *l*-amino acids, *l*-alanine and taurine (Martin and Shain, 1979). Since taurine, *l*-alanine, and GABA were mutually competitive inhibitors, and since 16 structural analogues of these three amino acids had similar inhibitory effects on the transport of all three substrates, it was concluded that all three entered the cell via a common transport system which had a different affinity for each. It follows that where both low- and high-affinity GABA-uptake systems exist, the low-affinity GABA influx may take place via transport through the high-affinity uptake system for *l*-alanine and taurine.

4.7.2. Neurotransmitter Synthesis and Release. In addition to transporting and concentrating exogenous neurotransmitters, glial cells are capable of synthesizing and releasing some transmitters. The first observation that suggested that glial cells can release a chemical transmitter was the detection of miniature endplate potentials (MEPPs) following the denervation of frog skeletal muscle (Katz and Miledi, 1959). After the nerve terminal degenerates following denervation, Schwann cells occupy the position previously held by the terminal. They then release quanta of ACh onto the muscle surface, producing the membrane depolarizations (Miledi and Slater, 1968). Since Schwann cells can synthesize low amounts of ACh (Heumann et al., 1981b), it is not necessary to hypothesize that Schwann cells acquire ACh through the uptake of ACh released by degenerating nerve terminals. The properties of the Schwann cell ACh-release mechanism are, however, very different from those of nerve terminals. In nerve terminals, membrane depolarization can be achieved by increasing exogenous K^+. Depolarization opens calcium channels, resulting in an influx of Ca^{2+} and release of ACh. In contrast, Schwann cells in high-K^+ medium do not release ACh (Miledi and Slater, 1968). Since the influx of Ca^{2+} into nerve terminals stimulates the release of ACh, it was asked if Ca^{2+} influx into Schwann cells releases ACh. When intracellular Ca^{2+} was increased by the use of several calcium ionophores, there was no increase in neurotransmitter release (Ito and Miledi, 1977). Similar conditions increased the rate of ACh release from axon terminals 100-fold. It follows that although Schwann cells can synthesize and release ACh, the mechanism of release is different from that of nerve terminals.

Glia can also synthesize and release the inhibitory neurotransmitter GABA. A low level of GABA synthesis appears to be a property of all cells derived from neuroectoderm. An examination of the free amino acid pools of many different cultured cell types showed that nerve, glia, and melanocytes all contained GABA, whereas none was detected in mesodermally derived cells (Schubert et al., 1975). In one group of approximately 20 CNS clonal cell lines of glia and nerve, the nerve cells contained between 6 and 75 residues of GABA per 1,000 free amino acids residues, whereas the glial cells generally contained only a few. Glutamic acid decarboxylase (GAD), the enzyme which synthesizes GABA from glutamic acid, was also detected in glia. The specific activity of

GAD was approximately 10-fold greater in stationary-phase C6 glioma cells than in exponentially dividing cells (Schrier and Thompson, 1974). These cells also spontaneously release GABA into the culture medium.

Evidence from other experimental preparations suggests that some types of glial cells can release GABA in a manner analogous to that of nerve terminals. When rat DRG are incubated with [^{3}H]GABA, only the glia concentrate isotope. When cells "loaded" with [^{3}H]GABA are incubated with increasing amounts of exogenous K^+, there is an enhancement of the rate of GABA release, reaching a maximum, fourfold stimulation at 64 mM K^+ (Minchin and Iversen, 1974). This K^+-stimulated increase is inhibited in Ca^{2+}-free medium, showing that GABA is released via a calcium-dependent mechanism similar to that of nerve terminals. In contrast to the results with DRG glia, potassium-stimulated release of GABA and CNS glia showed no calcium dependence (Sellstrom and Hamberger, 1977). Glia of rat SCG accumulated exogenous [^{3}H]GABA, and the rate of release from preloaded cells was increased approximately fourfold by 50 mM K^+ (Bowery et al., 1979). Ca^{2+} was not, however, required in the culture medium for the maximum evoked release. It can be concluded that some glial cells synthesize ACh and GABA and are capable of releasing these neurotransmitters on depolarization. The mechanism of release is not, however, always similar to that in nerve terminals. It is likely that additional neurotransmitters and neuropeptides will be found in glia in the near future.

Finally, an argument can be made for an additional role for neurotransmitters synthesized or concentrated by glial cells. Since nerve cells release potassium ions during the conduction of action potentials, these ions may depolarize glia, resulting in the release of transmitter. This glia-derived neurotransmitter may then alter the physiological properties of neighboring nerve cells in much the same manner as if it came from nerve cells. It follows that glia may modulate nerve cell activity by a process independent of neurotransmitter uptake and exogenous ion buffering.

4.7.3. *Neurotransmitter Receptors.*   Glia probably do not contain the spectrum of neurotransmitter receptors that is found on nerve cells. Although there is little published evidence for this conclusion from *in vivo* studies, it is supported by data from work with cultured cells. An examination of approximately 15 clonal nerve and glial cell lines derived from the rat CNS showed that the glia contained no detectable GABA or nicotinic ACh receptors (Napias et al., 1980; Schubert et al., 1974b). In contrast to the glia, five of the five nerve cell clones examined contained GABA receptors, and three contained nicotiniclike ACh receptors. The majority of the glial cell lines in this collection did contain typical β-adrenergic receptors, for the catecholamines NE and DA stimulated the intracellular accumulation of cyclic AMP between 3- and 130-fold (Schubert et al., 1976). Similar observations were made with the C6 glioma clone, in which NE caused a 100-fold increase in intracellular cyclic AMP (Gilman and Nirenberg, 1971; Oey, 1976). Because the response was blocked by β-adrenergic-receptor antagonists but not by α-adrenergic-receptor antagonists, the NE effect must have been mediated by a typical β-adrenergic receptor. The cyclic AMP response was also dependent on intracellular calcium, for when the calcium content of C6 was reduced fivefold by repeated exposure of the cells to a calcium chelator the cyclic AMP accumulation was reduced by 70% (Brostrom et al., 1979). This effect was reversed rapidly by the addition of extracellular calcium. NE and analogues of cyclic AMP also increased the specific activity of the glia-specific nucleoside-2',3'-cyclicphosphate 3'-nucleotidohydrolase, an enzyme which hydrolyzes 2',3'-cyclic AMP but not 3',5'-cyclic AMP (McMorris, 1977). In additional studies with the popular C6 clone, it was shown that the concentration of the β-adrenergic receptor varied as a function of the cell's growth cycle (Charlton and Venter, 1980). The highest concentration (9000 receptors/cell) was found during the S phase, and the receptor density was reduced to about 1500 receptors/cell during M, G1, and G2.

Finally, β-adrenergic agonists and cyclic AMP alter another property of glial cells—cell morphology—which we turn to next.

4.8 Morphology

When dibutyryl cyclic AMP is added to primary cultures of CNS glia or to cultures of most clonal glial cell lines, there is an increase in morphological complexity. The cells acquire multiple long processes, thought to be similar to those observed *in vivo* (Edstrom et al., 1974). These morphological changes are evident within a few hours and are usually reversible once the cyclic AMP analogue is removed. Agents which increase intracellular cyclic AMP, such as prostaglandin E_1, cholera toxin, and NE, cause morphological changes identical to those elicited by dibutyryl cyclic AMP. In C6 glioma cultures, there is an approximately 20-fold increase in GFAP-positive cells in morphologically differentiated stationary-phase cultures relative to the number in exponentially dividing cultures; dibutyryl cyclic AMP causes a similar increase in GFAP-positive cells and morphological complexity when added to dividing cultures (Raju et al., 1980). Thus, as in nerve cells, cyclic AMP mediates the morphological differentiation of glial cells. However, nerve cell lines from the rat CNS respond morphologically to both serum-free medium and cyclic AMP, whereas the morphology of CNS glia is altered only by cyclic AMP (Schubert et al., 1974a).

Neuronal and glial cells also differ in the manner in which they acquire their more complex morphologies. Nerve cells extend processes by growth behind an active growth cone (see Chapter 3). In contrast, studies have shown cultured glial cells acquire the majority of their morphological complexity by the retraction of an initially flattened cytoplasm (Steinbach and Schubert, 1975). Dibutyryl cyclic AMP (1 mM) was added to rat CNS glial lines that normally grew as flattened cells, and the morphological response was followed by time lapse microcinematography. Within 5 min after the addition of the cyclic nucleotide, the cytoplasm was partially retracted, leaving phase-dark processes behind. This retraction continued for several hours, producing the thin processes of the fully differentiated cell (Figure 4.4). The cells with this type of morphological response were termed "shrinkers." The cytoplasm appeared to flow inward toward the nucleus, and the cell body became more spherical. Usually the processes became progressively narrower and phase darker as the cytoplasm retracted, and respreading of parts of a process was rarely seen. All three glia examined in this study responded to cyclic AMP primarily by cytoplasmic retraction, whereas all five nerve cell lines extended processes exclusively by neurite outgrowth. These two modes of morphological differentiation were not mutu-

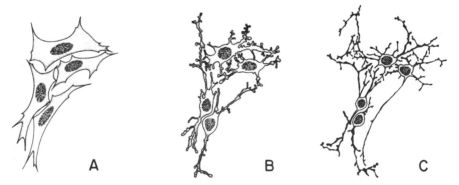

Figure 4.4. Cytoplasmic retraction by glial cells. B9 glial cells were plated at a low cell density. Two days later 1 mM dibutyryl cAMP was added and the morphological response of the cells followed by time lapse photography. Tracings were made of a group of cells from individual frames taken at the indicated times after the addition of the cyclic nucleotide. (A) 5 min. (B) 30 min. (C) 5 hr. From Steinbach and Schubert, 1975.

ally exclusive, for growth was seen at the ends of some glial processes previously formed by shrinkage, and all of the cells responded to dibutyryl cyclic AMP by becoming rounder in the area of the nucleus. As outlined in Chapter 3, most cell types show a morphological response to cyclic AMP. However, the only cell type besides glia known to develop processes by cytoplasmic shrinkage is the newt iris epithelial cell (Ortiz et al., 1973). In contrast, fibroblasts spread on the substratum in response to dibutyryl cyclic AMP, but do not acquire processes similar to those observed in glial cells.

—5—

Nerve–Muscle Interactions

5.1 Synaptic Specificity

A basic tenet of neurobiology is that individual neurons make synaptic connections with a limited subset of other neurons or target tissues such as muscle, and that this "wiring diagram" is highly specific with respect to the interacting cells. It has long been recognized that both the anatomical pattern of neuronal connections and their electrophysiological properties are very similar between individuals of the same species, and it has been argued that there is some inherent specificity in cellular migration, neurite outgrowth, and synaptic communication between pairs of cells. At least some of the patterns of connection are under strong genetic control. For example, the pattern of neurite outgrowth and innervation is the same in different grasshopper legs (Bentley and Candy, 1984), and the synaptic connections between cells in different individual nematodes of the same species are virtually identical (Sulston, 1984). The molecular basis for the formation of synaptic contacts between only certain cells is unknown, but a popular idea, put forth by Sperry in 1963, is that synaptic specificity is the result of a mutual affinity ("chemoaffinity") between the presynaptic growth cone or terminal and the postsynaptic cell. Sperry's chemoaffinity hypothesis predicted the existence of mutually complementary molecules which interact in much the same way as antigens and their antibodies. A natural extrapolation of this idea is that if it were possible to grow nerve cells in culture and assay the synaptic interactions between them in a pairwise manner, only limited pairs of cells should make functional synapses with each other. A major impetus in the decision of many laboratories to study cultured cells was the wish to test the chemoaffinity hypothesis experimentally with the hope of subsequently characterizing the molecules responsible for synaptic specificity.

In addition to the development of synaptic transmission, modification of the metabolism and physiology of both pre- and postsynaptic cells usually occurs as a consequence of synapse formation. For example, it is thought that the NGF synthesized by the postsynaptic cell is retrogradely transported through the axon of the sympathetic presynaptic cell to the cell body, where it induces enzymes and promotes cell survival. These types of interactions have been termed "trophic" or "inductive" functions (Guth, 1968; Harris, 1974), and it is likely that studies of cultured cells will play a major role in describing their molecular basis. To date, however, the information gleaned from studies of cell–cell interactions in culture has been largely limited to those between nerve and skeletal muscle. This is primarily because electrophysiological recording from large muscle fibers is much easier than recording from relatively small neurons. In addition, more information is available about the neuromuscular junction and the ACh receptor *in vivo* than about any other synapse. The following sections outline what is known about the neuromuscular junction in vertebrates and discuss the research on cultured cells.

5.2 The Neuromuscular Junction

The mammalian neuromuscular junction is a complex structure; it is shown schematically in Figure 5.1. In the mature junction there is extensive ultrastructural specialization in both the nerve and the muscle at the site of closest contact, and ACh sensitivity is localized to the immediate vicinity of the junction (Kuffler, 1943). Junctional folds in the muscle penetrate deep into the fiber. The ACh receptor accumulates at high density in the membrane at the tops of the folds, while AChE is localized

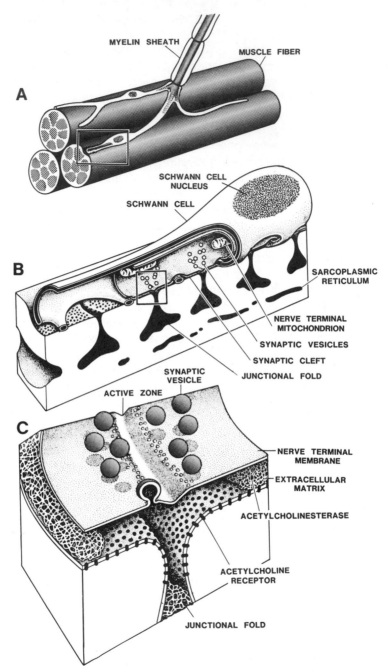

Figure 5.1. A schematic diagram of the frog neuromuscular junction. (A) Unmyelinated terminal branches of motor-nerve axon terminate in depressions in the membrane of muscle cells. (B) The nerve terminals are covered by Schwann cells on the top and sides while the synaptic specializations of the muscle are under the nerve. (C) Expanded view of the ACh release site. Redrawn from Lester (1977).

to the basement membrane in the cleft between nerve and muscle. A Schwann cell surrounds the nerve ending. The nerve terminal contains clear synaptic vesicles, mitochondria, and a limited number of different types of vesicles of unknown function.

ACh receptors function by binding ACh released from the nerve, resulting in the opening of a cation-specific channel in the membrane; this depolarizes the muscle membrane and initiates an action potential. The action potential causes the mobilization of calcium ions from the sarcoplasmic reticulum, resulting in muscle contraction. The receptor itself is a 250,000-dalton integral membrane protein made up of four polypeptide chains of known primary structure (see Changeux, 1981, for a review, and Karlin et al., 1984; and Numa et al., 1984 for discussions of molecular structure). The synthesis and degradation of muscle ACh receptors was discussed earlier (see Chapter 2).

Two techniques have been used to map the distribution of ACh receptors over the cell surface—iontophoresis and the binding of radioactive or fluorescently labeled α-bungarotoxin (αBT), a snake neurotoxin. The iontophoresis technique consists of releasing ACh by passing current through an ACh-filled electrode near the cell surface and assaying its effect on muscle membrane potential with an intracellular electrode (see Figures 2.3 and 5.4). The higher the density of ACh receptors, the greater is the depolarization in the muscle per unit current passed through the ACh electrode. The spatial resolution of this technique is about 5 μm. In contrast to the iontophoresis method, in the labeled toxin method αBT binds irreversibly to the receptor and can be coupled to a fluorescent or radioactive molecule to examine receptor distribution. The resolution of this technique is about 0.2 μm for fluorescent and 2 μm for [^{125}I]toxin using light microscopy. When coupled to electron-dense molecules, αBT can be used to map ACh receptors at the electron-microscopic level. Both techniques have been used extensively *in vivo* and with cultured cells.

The sequence of developmental events occurring at the rat diaphragm neuromuscular junction is shown in Figure 5.2. It is likely that a similar series of events occurs in other rat muscles. Nerve fibers are present in the mesodermal tissue before cell fusion is initiated. Myotube formation is initiated at about the 14th day of gestation, and nerves make functional synaptic contacts with the myotubes shortly thereafter. The distribution of receptors initially is uniform and the amount of receptor is low (about 150 sites/μm^2) as defined by fluorescent and ^{125}I-labeled αBT (Bevan and Steinbach, 1977; Harris, 1981). By embryonic day 15 aggregates of receptor are seen at junctions, but they consist of a loose distribution of small regions (called "speckles") of high ACh-receptor density.

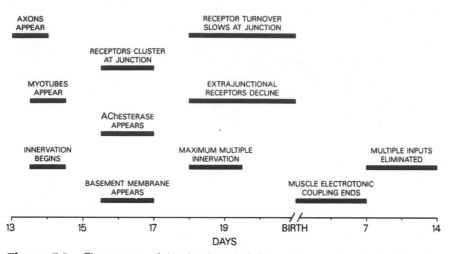

Figure 5.2. Time course of the development of the rat neuromuscular junction. Redrawn from Dennis et al. (1981).

Essentially nothing is known about speckle formation. Speckles may represent a partitioning of receptors in lipid microdomains or reflect interactions with the cytoskeleton. The speckles are soon replaced by a tightly clustered and well-defined plaque under the terminals (Figure 5.3) (Steinbach, 1981).

The initial nerve–muscle contact does not, however, have the specialization of the mature junction. At 16 days' gestation there are few synaptic vesicles, no postsynaptic membrane folds, and only traces of a basal lamina between nerve and muscle (Kelly and Zacks, 1969). The ACh-receptor plaques last through gestation and are found up to 10 days after birth. They are eventually transformed into the adult junctional aggregate—the only type seen 3 weeks post-partum.

In addition to the morphological differences the various types of receptor aggregates differ in their rates of turnover. The turnover of the initial loose aggregate is rapid and similar to that of extrajunctional ACh receptors. The turnover of junctional ACh receptors is very slow, with a half-time of about 16 days (Bevan and Steinbach, 1983). The ACh receptors in plaques and the adult junction are similar in metabolic stability, but they differ in their functional properties. For example, there is a postnatal decrease in ACh-channel open time (Michler and Sakmann, 1980) and turnover rate (Burden, 1977; Reiness and Weinberg, 1981). As the neuromuscular junction matures and the turnover of junctional ACh receptors declines, there is a decrease in the density of extrajunctional

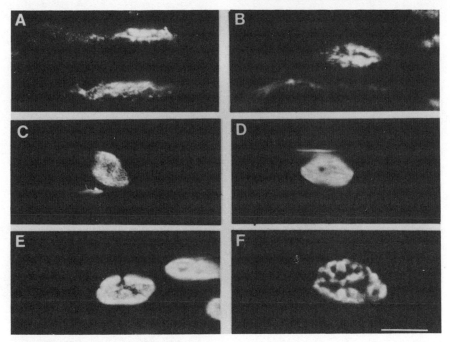

Figure 5.3.   Sequence of ACh-receptor aggregation in the developing rat neuromuscular junction. Junctional ACh receptor aggregates were stained with rhodamine-αBT in animals of different ages. (A) Loose aggregates from a 16-day embryo. (B) A tighter aggregate on another fiber from a 16-day embryo. (C) A plaque from a 20-day embryo. (D and E) Transitional shapes on fibers from a 7-day postnatal animal. (F) Adult neuromuscular junction of an 18-day-old animal. From Dr. J. H. Steinbach.

ACh receptors. Extrajunctional ACh-receptor density declines most rapidly prenatally between embryonic days 15 and 21 (birth), and then decreases more slowly until the adult levels are reached about 2 weeks postnatally (Bevan and Steinbach, 1983). Finally, most muscle fibers in the rat are multiply innervated by day 17, but the superfluous synapses are eliminated during the first 2 weeks after birth (Bennett and Pettigrew, 1974).

In normal mammalian muscle the majority of the ACh receptors are located at the neuromuscular junction. The density of extrajunctional ACh receptors is approximately 1000-fold lower than at the junction. Over 70% of the total receptor content is at the neuromuscular junction, on 0.1% of the muscle surface. The junctional and extrajunctional ACh receptors also differ in the following ways. (1) ACh receptors at the junction have a half-life in the membrane of 13 days; extrajunctional receptors have a half-life of about 22 hr (Berg and Hall, 1975a; Chang and Huang, 1975; Bevan and Steinbach, 1983). (2) Junctional receptors are immobile in the plane of the membrane; extrajunctional receptors are mobile (Axelrod et al., 1976). (3) The junctional and extrajunctional receptors have different isoelectric points (Brockes and Hall, 1975) and (4) biophysical properties (Neher and Sakmann, 1976). (5) The antigenic determinants are also distinguishable (Weinberg and Hall, 1979). Although these experiments suggest that junctional and extrajunctional molecules may be products of different genes, all of the data are compatible with post-translational modifications of the receptor. Recent data showing that there is only one ACh-receptor α-chain gene argue against the multiple-receptor concept (Numa et al., 1983). Evidence to be presented later shows that junctional receptors can be recruited from extrajunctional pools.

If the motor nerves of a skeletal muscle are sectioned, there is a profound change in the differentiated state of the muscle cells (see Harris, 1974, for a review). In addition to muscle atrophy, these changes include alterations in membrane electrical properties, energy metabolism, protein synthesis, and the distribution of ACh receptors on the muscle surface. The extrajunctional receptors increase several-hundredfold, primarily through the synthesis of new receptors (Devreotes and Fambrough, 1976; Brockes and Hall, 1975). The increase in extrajunctional receptors is controlled at least partially by nerve activity, as denervation supersensitivity can be blocked by direct muscle stimulation and, conversely, the inhibition of nerve-induced stimulation can lead to muscle changes characteristic of denervation (Lømo and Rosenthal, 1972). Upon reinnervation there is a gradual withdrawal of ACh sensitivity to the endplate region and a restoration of the normal muscle phenotype.

5.3 ACh Receptor Clusters on Muscle Cells

As outlined in Chapter 2, cultured skeletal muscle cells have ACh receptors on their surface. ACh receptors on cultured skeletal muscle are usually distributed diffusely over the surface of the cells, but in some culture systems receptors form clusters quite similar to those seen in rat embryos. ACh-receptor clusters (aggregates) have been found in the absence of nerve by electrophysiological, immunofluorescent, and autoradiographic techniques in primary cultures of chick, rat, and toad skeletal muscle (Fischbach and Cohen, 1973; Sytkowski et al., 1973; Anderson and Cohen, 1977). The fraction of total receptors which are clustered

varies tremendously between different preparations, but it is always relatively small (10% or less). The receptor aggregates are frequently localized to areas of cell–substratum adhesion (Bloch and Geiger, 1980).

How good is the analogy between receptor aggregates on cultured
skeletal muscle and those *in vivo*? Morphologically, the receptor aggregates in culture resemble those seen in the very early stages of embryonic development—the loose accumulation of receptors that precedes
the dense clusters at neuromuscular junctions (Bloch and Steinbach,
1983). Like the *in vivo* loose accumulations, the clustered *in vitro* receptors are rapidly metabolized, at the same rate as are the diffuse receptors (Schuetze et al., 1978). Attempts to induce the more stable
neuromuscular-junctionlike receptor aggregates in culture by the
cocultivation of nerve and muscle have met with mixed success. The
interaction between rat spinal cord neurons and skeletal muscle results
in generation of miniature endplate potentials (MEPPs) indicative of a
functional synaptic contact, but the ACh sensitivity of the muscle membrane under the nerve contact is indistinguishable from that of the extrajunctional areas (Kidokoro, 1980). In contrast, coculturing of chick spinal
neurons and muscle does result in increased ACh sensitivity on the
muscle at the point of nerve contact (Frank and Fischbach, 1979). The
rarity of these contacts, however, has prohibited more detailed studies of
the receptor at the nerve–muscle contact. Perhaps the best system for
studying nerve–muscle interactions in culture uses cells derived from
Xenopus tadpoles. We will describe this work, but first we will present a
discussion of the control of aggregate formation in cultured muscle and
a general discussion of functional synaptic contacts between cultured
cells.

The clustering of ACh receptors at nerve terminals suggests that a
signal may be released by the nerve. There have been several suggested candidates for the signal; most have been ruled out. ACh itself
has been one of the most extensively studied possibilities. However, one
of the first experiments using nerve–muscle cocultures ruled out a primary role for ACh in receptor clustering. When C1300 mouse neuroblastoma cells were plated on a monolayer of clonal L6 skeletal muscle
myotubes, an increase in ACh sensitivity was observed at the point of
nerve–muscle contact (Figure 5.4) (Harris et al., 1971). To rule out the
possibility that ACh release by the nerve cells induces receptor clustering, Steinbach and coworkers (1973) did three sets of experiments; the
results were as follows. (1) Localization still occurred when clonal variants of C1300 cells were used which were unable to synthesize ACh. (2)
Reagents which completely inhibit ACh synthesis by the nerve do not
alter the distribution of ACh receptors at the point of nerve–muscle con-

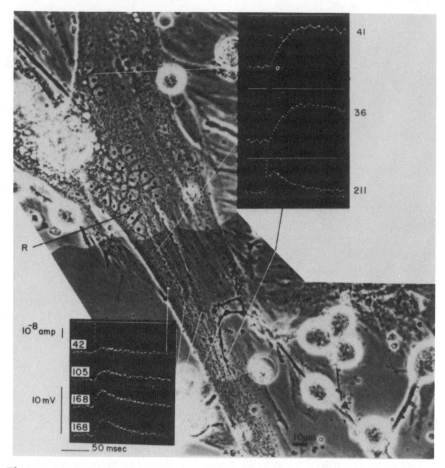

Figure 5.4. Localization of ACh sensitivity to the point of nerve contact. C1300 mouse neuroblastoma cells were cocultured with L6 myotubes for 4 days. The electrophysiological response of the muscle cells to the iontophoresis of ACh was determined by intracellular recording at site R. The number corresponding to each point refers to ACh sensitivity in mV/nC. R, From Harris et al. (1971).

tact. (3) Reagents which block ACh-receptor function have no effect on the interaction. Similar experiments have been done more recently with *Xenopus* cultures, with the same results (Anderson and Cohen, 1977). The above data do not, however, rule out the possibility that nerve releases something other than ACh that is responsible for receptor aggregation. This material cannot, however, be contained in synaptic vesicles, for high concentrations of Mg^{2+} ions inhibit the spontaneous release of material within ACh vesicles, but ACh-receptor accumulation on the muscle occurs as usual under these conditions (Y. Kidokoro, unpublished observations).

The possible role of spontaneous or nerve-induced electrical activity in regulating the distribution of ACh receptors was also tested in the C1300–L6 system (Steinbach, 1974). When muscle and nerve were co-cultured in the presence of high exogenous K^+, which depolarizes the cells and inactivates the action-potential mechanism, localization of ACh receptor to the point of nerve contact still occurred.

In summary, all of the data outlined above suggest that neither ACh nor ACh-receptor functioning nor electrical activity is required for ACh receptor to be localized to the point of nerve–muscle contact. It is likely that other mechanisms are responsible for ACh-receptor localization on muscle.

Cultured cells should provide ideal assays for the detection and isolation of such trophic factors. Extensive experimental activity has suggested that trophic factors exist, but none has yet been purified to homogeneity and chemically identified. Extracts prepared from spinal cord (Podleski et al., 1978) or the growth medium of nerve cells (Jessell et al., 1979; Bauer et al., 1981; Markelonis et al., 1982; Schaffner and Daniels, 1982) promote the formation of ACh-receptor aggregates on cultured cells. The active components in these preparations have very different molecular weights and physical properties, and their tissue distribution and specificity have not been examined extensively. Until these activities are characterized fully, their role at the neuromuscular junction will remain problematic. The use of cell culture systems as an assay is, however, the only way the hypothetical factors will be identified and isolated.

Unlike the situation with soluble factors, there is good evidence that insoluble extracellular matrix material is involved in the regulation of ACh-receptor clustering *in vivo*. When adult frog muscle is denervated and the muscle fibers in the denervated area destroyed, new myotubes are eventually formed within the scaffolding left by the original basement membrane. When the distribution of ACh receptors is assayed after muscle regrowth but in the continued absence of nerve, receptors are found clustered at the original site of innervation (Burden et al., 1979). If the nerve is allowed to regrow, it reinnervates the muscle at this original site, suggesting that something is deposited by the nerve on the basement membrane or on the surface of adjacent supportive cells which serves as a cue to direct the regrowth of the nerve and also to induce the localization of receptors. One candidate for this role is the group of macromolecules that is released by nerve cells which can become incorporated into extracellular matrix.

A protein which causes the localization of ACh receptors on cultured muscle cells has recently been isolated from the extracellular matrix of the electric organ of *Torpedo californica* (McMahon, 1984). The electric organ was chosen because it has a very high density of cholinergic synapses relative to that of skeletal muscle. Extraction of the extracellular matrix with high-salt or low-pH buffers solubilizes material which induces receptor aggregation in culture. This activity has been purified over 1000-fold and appears to be a protein of between 50,000 and 100,000 MW. There are also antigenic determinants of unknown structure that appear to be concentrated in the extracellular space surrounding the neuromuscular junction (Sanes and Hall, 1979). These observations indicate that the extracellular environment may play a major role in determining the distribution of cell surface molecules on muscle.

Several possible mechanisms for ACh-receptor cluster formation have been suggested; only the first has formally been ruled out. (1) Cluster formation is due to a specific decrease in the degradation rate of aggregated receptors relative to that of diffusely distributed receptors. (2) Junctional receptors are preferentially inserted into clusters. (3) There is a nucleation type of event involving direct receptor–receptor interactions. (4) Diffusely distributed receptors diffuse laterally into an area of the plasma membrane, but are retarded from leaving.

It is unlikely that the *reason* receptor aggregates form is low degradation rates for aggregated receptors. Receptor aggregates in the embryo and receptor clusters and noninnervated myotubes in culture have the same rate of turnover as diffuse receptors in these environments. It follows that differences in receptor stability *per se* are not responsible for cluster formation.

Similarly, a more rapid metabolic insertion of receptors into the cluster could not account for aggregate formation, because once a steady-state level of receptor density is reached the removal of receptors must be as rapid as insertion. It is possible, however, that once the high density is reached receptor insertion is reduced.

The possibility that direct receptor–receptor interactions cause receptor aggregates is also unlikely, because (1) during the *in vivo* development of rat muscle only one junctional aggregate is usually observed (Bevan and Steinbach, 1977), (2) in cultured rat myotubes receptor aggregates are preferentially localized to contacts with the substratum, and (3) small aggregates (speckles) appear to be organized into rows in the clusters on cultured rat myotubes (Bloch and Geiger, 1980).

Finally, an area of the membrane may limit the diffusion constant of mobile receptors by some sort of trap mechanism, resulting in their accumulation (Edwards and Frisch, 1976). It is known that the diffusion constant of receptors in clusters is very low compared to that of diffuse receptors. There are three alternatives which could account for the observed reduction in diffusion constants. (1) The passive diffusion of receptors into a localized region of membrane reduces their diffusion constant by means of interactions within the lipid bilayer. (2) There is a passive entrapment of the receptors by exogenous molecules, such as those in basement membrane. (3) The clustering results from specific interactions between the receptor and intracellular structures, such as the cytoskeleton.

The possibility that a local alteration in membrane composition or structure results in the accumulation of receptors is unlikely, for the lipid mobility in receptor clusters is the same as that in the other areas of the membrane (Axelrod 1978). In addition, it is unlikely that a regional alteration in membrane chemistry would specifically retard the receptor but not other molecules.

Since synaptic transmission precedes the formation of ACh-receptor clusters *in vivo,* it could be argued that the surface of the nerve ending or the extracellular molecules around the nerve trap laterally diffusing surface molecules in the muscle membrane. The diffusion trap must be specific for certain molecules, for not all of the surface proteins are localized to the extent of ACh receptors. Although ACh receptors do not diffuse within clusters, surface glycoproteins which bind some lectins diffuse freely within receptor aggregates (Axelrod et al., 1976; Axelrod, 1978). Evidence for the trap model was obtained by analysis of the redistribution of lectin receptors on dissociated *Xenopus* muscle cells following cell–cell contact (Chao et al., 1981). Initially, the receptors for soybean agglutinin (SBA) were uniformly distributed over the surface of the cells. However, when two cells were manipulated into contact, a high density of lectin receptors accumulated at the point of cell–cell contact within 10 min. This process was not dependent on metabolic energy, an observation consistent with the idea that the contact site serves as a trap for mobile SBA receptors. The experimental data agreed with the theoretical predictions of the trap model (Chao et al., 1981). ACh-receptor clustering can also be induced by isolated basement membrane material (Bloch and Steinbach, 1983), by pieces of degenerating nerve (Jones and Vrbova, 1974), and by positively charged latex beads (Peng et al., 1981). It should, however, be pointed out that some nerve cells can make functional synaptic contacts with muscle but are unable to induce ACh-receptor localization (Kidokoro and Yeh, 1982). *In toto,* these results clearly demonstrate that some extracellular macromolecules are able to induce the localization of mobile cell-surface molecules in a manner which is entirely consistent with a trap model.

The final alternative is that receptors interact with the cytoskeleton. The cytoskeleton is clearly involved in the patching, capping, and internalization of many cell-surface ligand–receptor complexes, and some aspects of ACh-receptor clustering are analogous to lectin-receptor capping in lymphocytes. For example, ACh-receptor clusters in rat myotubes, like lectin-induced caps, are dispersed by reagents which inhibit energy metabolism (Bloch, 1979). Unlike capping, however, colchicine blocks receptor clustering but promotes concanavalin-A-induced capping in lymphocytes. There is, however, extensive ultrastructural and morphological evidence that ACh receptors interact with the cytoskeleton. Ultrastructurally, postsynaptic densities at cholinergic synapses as well as receptor aggregates of cultured myotube are associated with cytoplasmic filamentous material (Ellisman et al., 1976). Some of this material is probably actin. Immunofluorescence techniques have been used on cultured myotubes to identify at least one cytoskeletal protein associated with ACh-receptor aggregates. Vinculin, an intracellular protein thought to be involved in cell–substratum adhesion, is associated with ACh-receptor clusters in rat myotubes (Bloch and Geiger, 1980). It could be argued that this association is fortuitous, because ACh-receptor aggregates are found primarily at points of cell–substratum adhesion in cultured rat myotubes. Virtually all contact sites display vinculin, but some are devoid of ACh receptors. Vinculin is, however, found near the postsynaptic membranes of neuromuscular junctions. Finally, the physical association of ACh receptors with the cytoskeleton can be shown directly by labeling myotubes with fluorescent or radioactive αBT and extracting the cells with a detergent, such as Triton X-100. The detergent removes the plasma-membrane lipids, releasing soluble intracellular molecules and leaving the cytoskeleton. When the distribution of ACh receptors is examined, most of the aggregated receptors are found to be retained with the cytoskeleton, whereas the majority of the diffuse receptors are lost (Prives et al., 1982). Most of the clustered ACh receptors remain associated with the IFs and microfilaments (Ben-Ze'ev et al., 1979).

In summary, extracellular matrix material may provide both the signal for induction of ACh-receptor clustering and a mechanism for trapping receptors which are mobile in the plane of the membrane. It is also likely, at least in mammalian skeletal muscle cells, that the cytoskeleton is involved in the assembly and maintenance of receptor aggregates. The actual molecules involved in the production of the signal for cluster location are unknown, and it has not been ruled out that the signal and the mechanism for clustering are the same.

5.4 Synapse Formation

There are numerous reports of synapse formation between nerve cells and between nerve and muscle cells in culture (see Nelson and Lieberman, 1981, for a review), and these studies will not be discussed here. Perhaps the greatest contribution of the work on cultured cells has been the insight provided about the nature of "synaptic specificity." Under normal conditions certain classes of neurons make synaptic connections with only a limited set of postsynaptic cells. It has generally been assumed that something is unique about the cells in each class. Perhaps synapse formation could occur only between selected subsets of pre- and postsynaptic cells. Such thinking led to the hypothesis that specific molecules on the two cells interact like antigens and antibodies to produce the synaptic specificity observed *in vivo* (Sperry, 1963). With the advent of cell culture the mechanisms of synaptic specificity could be addressed more directly, for the anatomical constraints present in whole animals are absent, and naked cell–cell interactions can be monitored directly. It became possible to ask what the minimal requirements for synapse formation are and how they relate to the proper wiring of the nervous system.

The examination of interactions between cultured cells has shown that only three requirements need be met to detect spontaneous synaptic activity (MEPPS) and evoked potentials in the postsynaptic cell. (1) The presynaptic cell must release a neurotransmitter. (2) The postsynaptic cell must have functional receptors for that transmitter. (3) The site of release must be sufficiently close to the postsynaptic receptors to generate a transmitter concentration high enough to activate a detectable number of the receptors.

The best studied pair of cells interacting in culture is nerve and skeletal muscle. This is because the postsynaptic muscle contains ACh receptors and is large, making electrophysiological recording relatively easy. Many neurons in clonal or primary cultures release ACh. The question was asked if there is any demonstrable specificity with respect to their ability to make functional synapses. Results of experiments using cholinergic cells from a variety of CNS and PNS sources indicate that all cells capable of synthesizing ACh generate MEPPS when in contact with muscle (Kidokoro et al., 1975; Kidokoro and Yeh, 1982). The establishment of synaptic interactions is very rapid; chemical transmission has been detected as soon as the main portion (palm) of the growth cone contacts the postsynaptic cell (Kidokoro and Yeh, 1982). If spinal neurons are plated upon cultures of *Xenopus* muscle, it is possible to follow the growth cone of the outgrowing neurite, record from the muscle cell that it is approaching, and stimulate the nerve cell body with an extracellular electrode. Although no synaptic potential is elicited when only the filopodia of the growth cone touch the muscle, when the palm of the growth cone overlaps the muscle membrane, nerve-evoked synaptic potentials are detected (Kidokoro and Yeh, 1982). These results show that the mobile growth cone is able to release neurotransmitter, and an analysis of endplate potentials suggests that transmitter release occurs in a quantal manner.

The apparent lack of specificity between pairs of cultured cells with respect to the formation of functional synaptic contacts is open to two interpretations: (1) If matched transmitter and receptor alone are sufficient to produce synapses, then a number of extrinsic developmental factors must be operating to generate the specificity observed *in vivo*. For example, the timing of nerve and target tissue formation may minimize chances for alternative interactions, and extracellular matrix material (which is less well organized in cultured cells) may be responsible for attracting or excluding ingrowing neurites. (2) Something in addition to neurotransmission *per se* is needed to establish a stable synapse *in vivo* which is not expressed in cultured cells. Synaptic activity may be necessary, but not sufficient. Perhaps the elusive "chemoaffinity" molecules are not made by cells in culture, or the cells do not make the trophic factors required for long-term interactions. In the latter case the lack of specificity in culture would not be a valid representation of the *in vivo* mechanisms. Since most other differentiated functions of cells are expressed in culture, the first alternative is more likely to be correct.

Finally, there are three classes of mechanisms which could account for the localization of nerve terminals and the release of ACh to points of high ACh-receptor density on the muscle. (1) The nerve could terminate at random points on the muscle; by probability some points would be associated with receptor clusters. (2) The nerve could detect pre-existing regions of high ACh-receptor density on the muscle and selectively release ACh at those points. (3) The nerve may induce or cause the formation of receptor clusters near the site of ACh release. Although the first two alternatives have not formally been ruled out, there is evidence against each. This information comes from studies on cocultures of *Xenopus* nerve and muscle. Support for the third alternative is also derived from the *Xenopus* system. We will now turn to this system.

5.5   The Development of the *Xenopus* Neuromuscular Junction

The sensitivity of innervated skeletal muscle to exogenously applied ACh is localized to the region of nerve contact (Kuffler, 1943). A long-standing problem has been the characterization of the biochemical mechanisms responsible for this developmental localization of ACh receptors at the neuromuscular junction. Although this question has been examined in cultures of rodent and chick muscle (see Section 5.3), the most productive experimental system has been that derived from the African clawed toad, *Xenopus laevis*. Cultures containing embryonic *Xenopus* spinal neurons and skeletal muscle myotomes were first used by Harrison (1910) to examine neurite outgrowth. In *Xenopus*, myotomes are the first muscle to be formed and are rapidly innervated by neurons in the spinal cord. Both the cord and muscle are easily accessible, they can be cultured at room temperature in relatively simple media, and the cellular interactions are highly visible. Another advantage is that the muscle cells do not fuse, but rather remain as large, flat cells that are accessible to histological and electrophysiological study. This cell culture system has been used to study the process of ACh receptor localization to areas of synaptic contacts.

If muscle cells from the *Xenopus* embryos are dissociated and placed in culture they adhere to the substratum and flatten. The distribution and relative densities of ACh receptors can be followed by staining the cells with fluorescent conjugates of α-BT, which binds specifically and irreversibly to ACh receptors. Within 1 day many small patches (or clusters) containing high densities of ACh receptors cover approximately 3% of the muscle surface (Anderson et al., 1977). These patches are up to 40 μm in length and lie mostly on the substratum-facing surface of the cell. Some of these patches are quite stable, lasting up to 4 days (Moody-Corbett and Cohen, 1982). AChE activity is normally found along with ACh receptors at the neuromuscular junction *in vivo*. It also codistributes with the patches of ACh receptors on the surface of noninnervated cultured muscle cells (Moody-Corbett and Cohen, 1981). Finally, limited areas of the plasma membranes have ultrastructures similar to that of the postsynaptic membrane of *in vivo* neuromuscular junctions (Weldon and Cohen, 1979). These similarities include (1) an increased electron density of the plasma membrane, (2) the appearance of a basal lamina on the extracellular surface, and (3) a high concentration of filaments under the dense plasma membrane. Although these ultrastructural specializations are similar to those of the *in vivo* synapse, they are relatively simple in comparison with the elaborate cytoarchitecture observed at the adult mammalian neuromuscular junction, where there is elaborate folding of the postsynaptic membrane (Figure 5.1). Although most cultured skeletal muscle cells form ACh-receptor clusters, no cells besides *Xenopus* myotomes have been studied which are able to form the postsynaptic specializations in the absence of nerve.

When spinal cord neurons are added to 1- or 2-day-old muscle cultures, the nerve cells rapidly extend neurites, and within 2 days sites of high ACh-receptor density are found associated with the points of nerve–muscle contact (Anderson et al., 1977). Electrical stimulation of the nerve cell bodies causes twitching of the nerve-contacted muscle cells. Since this muscle activation is blocked by αBT or curare, nicotinic chemical synapses must be present. Chemical synapses are, in fact, seen within 7 min after growth cone filopodia–muscle contact (Kidokoro and Yeh, 1982). Initially synaptic currents are of long duration and not prolonged by esterase inhibitors, but within 24 hr current duration decreases, currents become sensitive to esterase inhibitors, and a synaptic ultrastructure appears. The timing and sequence of events in this culture system are very similar to those observed *in vivo* (M. Cohen, 1980).

Before the accumulation of ACh receptors at the nerve, MEPPs are observed at low frequency. As ACh receptors cluster around the nerve, the frequency and sizes of the MEPPs increase (Kidokoro et al., 1980). This observation argues against the possibility that the nerve is selecting pre-existing sites of high receptor density. There is also cell-type specificity, for explants from tadpole spinal cord form synaptic contacts with the embryonic muscle cells, but do not cause ACh-receptor localization. Sympathetic and sensory neurons neither make synapses nor induce clustering (Cohen, 1980; Kidokoro et al., 1980). Finally, as in other culture systems and *in vivo,* the elimination of electrical activity in the nerve and muscle does not prevent the formation of the complete nerve–muscle junction in the *Xenopus* cultures.

The localization of ACh receptor to points along the nerve could be caused by the accumulation of pre-existing, mobile receptors at the points of nerve contact. There is ample evidence in support of this possibility (Anderson and Cohen, 1977; Moody-Corbett and Cohen, 1982). The effect of innervation on the distribution of pre-existing receptors was assayed by staining cultures with fluorescent αBT. The nerve cells were added in the presence of excess unlabeled toxin to block all newly appearing receptors. An examination of receptor distribution as a function of time indicated that (1) ACh receptors that were originally present become localized to the areas of nerve contact via their mobility in the plasma membrane, (2) although new ACh-receptor patches form on myocytes cultured alone, the innervation of the muscle apparently inhibits the formation of new patches, suggesting that nerve contact has a global effect on receptor metabolism, and (3) there is a net reduction in patches at sites away from the nerve, suggesting that receptor patches in these areas may contribute to the formation of patches along the nerve. A time-lapse analysis of receptor distribution following innervation has demonstrated that pre-existing receptors migrate in the plane of the muscle membrane and accumulate at the site of nerve contact (Y. Kidokoro, personal communication). These results show that nerve can directly alter the distribution and metabolism of ACh receptors at sites distant from the nerve–muscle contact. Similar conclusions have been arrived at from electrophysiological studies of these cells.

During the accumulation of ACh receptors at the nerve there is a fivefold increase in MEPP amplitude (Kidokoro et al., 1980). The increase in MEPP amplitude is due to an increase in the density of ACh receptors in the junctional regions (Kidokoro and Gruener, 1982). Quantitative autoradiographic techniques have demonstrated that there is about a ninefold difference between the density of ACh receptors in muscle background (104 receptors/μm^2) and that in the clusters (890/μm^2) (Kidokoro and Gruener, 1982).

Cultured *Xenopus* muscle cells are uniformly sensitive to iontophoretically applied ACh over their entire surface, with occasional points of higher sensitivity. In the presence of spinal cord neurons, there was a 50% increase in the overall background sensitivity of muscles to ACh (Gruener and Kidokoro, 1982). The background sensitivities of both innervated and noninnervated cells in the nerve–muscle cultures increased equally, and a similar increase in chemosensitivity was observed when muscle cells were grown in medium conditioned by the nerve cells. This increase in muscle sensitivity to ACh could be accounted for by (1) increased receptor number, (2) increased input resistance, or (3) a change in the ACh-receptor-channel properties such that more current flows each time a channel is activated. The first two alternatives have been ruled out by experiment, and evidence has been presented in support of the third. Electrophysiological measurements indicate that the membrane resistances are the same for the muscle cells grown under the two conditions, and quantitation of receptors with ^{125}I-labeled αBT shows that the background receptor densities are indistinguishable (Kidokoro and Gruener, 1982). However, an analysis of the kinetics of ACh-receptor channels of innervated muscle and muscle cultured in the presence of growth-conditioned medium showed that there is an increase in the mean channel open time for the receptors in all cultures exposed to nerve-conditioned media (Brehm et al., 1982).

There are two mechanisms that could account for the clustering of mobile receptors at the point of nerve–muscle contact. The first requires that the receptors be actively translocated via an energy-dependent cytoplasmic process. For example, the capping of certain ligand–receptor complexes is thought to occur through an interaction with cytoskeletal elements that attach to and draw the receptor complexes together. The second alternative is that an energy-independent mechanism exists which gathers freely translocating ACh receptors at certain sites on the plasma membrane. Evidence favoring the first alternative is the existence of apparent direct contacts between the postsynaptic membrane and underlying cytoplasmic filaments where the receptors are clustered (Heuser and Salpeter, 1979). Filamentous attachments between basement membrane and the postsynaptic membrane have also been demonstrated in *Xenopus*.

That randomly distributed molecules become localized to points of cell–cell contact by a trapping mechanism is supported by experimental work and on theoretical grounds. The *Xenopus* studies showed that pre-existing receptors can diffuse in the membrane and accumulate along the site of nerve contact. As discussed earlier, similar experiments have shown that SBA receptors on the surface of *Xenopus* muscle can rapidly (within 10 min) become localized to points of cell–cell contact by an energy-independent mechanism (Chao et al., 1981). Chao and his colleagues constructed a model by which a local region of the plasma membrane could serve as a "trap" for specific randomly diffusing molecules, derived equations which described the surface density distribution and lifetime of trappable molecules on the cell surface, and showed that the rate of trapping of SBA receptors fit the theoretically predicted values quite well. A similar model had been proposed earlier for the localization of ACh receptors (Edwards and Frisch, 1976). The mechanism responsible for the induction of the trap is unknown, but it could consist of an interaction of mobile molecules on the surface of one cell with ligands on an adjacent cell or structure. Since not all charged surface molecules can be trapped, the inducing agent must bind selectively to a subset of the cell's surface components. This mechanism suggests that pieces of cell debris, such as membrane fragments, could also induce the formation of receptor clusters in cultured cells. Some of the early published photographs of ACh-receptor "hot spots" on cultured skeletal muscle cells clearly show pieces of debris associated with receptor clusters.

Although receptor clusters and other postsynaptic specializations may form spontaneously in muscle cultures, a variety of conditions can induce their formation. These include various high- and low-molecular-weight cell extracts (see earlier discussion), basal lamina (Burden et al., 1979), silk threads (Jones and Vrbova, 1974), culture substrata (Bloch and Geiger, 1980), and positively charged latex beads (Peng and Cheng, 1982). In addition, positively charged Sepharose beads cause presynaptic specializations on cultured nerve cells, including synaptic vesicle clustering (Burry, 1980).

When latex beads coated with polylysine or PORN are placed in cultures of *Xenopus* muscle, ACh-receptor clusters localize to the point of contact between bead and muscle (Peng et al., 1981). Only positively charged beads were effective in inducing receptor localization, and, like neurites from spinal cord cells, the polylysine-coated beads caused the disappearance of clusters in non-contact areas. An extension of this work showed that essentially all of the postsynaptic specializations associated with cultured and adult *Xenopus* neuromuscular junctions could be induced by the beads (Peng and Cheng, 1982). These include (1) clusters of 11–12-nm intramembranous particles thought to be ACh receptors, (2) membrane invaginations extending 0.1 to more than 1 μm into the cell, (3) a 20–30-mm-thick membrane-associated cytoplasmic density, (4) 6–8-nm filaments (possibility of actin) apparently connected with the cytoplasmic density, and (5) a basal lamina usually found in the area between the beads and the muscle membrane.

Since it is unlikely that latex beads have any physiological significance in the *in vivo* development of the neuromuscular junction, what could account for the beads' ability to replace nerve in the induction of postsynaptic elements? The charged particle may create a local deformation of the macromolecular interactions in the adjacent plasma membrane, producing a signal which causes the cell to change its surface and submembrane structure at that point. This is a nonspecific mechanism which should be dependent on the charge density of the inducer. This does not, however, explain why there is nerve type-specificity in induction of localization. To account for the apparent biological specificity of receptor localization and the occurrence of some postsynaptic specializations in the absence of nerve or beads, the possibility should be considered that extracellular molecules or macromolecular complexes released into the culture medium bind to the beads. The beads may thereby acquire the characteristics of basal lamina. Large glycoprotein complexes called adherons are secreted by skeletal muscle cells, bind efficiently to various substrata, and are able to form extracellular matrices *de novo* (Schubert et al., 1983*a*; see also Chapter 6). This adsorption model might account for the fact that only polycationic beads work, because some extracellular muscle macromolecules involved in neurite extension bind only to positively charged surfaces (Collins, 1978*b*). The spontaneous appearance of aggregates in pure muscle cultures could be due to pieces of muscle membrane debris binding to the cell surface. This model predicts that extracellular molecules from spinal neurons and muscle itself would cause ACh-receptor localization, but extracellular molecules from sensory and sympathetic neurons would not.

In summary, the studies outlined above have set the minimal requirements for a functional synaptic interaction—simply that the presynaptic cell must release a neurotransmitter in close proximity to a cell which has the matching functional receptor. There is no example of a cell that synthesizes ACh and is not able to make functional synaptic contacts with skeletal muscle in culture. This observation necessitates an extension of the list of possible mechanisms responsible for the high degree of cell–cell specificity in the nervous system beyond the simple "lock and key" model proposed by Sperry (1963). Thus, extracellular constraints, such as basal lamina, timing of fiber outgrowth, and target availability must play a major role in development. One of the most obvious examples of trophic interactions at the neuromuscular junction, the formation of ACh-receptor clusters at the point of nerve contact, has been studied extensively in culture. The majority of the data indicate that most receptor clusters are formed by the aggregation of pre-existing receptors, probably by some sort of trapping mechanism induced by extracellular matrix material on the nerve membrane. Any antigenic or functional differences between ACh receptors are probably reflections of the local environment at the junction. There is currently no experimental result which necessitates postulating the synthesis of two different classes of receptor molecule by innervated muscle.

–6–

Extracellular Molecules and Adhesion

6.1 Extracellular Molecules

There is more to the mammalian cell than the material encompassed by the plasma membrane. Radiating from the lipid bilayer of essentially all cells is a poorly defined collection of glycoproteins, glycosaminoglycans, mucins, and other carbohydrate-rich molecules. Some of these molecules are extracellular extensions of proteins anchored in the membrane; many are more tenuously held by noncovalent interactions with the integral membrane proteins and lipids. The latter extracellular component, which tails off into the medium from the lipid bilayer, has been called the cell coat or glycocalyx (Cook, 1968). In a few types of cells part of this milieu is assembled into an ultrastructurally defined basement membrane. The basement membrane lies immediately outside the electron-dense layer comprising the cell coat, and consists of a more darkly staining matrix of collagen fibrils, glycoproteins, and glycosaminoglycans. Although the glycocalyx and basement membrane may not act as permeability barriers to small molecules, they may be barriers to viruses and antibodies, and to other cells. All cells release into their immediate environments not only "attached" extracellular material, but also molecules that are at least temporarily free in solution. These released molecules are derived either from classical secretory pathways or from the so-called "shedding" of membrane or glycocalyx components. Many of the released molecules reassociate with cells after a brief time in solution and may contribute to the assembly of such extracellular structures as the basement membrane.

Extracellular molecules have received only modest experimental attention during the last decade, particularly in the field of neurobiology. However, it is rapidly becoming clear that extracellular material is of critical importance in many aspects of cellular recognition, including synapse formation. Since the study of cultured cells has contributed enormously to this field, we will in this chapter discuss extracellular molecules and structures. We will outline the pathways of membrane and secretory protein elaboration first, and then discuss extracellular macromolecules from nerve and muscle. Lastly, we will describe the role of these molecules in cellular adhesion, emphasizing the adhesive interactions of myoblasts and neural retina cells.

6.2   The Synthesis of Secretory and Membrane Proteins

Cellular proteins are localized with a high degree of specificity to various subcellular and extracellular areas. For example, extracellular enzymes are not found free in the cytoplasm, and mitochondrial membrane proteins are not associated with the plasma membrane. What are the synthetic and transport pathways for the various kinds of membrane proteins, and what mechanism directs the traffic of the subset of proteins which is to be secreted?

The initial work on protein secretion was done by George Palade and his collaborators (Palade, 1975). They showed that proteins destined for secretion are synthesized on ribosomes associated with the rough endoplasmic reticulum (RER), an intracellular network of convoluted membranes containing ribosomes bound to the cytoplasmic face. The secretory proteins are then carried from the lumen of the RER by means of vesicles to another membrane system within the cell, the Golgi apparatus. From the Golgi apparatus vesicles the secretory proteins are transported to the surface, where the vesicle membranes fuse with the plasma membrane, releasing the vesicle contents to the outside (Figure 6.1).

The most difficult aspects of protein secretion from a conceptual point

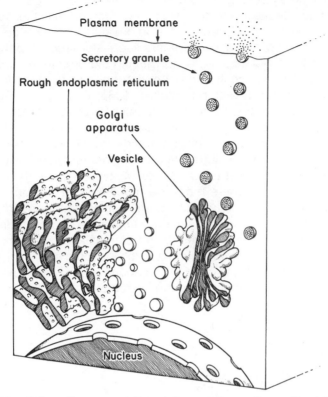

Figure 6.1. Schematic representation of the secretory pathway. The dotted vesicles are those derived from the Golgi apparatus en route to the plasma membrane. Redrawn from Wheater et al. (1979).

of view are the mechanisms which allow a water-soluble protein to pass through the hydrophobic membrane and the tagging of the protein to assure that it reaches the proper destination. Although a protein "pore" has not been isolated, it is clear that the membrane translocation takes place at the level of the RER membrane. In addition, the process of translocation is tightly coupled to protein synthesis. The resolution of the problem of selective protein transport came with the advent of the "signal hypothesis." In 1972, Milstein and coworkers showed that immunoglobulin light chains synthesized in a cell-free system from mouse myeloma mRNA were about 1500 daltons larger than the secreted protein. However, when membrane-bound myeloma ribosomes were used instead of the purified mRNA, only normal-sized light chains were detected; polysomes detached from the RER membrane again made the larger protein. Because peptide mapping showed that the additional amino acids were on the *N*-terminus, and because immunoglobulin light chains are secreted, these investigators suggested that the polypeptide extension could serve as a signal for getting the protein into the secretory pathway. This suggestion has been supported by observations in both animal cells and bacteria (Blobel, 1980; Emr et al., 1980). The signal hypothesis proposes that the *N*-terminal peptide extension (signal sequence) initiates binding of the translation complex to the RER membrane at a site where a pore is generated through which the growing peptide chain can be transferred vectorially as its synthesis proceeds (Figure 6.2). The signal sequence is then cleaved, leaving the protein in the lumen of the RER, where it can be processed for secretion as outlined above. The signal sequence usually contains polar residues near its *N*-terminus and the site of cleavage; it contains largely hydrophobic residues in between. Alternatives to the cotranslational mechanism of protein secretion outlined above must exist, for some secretory proteins are made without the transient signal sequence. The best example is ovalbumin, a major secretory protein of the chick oviduct (Gagnon et al.,

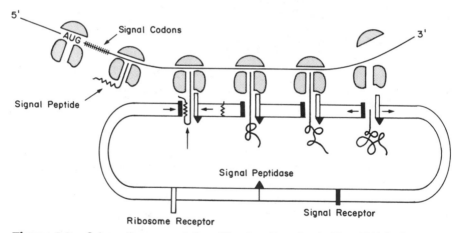

Figure 6.2. Schematic representation of the signal hypothesis. The mRNA for the secretory protein is represented by the line between 5′ and 3′. The AUG codon indicates the start of the protein message and is immediately followed by the signal sequence codons. As the protein is synthesized, the signal peptide binds to its receptor on the RER (solid bar), and the ribosome binds to a different receptor (open bar). These interactions result in the binding of the ribosome–nascent chain complex to the RER and the opening of a channel in the RER membrane. The protein is transported to the interior of the RER, and during transport the signal peptide sequence is enzymatically cleaved from the nascent chain by a signal peptidase. Redrawn from Blobel (1980).

1978). Although the *N*-terminal region of ovalbumin is not very hydrophobic, a hydrophobic region in the center of the molecule may serve as a signal. The presence of a signal sequence in the middle of the molecule does, however, pose some geometrical problems with respect to the protein being pushed by the ribosome through the membrane. Theoretical models for the spontaneous insertion of proteins into and across membranes have been proposed which may be relevant to proteins lacking signal peptides (Engelman and Steitz, 1981).

The signal hypothesis has been extended to include the assembly of integral membrane proteins (Wickner, 1980; Sabatini et al., 1982). Like secreted proteins, many membrane proteins have *N*-terminal signal peptides. To insert a protein into a membrane some mechanism is needed to stop the continued passage of the nascent chain across the lipid bilayer. This signal may be a unique amino acid sequence in the protein or a specific conformation adopted by the protein while being synthesized on membrane-bound ribosomes. Some membrane proteins also enter the membrane after their synthesis, for post-translational membrane assembly has been demonstrated. For example, mitochondrial membrane components are synthesized in the cytoplasm and then inserted into the mitochondrial membrane (Lallermayer and Neupert, 1976). As with secreted proteins synthesized on nonmembrane-bound ribosomes, there must be a sequence in the membrane polypeptide which is able to mediate its direct insertion into the lipid bilayer. Similar intramolecular signals may dictate the type of membrane (e.g., plasma, mitochondrial) into which the protein is inserted.

Once proteins are in the cisterna of the RER, some are destined to be secreted while others are directed to lysosomes. These lysosomal enzymes appear to be tagged for the lysosome by a mannose-6-phosphate group on a complex oligosaccharide which is added to the enzymes post-translationally (for a discussion, see Newark, 1979). Proteins containing this marker are sorted into lysosomes while in the Golgi apparatus. The mannose-6-phosphate group is missing from lysosomal enzymes in the human genetic disease called I-cell disease. Patients with this defect secrete their lysosomal enzymes. On the basis of these and similar observations, it is likely that carbohydrate moieties of glycoproteins tag proteins for transport to specific organelles following synthesis on the RER (Olden et al., 1982).

As adumbrated above, many secreted proteins are processed in a well-defined sequence of events terminating in the fusion of a secretory vesicle with the plasma membrane and the release of the vesicle contents. However, perhaps the majority of molecules which are released from cells do not follow the classical secretory pathway. These molecules are released from the cell as a consequence of membrane metabolism or "turnover." Warren and Glick (1968) were the first to note that the losses of proteins, lipid, and carbohydrate from fibroblast membranes were synchronous, with a turnover time of 2 or 3 days. The membranous origin of this class of extracellular molecules was further deduced from three experimental results. (1) Extracellular molecules comigrate on ion-exchange columns with molecules released from the cell surface by low amounts of trypsin (Doljanski and Kapeller, 1976). (2) The surfaces of cells were labeled with ^{125}I by the lactoperoxidase procedure, which iodinates tyrosine residues on proteins exposed to the extracellular enzyme. The cells were washed extensively to remove the enzyme and free iodine and then incubated in serum-free media for 8 hr. Concomitantly, cells were metabolically labeled with [^{3}H]leucine. The proteins released into the culture medium by both sets of cells were then compared by coelectrophoresis on SDS–acrylamide gels. The majority of the surface ^{125}I-labeled molecules comigrated with the leucine-labeled material, suggesting a common origin (Schubert, 1976). (3) A large number of surface antigens, including histocompatibility antigen H-2 and lymphocyte surface immunoglobulin, are released from cells into the culture medium (Doljanski and Kapeller, 1976).

The release of molecules from the cell surface demonstrated by the above experiments has been termed shedding to distinguish it from the classical secretory mechanisms of exocrine cells. Since the external plasma membrane consists primarily of carbohydrate-containing molecules, essentially all shed proteins are glycoproteins. Glycosylation is apparently not, however, required for protein shedding. The antibiotic tunicamycin inhibits the addition of core oligosaccharides to asparagine residues in proteins. It does not inhibit shedding (Damsky et al., 1979).

Cells release lipids as well as proteins into the culture medium (Blumenfeld et al., 1979). The ratios of various lipids in the culture medium are different from those of whole cells, showing that they do not originate simply from the lysis of cells. Once in the medium, the lipids become associated with the high- and low-density lipoproteins in the serum of the culture medium.

The most obvious alternative to shedding in accounting for extracellular molecules is the death of cells in culture and their subsequent lysis. That lysis contributes significantly to extracellular proteins has, however, been ruled out in clonal cell lines (Schubert et al., 1973; Doetshman and Jewett, 1981). The spectrum of proteins released into the medium of clonal cell lines is a minor subset of all cellular proteins. In addition, little DNA labeled with ^{3}H-TdR is released. Although cell death accounts for a minimal amount of released material in clonal cultures, the contamination of the culture medium with cell debris is much greater in primary cultures.

The mechanisms of cell-surface shedding have not clearly been defined. Cell-surface proteolytic activity may account partially for shedding, for naturally shed and trypsin-released molecules are similar. Extracellular proteases exist, and protease inhibitors partially block shedding in human erythrocytes (Bocci et al., 1979) and the release of plasminogen activator in transformed fibroblasts (O'Donnell-Tormey and Quigley, 1981). In addition, there may be sites on the surface of cells to which surface molecules must migrate before shedding can occur. These sites may be coated endocytotic pits (Brown et al., 1979), cell-surface caps (Karnovsky et al., 1972), or plasma membrane vesicles (Peterson and Rubin, 1969). If shedding occurs at defined sites on the cell surface, then inhibition of the lateral mobility of glycoproteins should inhibit the shedding process. When L6 myoblasts were incubated with a concentration of the multivalent lectin concanavalin A, which inhibits the mobility of some surface molecules, shedding of glycoproteins was also inhibited (Doetschman, 1980). This evidence is consistent with any of the three sites mentioned above being the release site.

Finally, it should be pointed out that the fundamental difference between cell-surface shedding and true secretion as defined by Palade (1975) may be minimal. Since most secreted and membrane proteins share a common pathway to the cell surface, which culminates in the fusion of the vesicle and plasma membranes, any loosely bound membrane proteins may be released along with the soluble vesicle contents. These could include proteins poorly inserted into the membrane or those with a low affinity for the lipid bilayer. Membrane proteins may differ from secretory proteins only in their relative affinities for the lipid bilayer. It is thus unlikely that there is an absolute distinction between macromolecular secretion and shedding.

The plasma membrane may be "turned over" or degraded by mechanisms other than shedding. These include: (1) internalization of a piece of membrane followed by lysosomal degradation of the entire unit; (2) aggregation of specific proteins into a high-density region called a patch or a cap, followed by internalization and degradation; (3) internalization and recycling into new membrane; and (4) shedding of a whole unit of membrane as a closed membrane vesicle. The shedding of plasma membrane vesicles can be induced by incubating cells at 4°C for 1 hr and then allowing the cells to warm up to 22°C. Within 2 hr most of the cells shed membrane vesicles and the cells become proportionally smaller (Liepins and Hillman, 1981). There is some evidence that each mode is used by some cell but that an individual cell may not employ all of them (Doyle and Baumann, 1979). However, essentially all cell types shed some membrane components.

A number of functions have been proposed for shed membrane molecules.

1. Shed tumor surface antigens present in the blood may act as blocking antigens and interfere with the immune response of the host, providing an escape mechanism for tumor cells (Kim et al., 1975). Molecules shed from the surface of normal cells play a role in inducing immunological tolerance during ontogeny. Normal cells also have an extracellular coat or glycocalyx which completely surrounds them, but this structure is apparently absent in most transformed cells. Finally, tumor cells shed membrane glycoproteins more rapidly than do normal cells.

2. Many serum proteins are derived from shed cell-surface molecules. Two common examples are the $beta_2$-microglobulins and cold-insoluble globulin, or fibronectin. Some cellular growth factors may also be released from cell surfaces. Table 6.1 lists some surface molecules which are released from cells and indicates whether or not they are present in serum.

3. Shed membrane components may be readsorbed and reinserted into the plasma membrane, contributing to membrane biogenesis (Peterson and Rubin, 1969). Some cell-derived serum proteins, such as fibronectin, can be readsorbed from serum onto cell surfaces and into extracellular matrices, suggesting that fibronectin in plasma may be a reservoir for cellular fibronectin (Oh et al., 1981).

4. Shed molecules may be involved in the assembly of the glycocalyx and basement membrane.

5. The spectrum of proteins released by a cell type may be used as a "fingerprint" for its identification (Jetten et al., 1979). It should be pointed out, however, that proteins released in exponentially dividing and stationary-phase cultures are quantitatively different from each other (Schubert, 1976).

6. Finally, macromolecules shed from the cell are involved in many aspects of cell–cell interactions, including cellular adhesion. One type of adhesion between the cell and substratum is mediated by extracellular molecules which bind to the surface of the culture dish. This collection of molecules has been termed substrate-attached material.

Table 6.1. Defined Extracellular Molecules

Molecule	Cell Type	Cell Surface	Serum	Medium	Reference
Collagen	Most	+	?	+	Hata et al. (1980)
Fibronectin	Mesoderm	+	+	+	Ruoslahti et al. (1973)
Beta$_2$-microglobulin	Leucocytes	+	+	+	Cejka et al. (1975)
Immunoglobulin (IgM)	Lymphocytes	+	+	+	Vitteta and Uhr (1972)
TL surface antigen	Lymphocytes	+	+	+	Yu and Cohen (1974)
Carcinoembryonic antigen	Some tumors and normal fetal colon	+	+	+	Sturgeon (1979)
AChE	Skeletal muscle	+	+	+	Rotundo and Fambrough (1980a)
IgA surface receptor	Epithelial cells	+	+	+	Kuhn and Kraehenbuhl (1979)
C3 complement component	Fibroblasts	?	+	+	Senger and Hynes (1978)
Alpha$_2$-macroglobulin	Fibroblasts	?	?	+	Mosher and Wing (1976)
Collagenase inhibitor	Smooth muscle	?	?	+	Nolan et al. (1978)
Plasminogen activator	Most	+	+	+	Kalderon (1979)
Glycosyltransferases	Fibroblasts	+	?	+	LaMont et al. (1977)
Glucose-regulated protein	Fibroblasts	+	?	+	McCormick et al. (1979)
Protease-nexin	Fibroblasts	?	?	+	Baker et al. (1980)
Transforming growth factors	Fibroblasts	?	?	+	Todaro et al. (1980)
Outer-segment molecules	Visual cells	+	+	?	O'Day and Young (1978)
Fertilized egg surface proteins	Egg	+	–	+	Johnson and Epel (1975)

6.3   Substrate-Attached Material

In addition to changing the chemical composition of their culture medium by the release of cellular material, cultured cells also alter the surface properties of the substratum on which they are grown. This observation was first made by Rosenberg in 1960. Using ellipsometry, an optical technique for measuring the thickness of thin films, he showed that cells cultured on glass leave a layer of material about 10 nm thick on the substratum after the cells are removed. This cell-derived film was called a "microexudate." Microexudates have also been characterized by the incorporation of isotopically labeled amino acids and carbohydrates, by electron microscopy, and by fluorescent antibody techniques. The microexudate has also been called the cell carpet or substrate-attached material (SAM). It should be noted, however, that molecules derived from serum also adhere to the substratum of culture dishes. Therefore, if cells are grown in serum-containing medium, the culture dish surface is coated with both cellular and serum-derived material.

Numerous biological functions have been ascribed to SAM. If cells are grown on tissue culture dishes and then removed by exposure to the calcium-chelating agent EGTA, the SAM remains on the substratum. When new cells are plated on the growth-conditioned surface of the culture dishes, they usually come to have different growth rates, adhesive characteristics, and cell morphologies from cells plated into virgin culture dishes. For example, when fibroblastlike cells are plated at low cell densities onto SAM prepared from exponentially dividing cells, growth is significantly enhanced (Yaoi and Kanaseki, 1972; L. Weiss et al., 1975). However, when cells are plated at the same low density onto substrata containing SAM from stationary-phase cultures, growth is inhibited relative to that in new dishes. These data may be a reflection of what takes place in normal cultures, for stationary-phase cells inhibit the growth of homologous cells seeded directly onto them (Eagle and Levine, 1967). They also suggest that the microexudate may be derived at least in part from the cell coat lying immediately exterior to the plasma membrane. This possibility has been supported further by an analysis of the chemical composition of SAM and experiments involving cell–substratum adhesive interactions.

The first experiments to examine the chemistry of SAM from fibroblast-like cells showed that SAM contained protein, carbohydrate, and small amounts of RNA (L. Weiss et al., 1975). By isotopic labeling of cells with precursors to carbohydrate ([^{3}H]glucosamine, [^{3}H]glucose), protein ([^{3}H]leucine, [^{35}S]methionine), and lipid (^{32}P]phosphate, [^{3}H]glycerol) a more detailed analysis of the composition of SAM was obtained. The vast majority of this work has been done with fibroblastlike cells; additional experiments with myoblasts and nerve cells will be discussed later in this chapter. Fibroblast SAM contains a number of identified proteins, plus at least a dozen proteins of unknown function (Culp et al., 1979). Fibronectin, myosin heavy chain, actin, and small amounts of collagen are found in SAM. In addition, isotopically labeled carbohydrate is incorporated into SAM-associated glycosaminoglycans (GAGs). Fibroblast SAM is highly enriched in GAGs relative to the whole cells— the GAG/protein ratio is at least an order of magnitude higher in the microexudate than in cells. Probably all classes of GAGs, including hyaluronic acid, heparin, heparan sulfate, and chondroitins are found in fibroblast SAM, although the relative amounts of each vary among cell lines. Although there are similarities among the GAGs and proteins of SAMs derived from different cells, there are also distinct differences, even within the class of rodent fibroblastlike cells. The molecular composition of SAM also depends on whether or not the cells are transformed. For example, the SAM of a tumorigenic variant of a normal skeletal muscle cell line lacks the chondroitin and collagen found in that of the parental cells (Schubert and LaCorbiere, 1980b).

The microexudates from some cells contain lipid in addition to proteins and carbohydrates. To determine the phospholipid content of SAM, Cathcart and Culp (1979) labeled fibroblastlike cells with [^{32}P]phosphate, extracted the total phospholipid in SAM, the cells, and partially purified membranes, and separated the individual phospholipids by thin layer chromatography. In 3T3 fibroblastlike cells, approximately 60% and 16% of the total cellular phospholipid are phosphatidylcholine (PC) and phosphatidylethanolamine (PE), respectively. SAM contains only 37% PC and 29% PE, suggesting that the lipid in the SAM is a unique subset of the total cellular phospholipid. Although cells release phospholipids into their culture medium (Peterson and Rubin, 1969), it is likely that the majority of thè phospholipid in 3T3-derived SAM is from pinched-off cellular "footpads" or attachment sites. The analysis of plasma-membrane-enriched fractions shows that their composition is more similar to that of whole cells than to that of SAM. If the phospholipid in this SAM preparation is in fact derived from footpad material, then there is probably a unique membrane fraction associated with the footpad adhesion sites. Cell-surface antigens and membrane virus receptors have also been demonstrated in microexudates (L. Weiss and Lachmann, 1964; Poste et al., 1973), but the adsorption from the growth-medium shed surface molecules could account for these data. Evidence that some of the components in the microexudate carpet are derived from pinched-off pieces of cells is limited to only a few fibroblastlike cell lines. In the majority of cell types SAM is probably derived from material released into the culture medium and then adsorbed to the surface of the culture dish. Where footpads are observed, SAM consists of a mixture of cellular and adsorbed molecules.

A knowledge of the origin of SAM is imperative for the proper interpretation of experimental evidence. There are two alternatives for the origin of the cellular microexudate; the evidence for each will be presented. (1) SAM is derived exclusively from pieces of cells which remain attached to the dish after cells have been removed by scraping or EGTA treatment. (2) The SAM is released from the cell in a soluble form and then adheres to the substratum or to material for which it has a high affinity already present on the culture dish. The first alternative would predict a punctate distribution of SAM corresponding to the cellular distribution, the second a more homogeneous distribution of material on the culture dish.

Fibroblastlike cells adhere to the substratum of the culture dish by slender projections from the cell body, called filopodia. They make contact with the substratum at the distal end of the filopodia by a flattened protrusion called the footpad. Electron micrographs of freeze-fractured material show that the internal membrane structure of footpads from mouse fibroblastlike cells is distinct from that of the rest of the plasma membrane. The density of intramembrane particles is about twice as great in the footpad region, and steroid components are lower in the footpads (Robinson and Karnovsky, 1980). If some fibroblastlike cells are exposed to 0.5 mM EGTA, the cell bodies become rounded and pull away from their substratum adhesion sites, leaving behind the footpads of the filopodia. Researchers studying the chemistry of the material remaining on the culture dish frequently assume that the material being examined is derived exclusively from the remaining footpads, and that the results can be interpreted in terms of the chemistry of these adhesion sites (see, for example, Culp et al., 1979). The justification for this approach comes mainly from autoradiographic and immunofluorescence studies. When cells are labeled with amino acid or carbohydrate precursors, the cells removed by EGTA, and the substratum autoradiographed, high densities of the residual isotope are found to be associated with areas which previously contained cells (Culp, 1975). This is to be expected, since the footpads left behind contain the highest *density* of macromolecular material on the substratum. However, isotope is also detected in areas devoid of cells, and the relative distribution of the isotope over the substratum and the contribution of each isotope to the total SAM have not yet been quantitated. Similar experiments using immunofluorescent techniques to assay the distribution of fibronectin showed that it also is localized to areas previously occupied by 3T3 fibroblastlike cells. These data show only that the areas which once contained cells have the highest density of material. They do not exclude the possibility that less dense deposits, perhaps consisting of different molecules, are distributed over the substratum not occupied by cells, nor do they rule out the possibility that this extracellular material mediates processes such as cell–substratum adhesion.

The second possible origin of SAM is that it is material that is released into the culture medium and then readheres to the dish surface. The following evidence shows that many macromolecules initially released into the culture medium are found in SAM. (1) Microexudate, as defined by ellipsometry (Rosenberg, 1960) and immunofluorescence (Bolund et al., 1970), is found uniformly distributed in the intercellular areas of subconfluent cultures. (2) When coverslips are placed at different distances away from monolayers of cells and the accumulation of macromolecules isotopically labeled with glucosamine and leucine is assayed after 5 hr, there is detectable protein and carbohydrate on the cell-free coverslips, with a gradient of accumulation which decreases by 50% 1 mm away from the cell layer (Yaoi and Kanaseki, 1972). Similar experiments with a rat nerve cell line showed that at 5 mm from the cell layer, about half as much SAM accumulates on the cell-free surface as in the SAM layer below the cells. This fractional distribution is not affected by the presence of serum in the culture medium (D. Schubert and M. LaCorbiere, unpublished). (3) Cells that grow exclusively in suspension culture, such as lymphoid cells, deposit as much protein and carbohydrate on the surface of the culture dish as do attached cells (Schubert, 1977). (4) Material from serum-free growth-conditioned medium that adheres to the surface of new culture dishes promotes cell–substratum adhesion (Moore, 1976; Schubert and LaCorbiere, 1980a,c). (5) SAM prepared from low-density cultures maximally stimulates the growth of cells replated on that surface (L. Weiss et al., 1975). It is therefore likely that all cells release material, probably derived from their surfaces, that is temporarily soluble in the culture medium and is able to adhere to the culture dish surface. The relative contributions of medium and footpad-derived material to the microexudate can be assayed with an inverted cell-free tissue culture dish placed in the medium above the attached cells being isotopically labeled. If the components of the culture dish and its "lid" surfaces are then both assayed following the removal of the cells, the difference between the cell substratum and the lid should be due to remnants of filopodia on the dish.

Several factors control the ultimate distribution of the released material. These include the solubility of the macromolecules and their relative affinities for the different accessible surfaces. For example, a molecule may have a high affinity for itself, a poorer affinity for the homologous cell surface, and little affinity for anything else. At equilibrium, these molecules would be found aggregated around the cells. They could be involved in the synthesis of basement membrane and the cell coat. Alternatively, some released molecules may have very poor affinities for their cells of origin and could become concentrated on other surfaces. Gradients away from cells may be generated by the release of nonspecifically adhesive molecules from cells, or by the release of molecules which are relatively insoluble in the extracellular milieu.

6.4 Protein Release from Muscle

Muscle cells release a variety of proteins into the culture medium, including collagen, fibronectin, actin, and AChE. Using Yaffe's L6 clonal skeletal muscle cell line, Schubert and coworkers (1973) showed that myoblasts release collagen and fibronectin. Since there are no contaminating cell types (e.g., fibroblasts) in clonal cultures, the cellular origin of these proteins was unambiguous. Fibronectin and collagen are also major constituents of basement membranes (see Yamada and Olden, 1978, for a review). Since the myotubes from the L6 line also possess a well-defined basement membrane (Schubert et al., 1973), the muscle cells alone must be able to synthesize and assemble the components of their own extracellular matrix. The possibility that fibroblasts were required for synthesis of the basement membrane surrounding myotubes has been ruled out formally by these experiments. Extracellular actin has also been identified in cultured skeletal muscle cells (Rubenstein et al., 1982). The function of this protein in the extracellular space is unknown.

In addition to releasing collagen, actin, and fibronectin, skeletal muscle cells also release large amounts of AChE, the enzyme that hydrolyzes ACh to choline and acetate, into the culture medium (Wilson et al., 1973). The release of AChE is energy dependent and temperature sensitive. Normally, between 90 and 95% of the enzyme produced by chick skeletal muscle cells is released into the culture medium (Rotundo and Fambrough, 1980a). If protein synthesis is inhibited by cycloheximide, there is no effect on AChE secretion until after 2.5 hr, which suggests that this is the transit time for the esterase from its site of synthesis to the extracellular space (Figure 6.3). The amounts of cell-surface, intracellular, and secreted AChE were determined by use of combinations of the irreversible esterase poison diisopropyl fluorophosphate (DFP) and reversible esterase inhibitors. About one-third of the total cellular enzyme

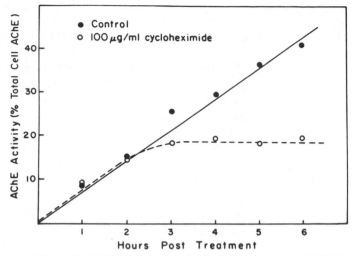

Figure 6.3. Effect of inhibiting protein synthesis on the secretion of AChE from chick skeletal muscle. Muscle cultures were incubated with or without 100 μg/ml cycloheximide, and the activity of AChE was determined in the culture medium at 1-hr intervals. The data are expressed as the secreted AChE as a percentage of the total amount of AChE present in control muscle cultures. From Rotundo and Fambrough (1980a).

is found on the surface, and once all of the AChE has been inactivated with DFP, the recovery of active surface AChE is about 3% of the previous total surface enzyme per hour. The first secreted enzyme was detected about 2.5 hr after DFP treatment, confirming the previously determined transit time. Cell-surface AChE is probably an integral membrane protein, because it is not removed by chelating agents, high salt concentrations, or low concentrations of trypsin. The degradation of the membrane AChE has a half-time of 50 hr. The cell-surface AChE is probably internalized and degraded, for it is not released into the medium. Since the ACh-receptor molecule, a known integral membrane protein, has a half-life on the cell surface of 19 hr, it can be concluded that the degradation rates or turnover of two membrane proteins in the same cell can be quite different. The observation that the transport time between synthesis and appearance on the cell surface is 2.5 hr for both AChE and ACh receptor in the chick myotubes supports Palade's suggestion that membrane and secretory proteins are transported to the cell surface via a common pathway (Palade, 1975). The mechanism by which AChE destined for the membrane is sorted from that to be secreted is, however, unknown. Additional evidence for the common pathway of secretory and membrane proteins was gleaned from experiments in which intracellular transport in myotubes was experimentally altered by a variety of reagents. In all cases the rate of ACh-receptor incorporation into membranes was identical to the rate of AChE secretion (Rotundo and Fambrough, 1980b).

Muscle cells also synthesize and release large amounts of GAGs (Ahrens et al., 1977). GAGs may be involved in the regulation of various morphogenetic events, including myogenesis (see Bernfield, 1980, for a review). For example, the continued presence of high concentrations of GAGs, particularly hyaluronic acid, may be required for the migration of cells. The interactions and differentiation of some cells are also affected by depletion of GAGs, both *in vivo* and in cell culture. Two examples involve skeletal muscle myogenesis and neural crest differentiation. During myogenesis, there is extensive degradation of extracellular hyaluronic acid after myoblast precursor cells cease migrating and begin to differentiate. It has been argued from this type of information that differentiation is specifically inhibited during cell migration by matrix hyaluronic acid, and that once this inhibitor is removed normal tissue differentiation can proceed. More direct evidence comes from studies of embryonic chick skeletal muscle myoblasts in culture. Within 24 hr after plating there is a rapid degradation of endogenous GAGs (Elson and Ingwall, 1980). This metabolic change occurs before the onset of myoblast fusion. In addition, hyaluronic acid, but not other GAGs, inhibits both myoblast fusion and the isozyme transition of creatine kinase normally associated with myogenesis. These results suggest that hyaluronic acid may play a regulatory role in the formation of skeletal muscle.

Neural crest cells are surrounded by an extensive extracellular matrix during the initial stages of their migration. However, the GAG composition of each area in which the different neural crest derivatives ultimately differentiate is unique. It is possible that these terminal environments directly influence the differentiation of the neural crest cells (Pintar, 1978).

6.5   Protein Secretion from Nerve and Glia

Nerve differentiation is usually associated with the movement of the growth cone over the surfaces of other cells or over extracellular matrix. The extracellular environment in the immediate vicinity of the nerve is developmentally critical. The study of extracellular neuronal molecules is, however, in its infancy. The existing work can be grouped into two categories. The first consists of examinations of the extracellular molecules as a whole and of their changes with neuronal and glial differentiation. Minimal emphasis has been placed on elucidating their functions. These studies will be discussed immediately below. The other group of studies involves the analysis of the subset of neuronally released molecules which is involved in cellular adhesion. It will be discussed in Sections 6.10 and 6.11.

When examining the effect of differentiation on the metabolism of any group of molecules, one must make sure that both experimental and control cultures have identical growth states. Many cellular properties, including the rates of synthesis of intracellular proteins (Garrels, 1979a) and of the proteins which are secreted by cells (Schubert, 1976), change dramatically between the exponential and stationary growth phases. Maintaining similar growth rates is particularly difficult with cultures of some nerve cells, in which the procedures used to induce differentiation generally inhibit cell division (see Chapter 3). This difficulty has been overcome in two ways. The first is to adjust conditions in the substrata on which the cells are grown so that they promote neurite extension but do not change the cell cycle time. The second is to use reagents to induce cell differentiation which allow a few days to elapse before cellular division is inhibited. For example, NGF ultimately inhibits cell division of responsive cells, but usually not until after 2–4 days of exposure.

When clones of the C1300 mouse neuroblastoma are grown in suspension and labeled with [^{3}H]glucosamine or [^{14}C]leucine, a number of glycoproteins released into the culture medium are detected by SDS–acrylamide gel electrophoresis. Predominant bands at about 87,000, 66,000, and 55,000 MW are observed (Schubert, 1973; Truding et al., 1975). However, when these cells are shifted to monolayer culture containing dibutyryl cyclic AMP or BUdR, both of which induce morphological differentiation in C1300 cells, the profile of released proteins changes, with an increase in the proportion of the 66,000-MW species (Truding et al., 1975). Although dibutyryl cyclic AMP rapidly inhibits cell division, BUdR does not. These results show that at least quantitative changes in the release of extracellular glycoproteins are associated with the differentiation of nerve cells. Similar results are obtained when C1300 cells are treated with dibutyryl cyclic AMP or BUdR and the surface molecules are labeled with [^{125}]I by the lactoperoxidase method. The differentiated cells release a quantitatively different spectrum of proteins from that released by the undifferentiated cells, and the released proteins change in a similar manner in the presence of either BUdR or dibutyryl cyclic AMP (Truding and Morell, 1977).

C1300 cells also secrete plasminogen activator into the growth medium (Laug et al., 1976). When cells are exposed to dibutyryl cyclic AMP or to other reagents which stimulate intracellular cyclic AMP synthesis, there is a 14-fold increase in plasminogen activator secretion. It is possible that this increase in extracellular protease associated with morphological nerve differentiation may have a role in partially breaking down the immediate environment of growth cones to allow less hindered neurite outgrowth. Plasminogen activator activity has been demonstrated directly in areas of growth cone activity (Krystosek and Seeds, 1981).

It is generally believed that nerve cells in the CNS are not surrounded by basal laminae, and that Schwann cells produce all of the extracellular matrix in the PNS. Extracellular matrix is, however, observed in the synaptic clefts of nerve–nerve synapses and at the neuromuscular junction. It is possible that nerve cells synthesize and secrete this material. That nerve cells can synthesize basement membrane components has been shown by the demonstration that C1300 nerve cells synthesize fibronectin, laminin, and type IV procollagen (Alitalo et al., 1980). Collagen secretion is only detected in the presence of ascorbic acid, which is a cofactor for the enzyme that hydroxylates collagen. The three basement membrane proteins are found in soluble form in the culture medium as well as in SAM deposited on the culture dish. Thus, C1300 nerve cells, like skeletal muscle myoblasts, are able to release matrix components into their immediate environment. These molecules, however, are apparently not assembled into a basal lamina.

NGF induces the differentiation of the clonal PC12 cell line in much the same way that dibutyryl cyclic AMP induces the differentiation of the C1300 clones (see Chapter 3). Changes in the syntheses of extracellular molecules in response to NGF have also been examined with this cell line (Schubert, 1978). To assay the effect of NGF on protein secretion, Schubert (1978) labeled cells for 6 hr with [³H]leucine in serum-free medium on days 1–4 after the addition of NGF. Control cultures without NGF were similarly labeled with [¹⁴C]leucine. The culture supernatants were removed, and the ³H- and ¹⁴C-labeled proteins from the same day were then mixed and coelectrophoresed on acrylamide gels. NGF caused a relative increase in the amounts of several secreted high-molecular-weight proteins; there was a slight decrease in the lower-molecular-weight peaks of radioactivity in the NGF-treated cells compared with those in control cultures. Cyclic AMP, which mediates some aspects of PC12 differentiation, had an effect on protein secretion very similar to that of NGF. Cyclic AMP and NGF also caused similar, but less dramatic, changes in the compositions of the SAM prepared from the cultures.

An examination of primary cultures of dissociated sympathetic neurons also shows that the spectrum of soluble extracellular proteins changes with the phenotype of the cell. As outlined previously (Section 3.8), dissociated neurons from neonatal rat SCG normally develop adrenergic characteristics when grown in the absence of other cell types, but acquire the ability to synthesize more ACh than catecholamines when cultured in the presence of conditioned medium from non-neural cells such as heart (for a review, see Patterson, 1978). An examination of the extracellular proteins synthesized by cultures which were either adrenergic (synthesized predominantly catecholamines) or cholinergic (synthesized predominantly ACh) showed differences between the two phenotypes (Sweadner, 1981). The conditioned medium that induced the cholinergic cell type altered the syntheses of four of the 18 extracellular proteins released by adrenergic cells. One protein was more heavily labeled in adrenergic cultures, while three were more abundant in cholinergic cells (Figure 6.4). The changes in expression of these proteins were between 3- and 10-fold.

Two things should be emphasized about the above studies. First, none of the data show unambiguously that a new protein is released following nerve differentiation. It can only be concluded that changes in the relative amounts of some extracellular proteins are associated with differentiation. Second, the spectrum of proteins released by each cell type appears to be unique. For example, the proteins released by the C6 glial line are dramatically different from those released by the C1300 mouse neuroblastoma (Schubert, 1973). It is therefore likely that cells can be classified phenotypically on the basis of their secreted proteins.

6.6 Cellular Adhesion

The ontogeny of the vertebrate nervous system can be divided into five phases—cell division, cell migration, nucleation (the coming together of distinct groups of cells), cellular differentiation, and ultimately the deaths of subsets of cells within the developing population. Cellular adhesion is intimately involved in each of these processes, with the possible exception of the last. For example, for cells to divide they must release the majority of their attachment sites to their neighboring cells or substratum. Once mitosis is complete, the daughter cells must readhere. The migration of glia and nerve from their origins in the ventricular zone to their more peripheral destinations requires a precise guidance mechanism. Various such mechanisms may be employed, including nerve cell migration along the extended filopodia of special glial cells (Sidman and Rakic, 1973) and through areas containing a high density of

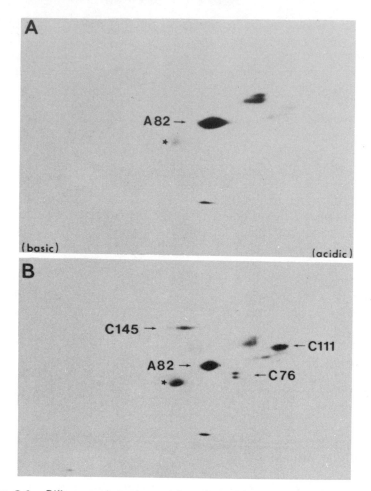

Figure 6.4. Differences between proteins released from adrenergic and cholinergic cultures. Cells grown under conditions which produced the adrenergic (A) or cholinergic (B) phenotype were labeled with [³H]leucine for 24 hr and the material released into the culture medium was electrophoresed on 2-D acrylamide gels. The numbers after the letters (A for adrenergic, C for cholinergic) indicate the protein molecular weight in kilodaltons. The star marks the faint nonmetabolic labeling of serum albumin in the culture medium. From Sweadner (1981).

extracellular matrix material (Pratt et al., 1975). Finally, the nucleation of similar cell types into a histotypically defined structure probably requires an accurate cell–cell recognition system, and neurite elaboration during the differentiation of neurons may follow cues generated by the adhesive properties of the surface on which they grow. It is clear from these few examples that a full description of the development of the nervous system must await the acquisition of a great deal of information about the adhesive properties of cells and growth cones. Recent progress in this area has come almost exclusively from studies of cultured cells.

There are basically four experimental paradigms used to study cellular adhesion. Each reflects a different aspect of the multiplicity of adhesive interactions available to most cells. The four classes of interaction are shown schematically in Figure 6.5. The first is cell sorting. Sorting usually involves the long-term (1–4 days) reassociation and segregation of a mixture of two tissue types that had been dissociated into single cells. The second is haptotaxis, a term derived from the Greek *haptein*, meaning contact, and *taktos*, meaning arrangement. This experimental paradigm acknowledges the preferential adhesion of cells to the most

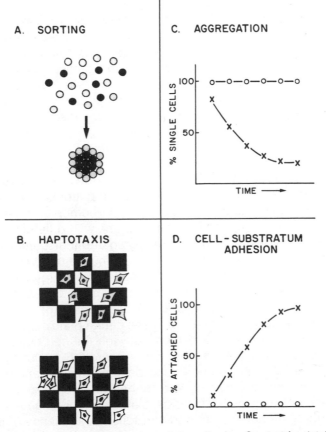

Figure 6.5. *In vitro* models for cellular adhesion. See text for details.

adhesive surface when they are grown on a grid made up of two or more types of substratum. The third paradigm involves the aggregation of single cells in suspension culture. These experiments usually measure the aggregation of cells with cells of the same type over periods of a few hours. Finally, when dissociated cells are plated onto culture dishes containing an adhesive substratum, the cells adhere to this surface as a function of time. Like aggregation experiments, cell–substratum adhesion experiments are usually short-term (1 hr). Both aggregation and cell–substratum adhesion experiments measure the kinetics of cellular adhesion and generate information regarding the number and affinities of adhesion-mediating molecules. Cell sorting and haptotaxis experiments measure equilibrium distributions of cells and frequently are interpreted in terms of thermodynamic parameters. These four aspects of cellular adhesion will be outlined directly below, and more specific examples relating to nerve and muscle will be discussed later.

6.6.1. Cell Sorting.   The first studies of cellular adhesion were cell-sorting experiments. In the early part of this century it was found that old sponges could be dissociated mechanically into single viable cells with ease. When the dissociated cells were placed together in sea water, they reassembled to generate a viable sponge. In 1940 these experiments were extended to vertebrates by Holtfreter and his colleagues (see Townes and Holtfreter, 1955, for a review). They observed that placing pieces of amphibian embryo in medium with a pH of 10 for about 5 min dissociated it into single cells. The cells were then placed in agar-coated culture vessels (to prevent cell–substratum adhesion) containing glucose–salt solutions at neutral pH. Over a period of 1–2 days the cells reaggregated, and in many cases histological patterns were produced that were very similar to those observed in the embryo. If neural plate and epidermis cells (derived from amphibians with different pigmentations) were mixed, the cells initially reaggregated indiscriminately, with no evidence of pattern formation. Within 1 hr they formed a single mass that initially was flat, but became rounded after 10 hr (Figure 6.6). After a few more hours, the epidermal and neural cells moved in opposite directions within the aggregate, and by 20 hr the surface of the aggregate contained exclusively epidermal cells. The neural plate cells were localized to the center and a cleft formed between the cell types; cells no longer adhered to cells of the other type (Figure 6.6C). During the next few days the epidermis expanded and became thinner, and the neural cells retracted from the surface and formed an internal cavity (Figure 6.6D). These results showed that dissociated vertebrate cells are capable of reaggregation in culture and can follow a sequence of develop-

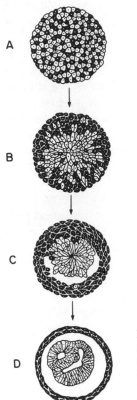

Figure 6.6. Cell sorting of embryonic neural plate and epidermis. Cells from the two embryonic tissues were dissociated, mixed, and allowed to reaggregate. Black cells are epidermis; white cells are from the neural plate. (A) 10 hr after mixing. (B) 13 hr. (C) 20 hr. (D) 2 days. Redrawn from Townes and Holtfreter (1955).

ment which roughly parallels in time that observed *in vivo*. Similar observations were made with numerous other tissues (Townes and Holtfreter, 1975). The experiment illustrated in Figure 6.6 also showed that when two different cell types are allowed to sort in culture, a spherical cell mass forms in which one tissue invariably lies internal to the other. This observation led Malcom Steinberg to develop his "differential adhesion hypothesis" to account for the types of adhesive interactions observed during development (Steinberg, 1970).

Single-cell suspensions from embryonic chick tissues are also able to aggregate and form relatively complete and morphologically differentiated organs (Moscona and Moscona, 1952; P. Weiss and Taylor, 1960). These results and those of Holtfreter demonstrate that separated cells possess the ability to "self organize" in the absence of the normal surrounding tissues. Therefore, the rules for histologically correct organization are inherent in the individual cells rather than imposed by something from the outside. In an extension of these experiments, cells from six distinct tissues of embryonic chicks were dissociated into single cells with trypsin, mixed two at a time in flasks on a gyrating shaker, and allowed to sort themselves (Steinberg, 1970). As with pairs of amphibian tissues, at the conclusion of the sorting process one cell type is always found to be surrounded by the other. Furthermore, the adhesive state of the starting material is unimportant, for results identical to those obtained with dissociated cells are obtained if blocks or sheets of undissociated tissue are pressed together and allowed to sort. If 6 tissues are studied pairwise, there are 15 possible combinations. Each tissue of a pair behaves in one of two ways. It either engulfs its partner or is itself engulfed by the other tissue. There is a well-defined hierarchy governing which tissue prefers the internal position. The hierarchy of embryonic chick tissues defined by Steinberg (1970) is epidermis > limb bud precartilage > eye pigmented epithelium > heart myocardium > neural tube > liver. Each tissue of this series tends to envelop any cell type preceding it when the two are mixed. For example, if epidermis were mixed with liver, epidermis cells would segregate internal to the liver cells.

What types of cellular interaction could account for these results? Two models have been proposed to explain the apparent specificity of adhesive interactions in the embryo. The first is that specific interactions, perhaps best thought of in terms of antigen–antibody binding, occur between pairs of cells. This model requires that there be qualitatively different adhesive molecules on the surfaces of different cell types. The alternative, the "differential adhesion hypothesis," states simply that qualitative differences in adhesive preference between pairs of cells are not necessary to explain the cell-sorting data. Instead, sorting is explained as a consequence of tissue-specific differences in the *relative* strengths of intercellular cohesiveness (Steinberg, 1970). Thus, in a binary tissue combination, the more cohesive cells would become enveloped by the less cohesive cells. The system reorganizes itself into a configuration which possesses a minimal total interfacial free energy. Although the differential adhesion hypothesis in no way contradicts or excludes the possibility of qualitative differences in intercellular adhesion, it does maintain that they are not necessary. For example, if specific interactions between pairs of cells were responsible for the hierarchy of segregation listed above, and if cell type A sorts out internal to B, and B internal to C, it would not be possible to predict accurately that cell type A would always sort out internal to C.

The physiological requirements for the differential adhesion hypothesis are quite simple—cells must adhere and they must be mobile. Since the cells are able to move within aggregates, the starting state (individual cells, slices, chunks) of the cellular mixture is not important; the final state can be viewed as one in thermodynamic equilibrium (i.e., a state in which the work of intercellular adhesions is maximized). The adhesive interactions between homologous cells need not be different in specificity, only in intensity. For example, if cell type A had a higher density of an adhesive molecule on its surface than B, then A would be relatively more adhesive than B and would lie internal to B in an aggregate. Therefore the frequency and relative densities of sites may be critical in some interactions.

The rules suggested by the differential adhesion hypothesis may be applied during early stages of embryogenesis, for example, when the ectoderm, mesoderm, and endoderm become precisely arranged during gastrulation. They may also be critical for neurite outgrowth during development and nerve regeneration, during which growth cones must "make decisions" about which surface to adhere to when they are given a binary choice. The differential adhesion hypothesis may not be applicable in later stages of development in which different tissues are not in direct contact or are separated by basement membrane. It should also be remembered that the adhesive properties of cells change as they differentiate.

Alternatives to the differential adhesion hypothesis have been proposed (see, for example, Harris, 1976). However, most criticism has been leveled at the proposed mechanisms of sorting rather than the more important findings that homologous cell types adhere preferentially to each other and that the more tightly adhering cells are always internalized in mixed populations. Adhesion should not be looked upon as an absolute phenomenon, only as a demonstration of the relative adhesiveness of cells to each other and to the available surfaces in the interacting system. The tenets of the differential adhesion hypothesis can be extended to cell–substratum adhesion and the study of haptotaxis.

6.6.2. *Haptotaxis.* Mouse fibroblasts plated upon a surface of cellulose acetate do not adhere. However, if the cellulose acetate is coated with evaporated palladium, the cells readily adhere to the substratum (Carter, 1967). By varying the amount of palladium deposited on the surface, one can prepare substrata with different adhesivenesses as well as with gradients of progressively increasing adhesiveness. When cells are plated upon an adhesive gradient or upon grids of two substrata (see Figure 6.5), they migrate, and ultimately reach a final position on the most adhesive surface. By presenting cells with two or more adhesive substrata, one can make a number of observations about cellular behavior (Harris, 1973). (1) When plated onto an adhesive gradient, cells always move up the gradient toward the more adhesive surface. (2) When plated onto a grid of two surfaces, cells move until they are on the more adhesive surface and then remain there. (3) In addition to adhering more readily to the more adhesive surface, cells spread to a greater extent on the more adhesive substratum. (4) If cells become overcrowded on the more adhesive surface, they tend to move into the less adhesive area. (5) The preference for adhesion and spreading correlates with the wettability of the surface. (6) Cells show an inhibition of movement on a surface to which they adhere strongly, but not on one to which they adhere poorly. (7) Cells cultured on less adhesive surfaces tend to overlap and pile up. (8) When the leading edge of a cell leaves the more adhesive surface and contacts a less adhesive area, its movement is immediately inhibited. This event is indistinguishable from contact inhibition of movement, in which cellular locomotion is temporarily inhibited when two cells come into contact.

The following hierarchies of preferred adhesiveness govern the behavior of all of the cell types so far tested on haptotactic substrata: cellulose acetate < metal < glass; Petri dishes (nonwettable polystyrene) < metal < tissue culture dishes (wettable polystyrene). For technical reasons adhesion cannot be directly compared between the two groups of surfaces. Thus, when cells are presented with a choice, they always accumulate on the more wettable substratum. It should be noted, however, that on biologically derived surfaces, not all cells have the same adhesive preferences (see, for example, Schubert and LaCorbiere, 1980c, 1982a). Finally, amoebae, which adhere to surfaces and are motile, served as a negative control, showing no adhesive preference when presented with grids of different adhesive surfaces.

Since there is a preferred relative accumulation of cells on different surfaces, it is possible to equate the haptotaxis results with those of Steinberg and his colleagues on differential adhesion in aggregates. In a manner directly analogous to cell–cell adhesion experiments, haptotaxis experiments show that cells have a tremendous "drive" to adhere to something, and when presented with a choice they will adhere to the surface which is more adhesive. These observations are quite important for the interpretation of data about nerve axon elongation and guidance, growth cone adhesion, and fasciculation or bundling together of neurites. The adhesive behaviors of fibroblasts and growth cones are surprisingly similar.

6.6.3. *Aggregation.* Cell-sorting and haptotaxis experiments measure the relatively long-term "equilibrium" adhesive preferences of cells. The kinetic parameters of cellular adhesion, which more directly reflect the number of adhesion sites and their affinities, can also be measured. Experimentally, the manipulations in cell aggregation experiments are quite similar to those in cell sorting. The major difference is the time involved—aggregation is usually measured over a few hours, while sorting requires several days. The first experiments to measure the specific aggregation of vertebrate cells were done by Moscona (1961). Moscona's major experimental innovation was the use of a rotary shaker to promote more uniform aggregation. Previously, cells had been allowed to settle out of suspension to the floor of the flask, which produced variable results because of the interaction of the cells with the substratum. By dissociating cells with chelating agents or trypsin and shaking the flasks during reaggregation, Moscona was able to form more uniform aggregates.

Two methods are generally employed to quantitate the extent of aggregation: (1) measuring the diameter of the aggregates (Moscona, 1961); or (2) measuring the disappearance of single cells from the suspension with electronic devices that can be adjusted to count only single cells. Typical experiments using the two methods are shown in Figures 6.7 and 6.8. Figure 6.7 shows the effect of an aggregation factor isolated from chick neural retina on the volumes of cellular aggregates formed from dissociated neural retina cells. Figure 6.8 shows the effect of an aggregating factor isolated from muscle on the disappearance of single cells from a cell suspension of rat myoblasts. The latter assay is used more frequently, primarily because of its ease and precision.

6.6.4. *Cell–Substratum Adhesion.* Of the four classes of cellular adhesion shown in Figure 6.5, perhaps the best understood is the adhesion of fibroblasts to culture dish substrata. The name fibroblasts is generally given to the group of cells from embryonic skin which is readily adapted to culture. Fibroblasts were one of the first cell types to be cultured, and primarily for this reason they were the cells of choice for most *in vitro* studies of viral and chemical transformation. Because transformation usually causes a decrease in the adhesion of cells to the culture dish substratum, a great deal of effort has been put into understanding the mechanisms of cell–substratum adhesion in normal fibroblasts and its alteration in tumorigenic derivatives. To date the evidence suggests that at least two extracellular proteins, collagen and fibronectin, along with GAGs, mediate cell–substratum adhesion in nontransformed cells. Inter-

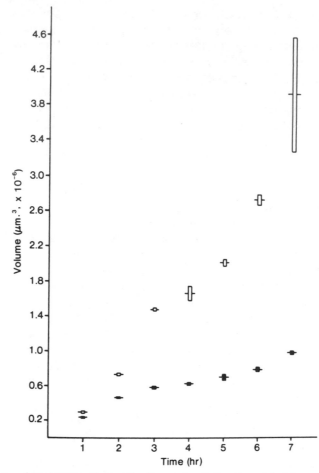

Figure 6.7. Aggregation of neural retina cells. Cells dissociated from 10-day chick neural retina were incubated with or without 0.2 μg/ml of a cell-aggregating protein on a rotary shaker at 38°C. At the indicated times the sizes of the aggregates were determined from photographs. Open bars indicate cells in presence of factor; solid bars indicate control cells. Mean volume of aggregate is plotted against time. The bars indicate the standard deviations for 100 aggregates. From Hausman and Moscona (1976).

actions between other molecules may be involved in the adhesion of transformed fibroblasts. The structures and syntheses of the three classes of molecules involved in fibroblast adhesion will be discussed first, followed by an examination of the interactions that lead to adhesion. This will also serve as a background for a description of the work describing muscle and nerve adhesion.

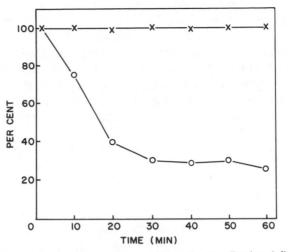

Figure 6.8. Aggregation of rat myoblasts. M3A cells, an adhesion-deficient variant of Yaffe's L6 skeletal muscle myoblast line which grows in suspension, were incubated alone or in the presence of 2 μg/ml of an adhesion-promoting glycoprotein complex isolated from the parental L6 muscle cells. Cells were incubated at 1×10^6 ml on a rotary shaker at 37°C. The disappearance of single cells from the suspension was monitored on a Coulter counter adjusted to count only single cells. ×, M3A cells alone; ○, M3A cells plus L6 adhesion-promoting factor. The ordinate represents the percentage of all cells that are single cells at the given times.

Collagen molecules are characterized by their high content of glycine (30%), proline (10%), and hydroxyproline (10%) and by their susceptibility to bacterial collagenases. Individual collagen molecules contain three polypeptide chains of about 95,000 daltons each, which interact to form a triple helix. There are at least five types of collagens, designated types I through V. Their subunits are thought to be derived from a group of eight or more genes. Type I collagen has two $\alpha1(I)$-chains and one $\alpha2(I)$-chain. Another form of type I collagen, named type I trimer, contains only $\alpha1(I)$-chains. Type I collagens are found mainly in skin, bone, and tendon. Type II collagen is localized to cartilage and is composed exclusively of $\alpha1(II)$-chains. Type III collagen has only $\alpha1(III)$-chains and is found primarily in blood vessels and skin. Type IV collagen contains two different types of polypeptide chain and is thought to be localized exclusively to basement membranes. Finally, type V collagen again contains two types of chain, $\alpha1(V)$ and $\alpha2(V)$, and is found in smooth muscle, bone, and skin.

The biosyntheses of types I, II, and III collagens are reasonably well understood (Fessler and Fessler, 1978). The individual collagen α-

chains are synthesized as molecules larger than the final product (Figure 6.9). These are called procollagens and they are more soluble than the α-chains. There are extension peptides of about 13,000 and 36,000 daltons at the amino and carboxy termini, respectively. During or immediately after translation, the pro-α-chains are modified by the hydroxylation of proline and lysyl residues and the glycosylation of some hydroxylysyl residues. Triple-helix formation and the formation of interchain disulfide bonds appears to occur intracellularly, and the assembled procollagen molecules are secreted. The extension peptides are removed from the secreted protein by specific proteases and the collagen chains are covalently cross-linked by the condensation of aldehydes formed from lysyl and hydroxylysyl residues.

Although some clones of cultured fibroblasts can synthesize completely assembled collagen molecules, most are unable to complete all of the required post-translational modifications because the enzymes required for removing extension peptides and for chain cross-linking are missing. Therefore, large amounts of procollagen accumulate in the culture medium (Layman et al., 1971). The amount of collagen synthesized by cultured cells varies between 0.001% and 10% of the total protein synthesis. Most fibroblast lines synthesize over 5% of their protein as collagen (Kleinman et al., 1981). The amount of collagen synthesized is also dependent on the amount of ascorbic acid in the culture medium. Ascorbic acid is a cofactor required for the hydroxylation of proline to hydroxyproline, and without hydroxyproline the conformation of collagen is defective. Since ascorbic acid lasts less than 12 hr in culture medium, cells must be fed ascorbic acid frequently for maximal collagen synthesis. Compared with normal cells, transformed or tumorigenic cells synthesize less collagen and perhaps lesser amounts of matrix components. Transformed fibroblasts generally are less adhesive to the substratum and have a less flattened morphology than their parent clones.

Cultured fibroblasts produce types I and III collagens, whereas smooth muscle cells make all forms except for type II. Type II collagen is made by initial explants of chondrocytes, but longer-term chondrocyte cultures make all of the isotypes except type IV. Finally, type IV collagens are made in culture primarily by epithelial cells, which is consistent with their role in basement membrane synthesis (Kleinman et al., 1981). The types of collagens synthesized by cultured cells are, however, somewhat variable and very dependent on culture conditions (Bissell, 1981).

Fibronectin, an abundant surface glycoprotein of fibroblasts, interacts with collagen to mediate the adhesion of some cell types (see Pearlstein

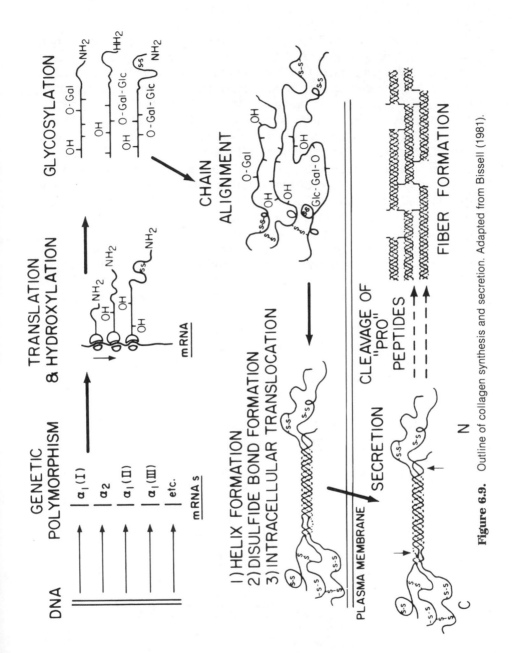

Figure 6.9. Outline of collagen synthesis and secretion. Adapted from Bissell (1981).

et al., 1980, for a review). Cold-insoluble globulin (CIG), which is closely related to fibronectin, was first described over 30 years ago as a precipitate which formed when blood clotted in the cold. The serum protein has two disulfide-bonded chains of about 220,000 MW each (Mosesson and Umfleet, 1970). The role of CIG in cell–substratum adhesion was discovered during the search for a serum factor required for the adhesion of myoblasts (Hauschka and White, 1972) and fibroblasts (Klebe, 1974). This factor was subsequently shown to be CIG (Pearlstein, 1976). Although CIG is frequently referred to as fibronectin, the blood-derived CIG and the cellular fibronectin are not identical. The blood molecule is slightly smaller than cell-surface fibronectin, and the cell form is about 50 times more active than CIG in restoring normal morphology to virally transformed cells (Yamada et al., 1976). It is possible that these differences are due to post-translational modifications of the protein, as the molecular basis for the differences is unknown.

The structure of fibronectin is represented in Figure 6.10. A number of fragments have been isolated from the native molecule by proteotypic cleavage, and several of these have biological functions. A 40,000-MW sequence near the N-terminus contains the collagen-binding region, whereas the large 160,000-MW C-terminal portion binds to the cell surface but not to collagen. Since some gangliosides inhibit the cell–fibronectin interaction but not the interaction between fibronectin and collagen, it has been argued that cell-surface glycolipids are the receptors for fibronectin (see Sekiguichi and Hakomori, 1980, and Kleinman et al., 1981, for reviews). Multiple binding sites for the GAG heparin have been identified. The major site is near the N-terminus, but heparin can also bind to the 160,000-MW component that interacts with the cell surface. Finally, actin binds to two of the three major domains shown in Figure 6.10.

It is thought that a series of interactions between the substratum,

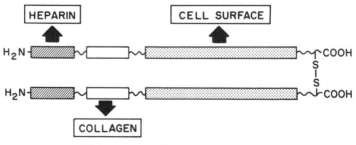

Figure 6.10. Structure of fibronectin.

collagen, fibronectin, and the cell surface are responsible for the initial adhesion and subsequent spreading of freshly plated fibroblasts and other cell types derived from mesoderm. Investigators frequently plate cultured cells on collagen-coated dishes to facilitate the initial adhesion of cells or to provide a suitably adhesive substratum for cells which do not synthesize sufficient quantities of collagen. Fibronectin binds to collagen at a defined position on the latter molecule (Kleinman et al., 1978). In the case of the $\alpha1(I)$ collagen chain, fibronectin interacts at a site within a 35 amino acid residue fragment near the collagen C-terminus. This region is atypical of the rest of the collagen molecule in that it lacks proline and hydroxyproline and its amino acid residues are hydrophobic. Fibronectin also binds to a variety of GAGs, including heparin and hyaluronic acid, and these GAGs appear to stabilize the interaction between collagen and fibronectin (Jilek and Hormann, 1979). In transformed mesodermally derived cells, interactions directly between GAGs or between GAGs and unknown molecules may replace the fibronectin–collagen interaction in mediating cellular adhesion (Schubert and La-Corbiere, 1980c).

Like collagen, fibronectin stimulates the adhesion of many mesodermally derived cells to culture dish substrata. It is frequently used in the absence of collagen to facilitate the adhesion and cell growth of anchorage-dependent cell lines. It should be noted that not all cells produce fibronectin, and that other molecules seem to fulfill the role of fibronectin in adhesion. For example, cartilage-producing chondrocytes do not make fibronectin, but they do secrete a lower-molecular-weight molecule called chondronectin, which promotes their adhesion to collagen (Kleinman et al., 1981).

Finally, two major groups of macromolecules which are also involved in cellular adhesion are the GAGs and the proteoglycans. GAGs are long, unbranched polysaccharide chains which contain numerous carboxylate and/or sulfate moieties. All GAGs except hyaluronic acid normally occur as proteoglycans, in which form they are attached by the reducing end of the terminal sugar to proteins. The GAGs are made of repeating disaccharide units; the seven classes of GAGs are listed in Table 6.2. The molecular weights of the individual glycosaminoglycan chains vary between 2000 and 20,000, and several hundred of these chains can be attached covalently to a single protein molecule to form a proteoglycan monomer (see Hardingham, 1981, for a review). The proteoglycans can, in turn, associate with hyaluronic acid via a specific "link protein" to form very large (350×10^6 MW) aggregates consisting of up to 200 proteoglycans and measuring 4 μm in length. Although the aggregation of proteoglycans with hyaluronic acid is most frequently

Table 6.2.   Composition of Mammalian Glycosaminoglycans

GAG	Hexuronic Acid	Hexosamine[a]	Sulfate
Hyaluronic acid	D-Glucuronic acid	D-Glucosamine	None
Chondroitin 4-sulfate	D-Glucuronic acid	D-Galactosamine	O-Sulfate
Chondroitin 6-sulfate	D-Glucuronic acid	D-Galactosamine	O-Sulfate
Dermatan sulfate	L-Iduronic acid or D-glucuronic acid	D-Galactosamine	O-Sulfate
Keratan sulfate	D-Galactose	D-Glucosamine	O-Sulfate
Heparan sulfate	D-Glucuronic acid or L-iduronic acid	D-Glucosamine	O-Sulfate and N-Sulfate
Heparin	D-Glucuronic acid or L-iduronic acid	D-Glucosamine	O-Sulfate and N-Sulfate

[a] B-acetylated, except when N-sulfated.

associated with the cartilage matrix, similar structures have been demonstrated in cultured glial cells (Norling et al., 1978).

The chemical composition of GAGs and proteoglycans gives them a number of properties unique among biological macromolecules. Their high charge density makes them very hydrophilic. In solution they are highly expanded and occupy a large volume, thus contributing to the resilience of such tissues as cartilage. The function of GAGs in noncartilaginous tissues, however, is relatively unknown.

6.7   The Intermolecular Forces Involved in Adhesion

Before considering specific experiments involving nerve and muscle adhesion, it may be worthwhile briefly to discuss the types of intermolecular forces that are thought to be responsible for cellular adhesion. Since cellular adhesion is probably mediated by receptor–receptor or receptor–ligand interactions, adhesions have been modeled after three fairly well-understood types of interaction–antigen–antibody, lectin–carbohydrate, and ligand–enzyme. Cells are highly negatively charged because of the ionized carboxylate groups of sialic acid on the cell surface. Electrostatic interactions are, however, minimally involved in most cell–cell and cell–substratum interactions (Bell, 1978). This conclusion can be drawn from theoretical considerations and the following experimental evidence. (1) Carboxyl groups of sialic acid are masked by positively charged ions in the culture medium under physiological conditions. (2)

Sialic acid can be removed from the cell surface without changing the cell's adhesive characteristics. (3) Cell–substratum adhesion varies with the wettability of the surface rather than with the net charge of the substratum. (4) Electrostatic forces are present on all surfaces and are probably nonspecific.

The other major group of forces that could influence adhesion is the electrodynamic or van der Waals forces. Van der Waals forces are caused by the mutual attraction between induced dipoles in molecules, and occur at a significantly longer range than do the repulsive electrostatic interactions. They are the predominant forces in most antigen–antibody binding. Figure 6.11 shows a theoretical energy-separation curve for cells which indicates that cells are attracted to each other until they come close enough to reach the secondary minimum (r_s) at about 7 nm (Bell, 1978). Most cells are separated by approximately this distance *in vivo*. It follows that although cells have highly charged surfaces, longer-range van der Waals forces probably mediate most cellular adhesions. Finally, it should be pointed out that adhesive forces may be sufficiently strong that an integral membrane protein subjected to a shear force may be pulled out of the membrane rather than break the adhesive bond. The high strength of adhesive bonding is due to the simultaneous bonding of many receptors. Although a single interaction may be reversible, the probability of many receptor–ligand complexes dissociating simultaneously is low. This multivalent nature of adhesive interactions also leads to a greater degree of specificity (Ehrlich, 1979)

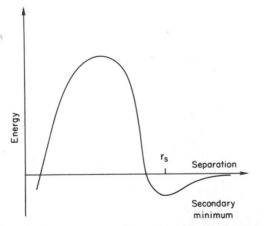

Figure 6.11.   Theoretical energy-separation curve, showing the energy due to electrostatic and electrodynamic forces plotted against the separation between two cell membranes. From Bell (1978).

and makes it likely that the properties, including specificity, of isolated monovalent adhesion molecules may be very different from those *in vivo*.

6.8   The Role of Substratum Collagen in Myogenesis

The interactions between the cell and its substratum have profound effects on cellular metabolism, DNA synthesis, division, and differentiation. In most untransformed cell lines, a surface on which to adhere and flatten is required for maximal rates of DNA and protein synthesis and of cell division to be attained (Folkman and Moscona, 1978; Ben-Ze'ev et al., 1980). Varying the extent and kind of adhesive contacts therefore generates a variety of regulatory responses (see Chapter 1 for a discussion).

One of the first discovered examples of the role of cell–substratum adhesion in a developmental process was that a collagen-coated substratum is required for myogenesis in chick cells. When the work on cultured embryonic chick skeletal muscle cells was in its infancy, it was found that the survival, division, and differentiation of myoblast clones was dependent on the use of culture medium in which high-density cells had been grown for several days. One observation suggested that the conditioned medium may alter the culture dish surface (Hauschka and Konigsberg, 1966). If conditioned medium was placed in new Petri dishes for 3 days, the medium removed, the dishes washed extensively with water, and new medium put into the dishes, the dishes were able to support the development of muscle clones as well as did dishes containing growth-conditioned medium. Since stationary-phase but not exponentially dividing chick fibroblasts were able to condition medium as well as muscle cultures could, and since only stationary-phase chick fibroblasts secrete collagen, Hauschka and Konigsberg reasoned that collagen was the material that was released into the culture medium and altered the Petri dish surface. They subsequently showed that a substratum prepared from rat-tail collagen was able to eliminate completely the requirement for conditioned medium, suggesting that collagen is the active ingredient in the growth-conditioned medium. Collagen types I through IV are all equally effective (Ketley et al., 1976). It should be pointed out, however, that rodent myoblast cultures do not depend on a collagen-coated substratum as chick myoblasts do. This is probably because rat myoblasts secrete large amounts of collagen (Schubert et al., 1973).

Since collagen stimulates cell–substratum adhesion by interacting with fibronectin, a number of laboratories have examined the distribution of fibronectin in muscle cultures. Using the peroxidase–antiperoxidase method to localize anti-fibronectin antibody binding to cultures of clonal L6 rat myoblasts, Furcht and colleagues (1978) observed two forms of fibronectin. Low-density myoblasts have fibronectin sparsely distributed on their surface and produce very little extracellular matrix. As the cultures become confluent, myoblasts become associated with an extensive extracellular filamentous matrix containing heavily staining fibronectin-containing globules along the filaments. By the time fusion occurs, there is a diffuse distribution of fibronectin on the cell surface of multinucleate myotubes. The presence of fibronectin in extracellular filaments suggests that it may interact with other molecules such as collagen and GAGs to produce the extracellular matrix structure.

6.9 Muscle Lectins

Lectins are proteins which agglutinate erythrocytes (hemagglutinate) via interactions with carbohydrate moieties on the cell surface. The agglutination activity is specifically inhibitable by simple carbohydrates. Although this type of activity was first isolated from plants, predominantly from legumes, lectins have also been demonstrated in many vertebrate tissues. Since lectins are able to induce cellular aggregation, they may play a role in cell–cell adhesion *in vivo*. Extracts of skeletal muscle contain carbohydrate-binding proteins which can be assayed as saccharide-inhibitable hemagglutinins. At least three distinguishable lectins have been purified from muscle. Although there is no evidence that these activities are directly involved in myogenesis, they may mediate some cell–matrix interactions, or they may be involved directly in the synthesis of myotube basement membrane. The first lectin, appropriately called chicken-lactose-lectin-I, was purified on the basis of its ability to agglutinate fixed trypsinized rabbit erythrocytes; agglutination was inhibited by low concentrations of lactose, but not by most other carbohydrates (Nowak et al., 1977). Chicken-lactose-lectin-I activity is low in extracts of 8-day embryonic chick pectoral muscle, increases about 10-fold by day 16, and then declines. Chicken-lactose-lectin-I has been purified by affinity chromatography. It has a molecular weight of 30,000 daltons, with two subunits of 15,000 daltons each (Nowak et al., 1977). Lactose-inhibitable lectins similar but probably not identical to Lectin I have been found in a variety of tissues, including liver, brain, and intestine (Beyer et al., 1980).

Chicken-heparin-lectin does not agglutinate the same type of erythrocyte as does chicken-lactose-lectin-I but it does agglutinate aged or alcohol-washed erythrocytes. The specificity of sugar inhibition of the heparin lectin is also distinct from that of the lactose lectin (Ceri et al., 1979). Chicken-heparin-lectin-induced agglutination is blocked by *N*-acetyl-D-galactosamine, but not by lactose. Agglutination is also inhibited by heparin, heparan sulfate, and chondroitin-2-sulfate (dermatan sulfate), but not by hyaluronic acid, chondroitin-4-sulfate, or chondroitin-6-sulfate. The heparin lectin is found in cells and the growth-conditioned medium of cultured chick myoblasts. Figure 6.12 shows that chicken-heparin-lectin activity appears in the medium only after the initiation of myoblast fusion. Cultured myoblasts secrete such basement membrane components as fibronectin, collagen, and GAGs, but they do not possess a basement membrane. Once cultured myoblasts have fused, the multinucleate myotubes frequently become surrounded by basement membrane (Schubert et al., 1973). It is possible that the release of chicken-heparin-lectin after fusion is involved in the assembly of the basement membrane. Finally, a potent inhibitor of the lectin-induced agglutination, identified as dermatan sulfate, has been extracted from the SAM synthesized by the cultures (Ceri et al., 1981). This datum shows again that the lectin can interact with GAGs in the immediate extracellular environment.

The last lectin that has been purified is from the clonal L6 rat skeletal muscle myoblast line (Podleski and Greenberg, 1980). This lectin, termed electrolectin because it was first characterized in the electric

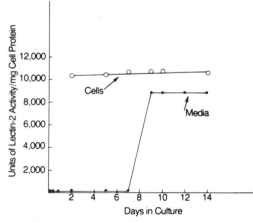

Figure 6.12. Appearance of chicken-heparin-lectin in the medium of chick muscle cultures. The appearance of chicken-heparin-lectin activity in chick primary skeletal muscle cultures was followed as a function of time. Cell fusion began at day 6. From Ceri et al. (1979).

organs of eels, has an apparent molecular weight of 13,000 and is probably similar to the lactose-inhibitable Lectin I activity described by Nowak and colleagues (1977). Antibodies prepared against electrolectin have been used to demonstrate that the lectin occurs on the cell surfaces of myoblasts from both the clonal rat L6 line and rat skeletal muscle primary cultures. The amount of electrolectin associated with myotubes is much less than that found on myoblasts.

The possible roles of any of the above lectins in myogenesis remain to be defined.

6.10 Muscle Adherons

The above data show that the collagen–fibronectin system is involved in myoblast adhesion and that collagen, fibronectin, and at least one muscle lectin are found extracellularly. Studies with the clonal L6 skeletal muscle myoblast line have demonstrated the existence of an extracellular particle that contains collagen, fibronectin, and GAGs. This particle mediates cell–cell and cell–substratum adhesion of myoblasts, and has been called an adheron (Schubert and LaCorbiere, 1982a). Similar adhesion-mediating particles have been found in most cell lines that grow attached to the surface of culture dishes.

L6 skeletal muscle cells release into the culture medium molecules that adhere to the surfaces of Petri dishes and promote cell–substratum adhesion of myoblasts. L6 myoblasts normally do not adhere to plastic Petri dishes. However, if the growth-conditioned medium from L6 myoblasts is placed on new Petri dishes for 18 hr, material is adsorbed onto the surface of the Petri dishes which promotes the initial rate of myoblast adhesion (Figure 6.13). One can follow adhesion easily by isotopically labeling the cells with a tritiated amino acid, pipetting the labeled cells into culture dishes, aspirating the unattached cells at various times, and counting the amount of radioactivity remaining associated with the dish. The data are then plotted as the percentage of all input cells that adhere at each time point (Figure 6.13).

When the growth-conditioned medium of L6 cells was centrifuged for 3 hr at 100,000 g, *all* of the adhesion-promoting activity was found in the pellet, suggesting a high molecular weight. About 40% of the total glucosamine and about 5% of the total amino acid incorporated into extracellular macromolecules is pelleted. If the adhesion-promoting material is a large molecule or macromolecular complex, it should sediment into a sucrose gradient, and the adhesion-mediating activity should be associated with this entity. To characterize the particulate material, Schubert

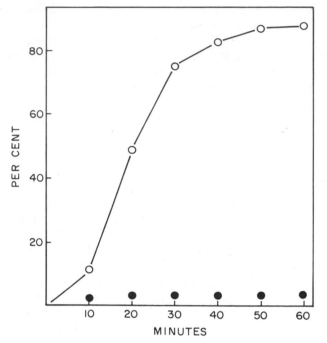

Figure 6.13.   Adhesion of L6 cells to SAM prepared from L6 growth-conditioned medium. Exponentially growing L6 myoblasts were incubated overnight in serum-free HEPES medium. The growth-conditioned medium was then placed in 35-mm Petri dishes overnight at 37°C. Exponentially dividing L6 cells were isotopically labeled, dissociated with EGTA, washed, and plated at 3×10^5 cells/dish into the SAM-coated dishes containing 2 ml HEPES medium and 0.2% bovine serum albumin. The data are presented as the percentage of all input cells that were adherent at each time point. ●, L6 cells plated on new Petri dishes; ○, L6 cells plated on Petri dishes incubated overnight with L6-conditioned medium.

and LaCorbiere (1980a) sedimented it in 5–20% sucrose gradients. The glucosamine-labeled material defined a peak that sedimented about one-third of the way into the gradient and had a sedimentation value of 16S relative to ribosomal RNA markers (Figure 6.14). When the ability of each gradient fraction of the unlabeled material to mediate cell–substratum adhesion was assayed, it was found that only the gradient fractions that coincided with the presence of the isotope effected an increase in the adhesion of myoblasts. In addition to stimulating myoblast adhesion, the particles also stimulated myoblast aggregation and *inhibited* the adhesion of a clone of sympathetic nerve cells to plastic Petri dishes (Figure 6.14). The latter two activities will be discussed in detail later.

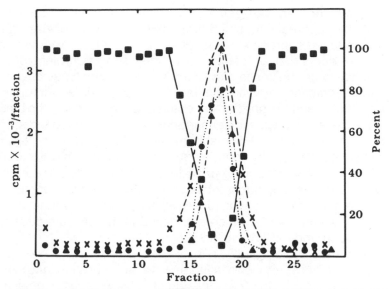

Figure 6.14. Cosedimentation of L6 GAGs and adhesion activities in sucrose gradients. L6 cells were labeled with [³H]glucosamine in serum-free medium or were incubated in serum-free medium alone for 18 hr. The media were then centrifuged at 100,000*g* for 2 hr. The two 100,000*g* pellets were then centrifuged into calcium-free 5–20% linear sucrose gradients and fractionated. The isotope data (×) are presented as cpm per fraction. Each gradient fraction with unlabeled cellular material was divided into three aliquots. One was diluted into HEPES medium to determine its effect on myoblast agglutination. These data are presented as the percentage of maximal aggregation at 40 min (▲). The remaining two aliquots were each diluted into HEPES medium and incubated in Petri dishes overnight at 37°C. The following day the adhesions of myoblasts (●) and the sympathetic nerve cell line PC12 (■) to the substratum were determined; these data are presented as the percentage of all input cells that adhered at 50 min. The top of the gradient is at the right.

The 16S skeletal muscle adheron contains one mole each of collagen and fibronectin, and four distinguishable proteoglycans (Schubert et al., 1983a). A few other proteins are associated with the particles, and are bound in nonstoichiometric amounts. GAGs account for about 30% of the particle mass, and 70% is protein. No nucleic acids or phospholipids are detectable. As viewed by electron microscopy, the L6 adherons are spheres approximately 20 nm in diameter.

The next question is how skeletal muscle adherons interact with cells and with themselves to produce adhesions between cells and between cells and adheron-coated surfaces. Two classes of adheron interactions can be distinguished experimentally by their dependences on exogenous calcium ion concentration. The first type is that between the particle and the cell surface or the culture dish. L6 adherons bind to essentially all surfaces and all cell types in the absence of calcium. In contrast, the aggregation of L6 adherons with themselves, the L6-adheron-induced aggregation of myoblasts, and cell–substratum adhesion are all dependent on calcium (Schubert and LaCorbiere, 1982b). The calcium dose–response curves for the latter three types of adheron mediated interactions are very similar. Thus, the particle–particle interactions which mediate adhesion are calcium dependent and the binding of the adherons to cells and other surfaces is calcium independent. This adhesion mechanism is shown schematically in Figure 6.15. It is very similar to that employed by sponge aggregation factors, in which there is a calcium-independent binding of aggregation factors to surface receptors and a calcium-dependent interaction between factors on adjacent cells that promotes aggregation (Jumblatt et al., 1980). A variant cell line has been isolated from L6 which is blocked in the calcium-dependent step (Schubert and LaCorbiere, 1982b).

If adherons are biologically relevant, then they should show some adhesive specificity. As shown in Figure 6.14, adherons from the skeletal muscle cell line L6 inhibit the adhesion of PC12 sympathetic-nervelike cells to plastic substrata. The PC12 cells normally rapidly adhere to Petri dish surfaces, but if the dishes are coated with adherons isolated from skeletal muscle adhesion is inhibited. This may be a reflection of the fact that sympathetic nerves normally innervate smooth muscle but not skeletal muscle. If there is some adhesive specificity inherent to the adherons,

PARTICLE MEDIATED ADHESION

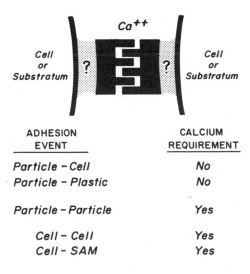

ADHESION EVENT	CALCIUM REQUIREMENT
Particle – Cell	No
Particle – Plastic	No
Particle – Particle	Yes
Cell – Cell	Yes
Cell – SAM	Yes

Figure 6.15. A model for adheron-induced adhesion of L6 skeletal muscle myoblasts.

then adherons from smooth muscle should promote the cell–substratum adhesion of PC12 cells and adherons from skeletal muscle should inhibit PC12 adhesion. Adherons from four cell lines or from primary cultures of each muscle type have been examined. Those from smooth muscle all promote PC12 adhesion, whereas adherons from any of the skeletal muscle preparations inhibit adhesion (Figure 6.16). In addition, the adherons from the two muscle classes have the following structural differences (Schubert and LaCorbiere, 1982a): (1) The particles from skeletal muscle have a sedimentation value of 16S, but those from smooth muscle are 12S. (2) The GAG compositions of the smooth and skeletal muscle adherons are also distinct. Those of smooth muscle contain no detectable hyaluronic acid, whereas the skeletal muscle cell lines all contain this GAG. In addition, all of the smooth muscle particles

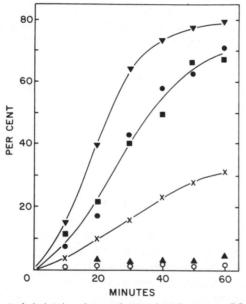

Figure 6.16. Effect of skeletal and smooth muscle adherons on PC12 cell–substratum adhesion. Experimentally growing skeletal or smooth muscle cells were incubated overnight in serum-free HEPES medium. The growth-conditioned medium was then placed in 35-mm Petri dishes overnight and the coated dishes were washed. Exponentially dividing PC12 cells were isotopically labeled, washed, and plated into the coated dishes containing 2 ml HEPES medium and 0.3% bovine serum albumin. The data are presented as the percentage of all input cells that adhered at each time point. ×, Petri dish alone; ▼, BC₃HI (smooth muscle) SAM; ●, A10 (smooth muscle) SAM; ■, A7 (smooth muscle) SAM; ▲, B44 (skeletal muscle) SAM; O-O-O, L6 (skeletal muscle) SAM. From Schubert and LaCorbiere (1982a).

contain an unidentified GAG-like carbohydrate. (3) Antibodies prepared against skeletal muscle particles inhibit skeletal muscle myoblast adhesion and therefore fusion, but do not block the adhesion of PC12 cells to smooth muscle adherons. However, the protein contents of skeletal and smooth muscle adherons are similar, for both contain collagen and fibronectin as their major constituents. Therefore the only systematic differences in composition between smooth and skeletal muscle adherons are the lack of hyaluronic acid and the presence of an unknown carbohydrate in the smooth muscle particles. It must also be stressed that only skeletal muscle adherons inhibit the adhesion of PC12 cells. Adherons from smooth muscle cells and from all other cell types tested promote PC12 adhesion to varying extents. This shows that the properties of adherons from skeletal muscle are distinct from those from all other types of cells tested with respect to the adhesion of PC12 cells. Skeletal muscle adherons do, however, promote the adhesion of other cell types, such as skeletal muscle myoblasts.

The adhesion-modulating activity described above could influence the innervation of embryonic muscles in two ways. Since the adheron is released in a soluble form from cells, it may diffuse a small distance away from the muscle and attach to the surface of surrounding cells. In the case of smooth muscle cells, adherons may form a limited adhesion gradient to guide incoming sympathetic axons. This substratum would have to be more adhesive for sympathetic nerve cell growth cones than the substratum generated by neighboring tissues. In contrast, particles from skeletal muscle could establish an inhibitory gradient for sympathetic nerve around that tissue. For this extrapolation from *in vitro* data to the developing animal to be valid, the adhesive properties of whole cells must be reflected in the adhesion of the growth cone. There is evidence for this, for the adhesive properties of sympathetic nerve cell bodies and the directional turning responses of growth cones are affected in a similar way by exogenous NGF, calcium ions, and cyclic nucleotides (Schubert et al., 1978; Gundersen and Barrett, 1980). Once the axon terminals are near the smooth muscle target, a high density of adhesion-promoting particles either on the surface of the cell or in its extracellular matrix may physically stabilize the nerve–muscle contact.

Glycoprotein particles similar in size to L6 adherons exist in basement membrane. After osmium-fixed tissues were stained with ruthenium red, particles were observed in a regularly spaced pattern in the basal lamina (Hay, 1978). These particles were sensitive to chondroitinase, showing that they contained GAGs. A 2-D symmetrical array of particles was also found in the basal lamina of mammary epithelium (Gordon and Bernfield, 1980). More recently, an ultrastructural analysis of the extracellular matrix synthesized by endothelial cells showed that it consists of 30-nm spherical "nodes" which are stacked in parallel arrays and separated by approximately 100 nm (MacCallum et al., 1982). It is possible that the particulate material released by skeletal and smooth muscle generates similar structures.

6.11 Adhesion Factors from the Chick Neural Retina

In addition to the work outlined above on the role of collagen and fibronectin in the adhesion of mesodermally derived cells, a great amount of effort has been put into purifying material from growth-conditioned media that promotes the cell–cell aggregation of embryonic chick neural retina cells. This has been a popular experimental system because embryonic retinal cells are easily dissociated into single cells and remain viable for extended periods of time in culture. Neural retina tissue is, however, extremely heterogeneous, for it contains at least 20 distinguishable cell types. A standardized aggregation assay, in which trypsin-dissociated cells are allowed to reaggregate on a rotary shaker (Figure 6.7), can be used to show that growth-conditioned medium from 10-day chick embryo neural retina specifically aggregates 10-day embryonic neural retina cells (Lilien and Moscona, 1967). Growth-conditioned media from embryonic heart, liver, and limb buds have no demonstrable effect on aggregation of neural retina cells, nor does neural-retina-conditioned medium aggregate the other cell types (Lilien, 1968).

On the basis of his aggregation assay, Moscona and colleagues have purified a protein from growth-conditioned medium and neural retina cell membranes that specifically aggregates embryonic chick neural retina cells (McClay and Moscona, 1974; Hausman and Moscona, 1975, 1976). The growth-conditioned medium of 10-day chick neural retina can be concentrated and an aggregation activity purified by isoelectric focusing and column chromatography. The aggregation-promoting activity resides in a glycoprotein (15% carbohydrate) with a molecular weight of approximately 50,000. A protein of similar size and specificity can be extracted with butanol from purified neural retina membranes (Hausman and Moscona, 1976). The 50,000-MW protein, called cognin, aggregates cells maximally at concentrations between 0.2 and 1 µg/ml. Cognin is able to bind to cells at 4°C, but not to aggregate them. When the cells are washed and the temperature is raised to 37°C, the cells aggregate. Cognin has not been extractable from membranes of embryonic neural retina older than 13 days—the time when the morphogenetic cell organization of this tissue is completed. Nor can the older cells be aggregated by cognin from younger cells.

Antibodies have been prepared against retina cognin and used to examine its tissue distribution and the age dependence of expression on the cell surface (Ben-Shaul et al., 1979; Hausman and Moscona, 1979). By use of both complement-mediated lysis and fluorescent antibody staining techniques, cognin is demonstrable on 6-, 8-, and 10-day embryonic neural retina, is lower on 12-day cells, and is not present on cells past day 13. Cognin is, however, detected in other parts of the nervous system, but not on the surface of non-neuronal cells. The antiserum against cognin inhibits neural retina aggregation.

In contrast to Moscona and coworkers, who isolated a molecule from neural retina on the basis of its ability to promote cellular aggregation, Edelman and his colleagues have isolated a molecule thought to be involved in neural retina cell aggregation on the basis of its ability to neutralize the adhesion-inhibiting effect of an antibody against whole retina cells which inhibited cellular aggregation. The immunological technique employed in these studies to isolate potential adhesion-mediating molecules was initially devised by Huesgen and Gerisch (1975) for the isolation of similar molecules from slime mold. Antisera prepared against either whole cells or purified plasma membrane usually inhibit cell−cell aggregation and cell−substratum adhesion. For example, antisera prepared against live nerve cells cause attached and spread cells to become rounded and detach from the substratum, and inhibit the adhesion of freshly plated cells (Stallcup, 1977b; Hsieh and Sueoka, 1980). Alterations in cellular adhesiveness are also caused by antisera prepared against material shed into serum-free medium (Damsky et al., 1981).

If a molecule is involved in the cellular function (e.g., adhesion) which is blocked by the antiserum, this molecule should neutralize the inhibitory effect of the antiserum. This neutralization assay can serve as the basis for the purification of the molecule involved in the interaction being studied. The only requirements are that the cells adhere to the substratum or spontaneously aggregate and that the rate-limiting step in the adhesion involve a small number of molecules. In chick neural retina these criteria are met, and a protein thought to be involved in adhesion has been isolated (Thiery et al., 1977). When cells are dissociated with trypsin, many of their surface molecules are destroyed. Therefore chick neural retina cells are trypsinized and maintained for 20 hr in a low-calcium medium to allow the recovery of trypsin-sensitive surface molecules. When the cells are concentrated in normal media, they spontaneously aggregate. If, however, monovalent Fab′ fragments of an antibody prepared against whole cells are present, aggregation is inhibited. When growth-conditioned medium is present, the effect of the antibody is neutralized and the cells aggregate. The neutralizing activity in the supernatant has been purified by ion exchange chromatography followed by a sizing column followed by acrylamide gel electrophoresis in the presence of SDS. Three proteins were isolated from conditioned medium which neutralized the inhibition by antibody of cell aggregation (Thiery et al., 1977). These proteins of 140,000, 120,000, and 65,000 MW were pooled and antibody was prepared against them. This antiserum specifically precipitates the three proteins and inhibits cell aggregation. The antiserum also precipitates a protein of 140,000 MW from detergent extracts of neural retinal cell membranes. Precipitation is inhibited by all three supernatant proteins, suggesting that the larger surface molecule is released into the culture medium and partially degraded. The antibody against the 140,000-MW surface component, which is called "cell adhesion molecule" (CAM), has been used in a series of experiments designed to explore the function of CAM in the nervous system.

By using anti-CAM antibody and the peroxidase–antiperoxidase procedure to visualize the binding of antibody to cells, Rutishauser and coworkers (1978a, b) have shown that CAM is distributed fairly uniformly over the majority of neural retina cells. Like cognin, CAM is less abundant on the surface of embryonic neural retina cells more than 14 days old. When frozen sections of various chick tissues are stained with anti-CAM antibodies, all areas of the nervous system are heavily stained, but very little CAM antigen is detected on liver. Finally, monovalent Fab' fragments prepared from antibodies against CAM inhibit the sorting of cells and neurites in large aggregates of neural retina. Neural retina tissue from 6-day-old embryos differentiates over 3 days in organ culture in a manner roughly paralleling that of normal retina. If explants are placed in the presence of anti-CAM Fab' fragments, the normal development of the histological layers is disrupted (Buskirk et al., 1980). Anti-CAM antibodies also alter the appearance of normal histotypic patterns in neural retina cell aggregates formed from single cells (Rutishauser et al., 1978b). These results suggest that CAM is intimately involved in some of the cellular interactions responsible for the development of neural retina.

The effect of anti-CAM Fab' on neurite interactions in retina aggregates has been extended to cultures of DRG from 10-day-old chick embryos (Rutishauser et al., 1978a; Rutishauser and Edelman, 1980). Under normal growth conditions explants of DRG form a halo of neurites extending from the central ganglion (see Figure 1.1). The majority of the neurites form fascicles or bundles of neurites wrapped around each other. In the presence of monovalent anti-CAM, the morphology of the neurites is dramatically altered, but the diameter of the halo is not changed. The neurites formed in the presence of antibody are much more randomly oriented, fewer fascicles are present, and the number of single neurites is increased. The process is reversible on removal of antibody. It was concluded that anti-CAM inhibits the side-to-side adhesions between neurites, but not the interaction between the growth cone and the substratum.

Although it is likely that CAM plays some role in nerve cell adhesion and the formation of fascicles, it cannot yet be concluded that this molecule is uniquely responsible for these or any other adhesive interactions within the nervous system. Since it has not been demonstrated that CAM can act as a direct ligand in the formation of adhesive bonds between two cells, it may be a part of a larger assembly involved in the adhesion process. For example, a divalent linker, such as an antibody, may bridge two identical receptors on two different cells (e.g., R_1–L–R_1). Alternatively, adhesion may occur directly via two identical or nonidentical receptors (R_1–R_1 or R_1–R_2). Antibodies prepared against any one of the receptors or the linker would inhibit adhesion. Since cognin directly aggregates cells, whereas CAM apparently does not, cognin could be the linker (L) in the first model, and CAM could be a surface receptor. The problem of retina cell adhesion is, however, even more complex.

During the initial purifications of both cognin and CAM, data were obtained that suggested that these proteins exist in the culture medium in higher-molecular-weight forms. Material containing CAM has a molecular weight of about 400,000 daltons on a sizing column during the initial stages of purification (Thiery et al., 1977), and cognin is pelleted by high-speed centrifugation (McClay and Moscona, 1974). Since the biological activity of cognin is similar to that of muscle adherons, it was asked if the adhesion-promoting activities released into the culture medium by chick neural retina cells are associated with glycoprotein complexes similar to those isolated from skeletal muscle. Schubert and colleagues (1983) showed that an extracellular particle (adheron) containing proteoglycans and proteins promotes cell–substratum and cell–cell adhesion of chick neural retina cells, that the adheron-mediated adhesion is cell-type specific, and that the specific activity of the adhesion-mediating complex varies with embryonic age.

As with the adhesion-promoting activity in the culture medium of muscle cells, all of the biological activity in the growth-conditioned medium of 10-day chick neural retina cells is removed by high-speed centrifugation. When the pelleted material is centrifuged into a sucrose gradient, the cell–substratum- and cell–cell-adhesion-promoting activity cosediments with the pelleted protein and carbohydrate in a symmetrical 12S peak. By both negative staining and rotary shadowing techniques, electron microscopy revealed that the majority of the 12S material is spherical, although some particles have a pinwheel configuration (Figure 6.17). The pinwheel shape of some neural retina adherons is amazingly similar to the configuration of sponge aggregation factors (Henkart et al., 1973). The spherical neural retina adherons have a mean diameter of 15.4 ± 2.0 nm.

Although the functional and biochemical properties of chick neural retina adherons are similar to those of skeletal muscle in many respects, there are a number of differences. The following traits are shared. (1) Both neural retina and skeletal muscle adherons contain all of the adhesion-mediating activities in their respective growth-conditioned media as defined by short-term (1 hr) cell–cell and cell–substratum adhesion assays. (2) L6 and neural retina particles are relatively homogeneous spheres about 20 nm and 15 nm in diameter, respectively. (3) Both adhere nonspecifically to a variety of surfaces. (4) Both adherons can be incorporated into extracellular matrix material and are localized primarily to the tissue of origin. (5) Monovalent Fab′ fragments prepared against L6 skeletal muscle or neural retina adherons inhibit both the spontaneous cell aggregation and the particle–cell adhesion of the cell

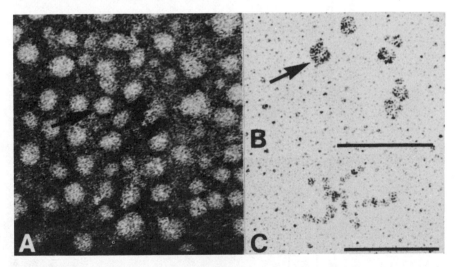

Figure 6.17.   Electron micrographs of neural retina adherons. Adherons were either negatively stained with uranyl formate (A) or sprayed and rotary shadowed with platinum (B and C). In some preparations the particles appear to generate "pinwheels" (C). Arrows indicate individual adherons. Bars are equal to 100 nm. From Schubert et al. (1983).

Table 6.3. Comparison of Cognin, CAM, and Neural Retina Adheron

Property	Cognin	CAM	Adheron[a]	References
Cellular origin	Chick neural retina	Chick neural retina	Chick neural retina	McClay and Moscona (1974); Thiery et al. (1977)
Source of purified molecule	Growth-conditioned medium, cell surface	Growth-conditioned medium, cell surface	Growth-conditioned medium	McClay and Moscona (1974); Hausman and Moscona (1976); Thiery et al. (1977)
Cellular localization	Plasma membrane	Plasma membrane	Membrane and extracellular matrix	Hausman and Moscona (1979); Rutishauser et al. (1978a,b; LaCorbiere (unpublished data)
Binds to cell surface	Yes	?	Yes	Hausman and Moscona (1979)
Stimulates aggregation	Yes	No	Yes	Hausman and Moscona (1975)
Ability of antiserum against factor to inhibit aggregation	Yes	Yes	Yes	Thiery et al. (1977); Hausman and Moscona (1979)

Cell-type distribution	Most neural cells; not on liver, heart, red cells, or kidney	Most neural cells; much less on liver	Neural cells only	Rutishauser et al. (1978a,b); Hausman and Moscona, 1979; LaCorbiere (unpublished data)
Number of molecules per neural retina cell	2–3×10^4	?	?	Hausman and Moscona (1979)
Time course of appearance of activity	Maximum at 8–13 days, then declines	Maximum at 8–13 days, then declines	Maximum at 12 days	Hausman and Moscona (1976); Rutishauser et al. (1978a,b)
Molecular weight	50,000	140,000	Approximately 500,000	McClay and Moscona, (1974); Thiery et al. (1977)
Evidence for association with higher-molecular-weight complex	Yes	Yes	Yes	Hausman and Moscona (1975); Thiery et al. (1977)

[a] All adheron data are from Schubert et al., (1983).

type from which the antigen (adheron) is derived, but not of the other cell type. (6) Finally, the two adherons contain several proteoglycans and proteins. It should be noted, however, that the chick neural retina adherons contain many more proteins than adherons prepared from clonal cell lines (Schubert et al., 1983). This is probably because the neural retina contains at least 20 distinguishable cell types. Included among the proteins are those with molecular weights near 50,000 and 140,000. Whether or not these are cognin and CAM, respectively, remains to be determined.

There are also several differences between the two classes of adherons. (1) The cell–cell and cell–substratum adhesions mediated by skeletal muscle adherons are calcium dependent, while those mediated by neural retina particles are calcium independent. (2) The principle interaction in the adhesion between skeletal muscle myoblasts and particle-coated substrata involves collagen and fibronectin, whereas the neural-retina-stimulated adhesion probably involves an interaction between a protein in the adheron and a heparan sulfate proteoglycan on the cell surface. (3) The biological activity and composition of the skeletal muscle adheron remains constant throughout myogenesis, whereas the activity of the neuronal retina particle varies with the developmental age of the retina (Schubert et al., 1983).

A systematic comparison of the properties of cognin, CAM, and neural retina adherons is presented in Table 6.3. Although the macromolecules are functionally related, there are no data on their structural relationship. The most puzzling aspect of the possible relationship between cognin and CAM is that the lower-molecular-weight cognin is able to cause aggregation, whereas the larger CAM apparently does not. This rules out the possibility that cognin is a breakdown product of CAM, and further, suggests that if indeed they are part of the same adhesion mechanism, cognin may be a divalent linker and CAM a cell receptor molecule. The possibility that CAM and/or cognin are incorporated into neural retina adherons has not been ruled out.

The widespread distribution of both cognin and CAM in neuronal tissues does not preclude specific adhesive functions for these surface molecules during development, for Steinberg's differential adhesion hypothesis (Steinberg, 1970) clearly shows that quantitative changes in the expression or cell-surface distribution of a single adhesive determinant can lead to selective adhesion between different cell types. It is also clear that molecules in addition to or instead of cognin and CAM could determine the highly specific ordering of cells within the nervous system. The rate-limiting factor in the identification of these molecules is the development of suitable assays.

The association of a large proportion of the extracellular adhesion-mediating activities in both nerve and muscle with a macromolecular complex may reflect a need to maintain several biological activities and structural requirements in juxtaposition. For example, both skeletal muscle adherons and neural retina adherons are associated with extracellular matrix material *in vivo* and in cultured cells. Since many *in vivo* adhesions which had previously been viewed as direct cell–cell contacts may in fact be cell–matrix interactions, the various adhesion activities may of necessity be associated with the structural components of the matrix. It has been shown that L6 skeletal muscle adherons can be incorporated directly into extracellular matrix (Schubert et al., 1984). The extracellular glue may therefore be of more importance in the construction of tissues than the integral membrane proteins. The packaging of the adhesion-mediating, structural, and perhaps other biological activities into adherons may be needed to organize the matrix as well as to limit the distribution of activities within the tissue.

References

Abbott, J., Schiltz, J., Dienstman, S., and Holtzer, H. 1974. The phenotypic complexity of myogenic clones. *Proc. Natl. Acad. Sci. USA* **71,** 1506–1510.

Abercrombie, M. and Heaysman, J. 1954. Observations on the social behavior of cells in tissue culture. *Exp. Cell Res.* **6,** 293–306.

Abercrombie, M., Heaysman, J., and Pegram, S. 1970. The locomotion of fibroblasts in culture I. Movements of the leading edge. *Exp. Cell Res.* **59,** 393–398.

Adhya, S. and Garges, S. 1982. How cyclic AMP and its receptor protein act in *Escherichia coli*. *Cell* **29,** 287–289.

Adler, R., Manthrope, M., Skaper, S., and Varon, S. 1981. Polyornithine-attached neurite-promoting factors. Culture sources and responsive neurons. *Brain Res.* **206,** 129–144.

Adler, R. and Varon, S. 1980. Cholinergic neurotrophic factors V. Segregation of survival and neurite-promoting activities in heart-conditioned media. *Brain Res.* **188,** 437–448.

Aggeler, J., Kapp, L. N., Scheffer, C., Tseng, G., and Werb, Z. 1982. Regulation of protein secretion in Chinese hamster ovary cells by cell cycle position and cell density. *Exp. Cell Res.* **139,** 275–283.

Agy, P. C., Shipley, G. D., and Ham, R. G. 1981. Protein-free medium for C-1300 mouse neuroblastoma cells. *In Vitro* **17,** 671–680.

Ahrens, P. B., Solursh, M., Meier, S. 1977. The synthesis and localization of glycosaminoglycans in striated-muscle differentiating in cell-culture. *J. Exp. Zool.* **202,** 375–388.

Albrecht-Buehler, G. 1977. Daughter 3T3 cells: Are they mirror images of each other? *J. Cell Biol.* **72,** 595–603.

Al-Ghaith, L. K. and Lewis, J. H. 1982. Pioneer growth cones in virgin mesenchyme: an electron-microscope study in the developing chick wing. *J. Embryol. Exp. Morphol.* **68,** 149–160.

Alitalo, K., Kurkinen, M., Vaheri, A., Virtanen, I., Rohde, H., and Timpl, R. 1980. Basal lamina glycoproteins are produced by neuroblastoma cells. *Nature* **287,** 465–466.

Anderson, M. J. and Cohen, M. W. 1977. Nerve-induced and spontaneous redistribution of acetylcholine receptors on cultured muscle cells. *J. Physiol. (London)* **268,** 757–773.

Anderson, M. J., Cohen, M. W. and Zorychta, E. 1977. Effects of innervation on the distribution of acetylcholine receptors on cultured muscle cells. *J. Physiol. (London)* **268,** 731–756.

Angeletti, R. H. and Bradshaw, R. A. 1971. Nerve growth factor from mouse submaxillary gland: amino acid sequence. *Proc. Natl. Acad. Sci. USA* **68,** 2417–2420.

Anglister, L., Farber, I. C., Shahar, A., and Grinvald, A. 1982. Localization of voltage-sensitive calcium channels along developing neurites: their possible role in regulating neurite elongation. *Dev. Biol.* **94,** 351–365.

Appenzeller, O. and Palmer, G. 1972. The cyclic AMP (adenosine 3',5'-phosphate) content of sciatic nerve. Changes after nerve crush. *Brain Res.* **42,** 521–524.

Arner, L. S. and Stallcup, W. B. 1981. Rubidium efflux from neural cell lines through voltage-dependent potassium channels. *Dev. Biol.* **83,** 138–145.

Augusti-Tocco, G. and Sato, G. 1969. Establishment of functional clonal lines of neurons from mouse neuroblastoma. *Proc. Natl. Acad. Sci. USA* **64,** 311–315.

Axelrod, D. 1978. Control of acetylcholine receptor mobility and distribution in cultured muscle membranes. *Biochim. Biophys. Acta* **511,** 23–38.

Axelrod, D., Ravdin, P., Koppel, D. E., Schlessinger, J., Webb, W. W., Elson, E. L., and Podleski, T. R. 1976. Lateral motion of fluorescently labeled acetylcholine receptors in membranes of developing muscle fibers. *Proc. Natl. Acad. Sci. USA* **73,** 4594–4602.

Bachrach, U. 1975. Cyclic-AMP-mediated induction of ornithine decarboxylase of glioma and neuroblastoma cells. *Proc. Natl. Acad. Sci. USA* **72,** 3087–3091.

Baker, J. B., Low, D. A., Simmer, R. L., and Cunningham, D. 1980. Proteasenexin: a cellular component that links thrombin and plasminogen activator and mediates their binding to cells. *Cell* **21,** 37–45.

Ball, E. H. and Singer, S. J. 1981. Association of microtubules and intermediate filaments in normal fibroblasts and its disruption upon transformation by a temperature-sensitive mutant of Rous sarcoma virus. *Proc. Natl. Acad. Sci. USA* **78,** 6986–6990.

Bandman, E., Matsuda, R., and Strohman, R. C. 1982. Developmental appearance of myosin heavy and light chain isoforms *in vivo* and *in vitro* in chicken skeletal muscle. *Dev. Biol.* **93,** 508–518.

Barnes, D. and Sato, G. 1980. Serum-free cell culture: a unifying approach. *Cell* **22,** 649–655.

Bauer, H. C., Daniels, M. P., Pudimat, P. A., Jacques, L., Sugiyama, H., and Christian, C. N. 1981. Characterization and partial purification of a neuronal factor which increases acetylcholine receptor aggregation on cultured muscle cells. *Brain Res.* **209,** 395–405.

Begg, D. A., Rodewald, R., and Rebhum, L. I. 1979. The visualization of actin filament polarity in thin sections. *J. Cell Biol.* **79,** 846–852.

Bell, G. 1978. Models for the specific adhesion of cells to cells. *Science* **200,** 618–627.

Benda, P., Lightbody, J., Sato, G., Levine, L., and Sweet, W. 1968. Differentiated rat glial cell strain in tissue culture. *Science* **161,** 370–371.

Bennett, M. R. and Pettigrew, A. G. 1974. The formation of synapses in striated muscle during development. *J. Physiol.* (*London*) **241,** 515–545.

Benoff, S. and Nadal-Ginard, B. 1979. Most myosin heavy chain mRNA in L₆E₉ rat myotubes has a short poly(A) tail. *Proc. Natl. Acad. Sci. USA* **76,** 1853–1857.

Ben-Shaul Y., Hausman, R. E., and Moscona, A. A. 1979. Visualization of a cell surface glycoprotein, the retina cognin, on embryonic cells by immuno-latex labeling and scanning electron microscopy. *Dev. Biol.* **72,** 89–101.

Bentley, D. and Candy, M. 1984. Pathways of pathfinding by peripheral pioneer neurons in grasshoppers. *Cold Spring Harbor Symp. Quant. Biol.* **48,** in press.

Ben-Ze'ev, A., Duerr, A., Solomon, F., and Penman, S. 1979. The outer boundary of the cytoskeleton: a lamina derived from plasma membrane proteins. *Cell* **17,** 859–865.

Ben-Ze'ev, A., Farmer, S. R., and Penman, S. 1980. Protein synthesis requires cell-surface contact while nuclear events respond to cell shape in anchorage-dependent fibroblasts. *Cell* **21,** 365–372.

Berg, D. K. and Hall, Z. W. 1975a. Loss of α-bungarotoxin from junctional and extrajunctional acetylcholine receptors in rat diaphragm muscle *in vivo* and in organ culture. *J. Physiol.* (*Lond.*) **252,** 771–789.

Berg, D. K. and Hall, Z. W. 1975b. Increased extrajunctional acetylcholine sensitivity produced by chronic post-synaptic neuromuscular blockade. *J. Physiol.* (*London*) **244,** 659–676.

Berger, E. A. and Shooter, E. M. 1977. Evidence for pro-β-nerve growth factor, a biosynthetic precursor to β-nerve growth factor. *Proc. Natl. Acad. Sci. USA* **74,** 3647–3651.

Bernfield, M. R. 1980. Organization and remodeling of the extracellular matrix. In

Morphogenesis and Pattern Formation: Implications for Normal and Abnormal Development, L. L. Brinkley, B. M. Carlsson, and T. G. Connelly, Eds., Raven Press, New York, pp. 139–162.

Betz, W. and Sakmann, B. 1973. Effects of proteolytic enzymes on function and structure of frog neuromuscular junctions. *J. Physiol. (London)* **230,** 673–688.

Bevan, S. and Steinbach, J. H. 1977. The distribution of α-bungarotoxin binding sites on mammalian skeletal muscle developing *in vivo. J. Physiol. (London)* **267,** 195–213.

Bevan, S. and Steinbach, J. H. 1983. Denervation increases the degradation rate of acetylcholine receptors at endplates *in vivo* and *in vitro. J. Physiol. (London)* **336,** 159–177.

Beyer, E. C., Zweig, S. E., and Barondes, S. H. 1980. Two lactose binding lectins from chicken tissues. *J. Biol. Chem.* **255,** 4236–4239.

Bhattacharyya, B. and Wolff, J. 1975. Membrane-bound tubulin in brain and thyroid tissue. *J. Biol. Chem.* **250,** 7639–7646.

Bignami, A., Eng., L. F., Dahl, D., and Uyeda, C. T. 1972. Localization of the glial fibrillary acidic protein in astrocytes by immunofluorescence. *Brain Res.* **43,** 429–435.

Birks, R. I. 1974. The relationship of transmitter release and storage to fine structure in a sympathetic ganglion. *J. Neurocytol.* **3,** 133–160.

Bischoff, R. 1979. Tissue culture studies on the origin of myogenic cells during muscle regeneration in rat. In *Muscle Regeneration,* A. Mauro, Ed., Raven Press, New York, pp. 13–29.

Bischoff, R. and Holtzer, H. 1969. Mitosis and the processes of differentiation of myogenic cells *in vitro. J. Cell Biol.* **41,** 188–200.

Bissell, M. J. 1981. The differentiated state of normal and malignant cells or how to define a "normal" cell in culture. *Int. Rev. Cytol.* **70,** 27–100.

Biswas, D. K., Abdullah, K. T., and Brennessel, B. A. 1979. On the mechanism of 5-bromodeoxyuridine induction of prolactin synthesis in rat pituitary cells. *J. Cell Biol.* **81,** 1–9.

Blobel, G. 1980. Intracellular protein topogenesis. *Proc. Natl. Acad. Sci. USA* **77,** 1496–1500.

Bloch, R. 1979. Dispersal and reformation of acetylcholine receptor clusters of cultured rat myotubes treated with inhibitors of energy metabolism. *J. Cell Biol.* **82,** 626–643.

Bloch, R. J. and Geiger, B. 1980. The localization of acetylcholine receptor clusters in areas of cell–substrate contact in cultures of rat myotubes. *Cell* **21,** 25–35.

Bloch, R. J. and Steinbach, J. H. 1983. The distribution of acetylcholine receptors on vertebrate skeletal muscle cells. In *Receptors in Biological Systems,* R. M. Gorczynski and G. B. Price, Eds., Marcel Dekker, New York, in press.

Bloom, E. M. and Black, I. B. 1979. Metabolic requirements for differentiation of embryonic sympathetic ganglia cultured in the absence of exogenous nerve growth factor. *Dev. Biol.* **68,** 568–578.

Bloom, G. S. and Lockwood, A. H. 1980. Redistribution of myosin during morphological reversion of Chinese hamster ovary cells induced by db-cAMP. *Exp. Cell Res.* **129**, 31–45.

Blumenfeld, O. O., Schwartz, E., and Adamany, A. A. 1979. Efflux of phospholipids from cultured aortic smooth muscle cells. *J. Biol. Chem.* **254**, 7183–7190.

Bocci, V., Pessina, G. P., and Paulesu, L. 1979. The role of protease inhibitors and albumin on the membrane shedding of human erythrocytes. *Life Sci.* **24**, 1351–1356.

Bolund, L., Darzynkiewicz, Z., and Ringertz, N. R. 1970. Cell concentration and the staining properties of nuclear deoxyribonucleoprotein. *Exp. Cell Res.* **62**, 76–89.

Boonstra, J., Van Der Saag, P. T., Moolenaar, W. H., and DeLatt, S. W. 1981. Rapid effects of nerve growth factor on the Na^+,K^+-pump in rat pheochromocytoma cells. *Exp. Cell Res.* **131**, 452–454.

Bothwell, M. A. and Shooter, E. M. 1978. Thermodynamics of interaction of the subunits of 7S nerve growth factor. *J. Biol. Chem.* **253**, 8458–8464.

Bottenstein, J. E. 1981. Differentiated properties of neuronal cell lines. In *Functionally Differentiated Cell Lines,* G. Sato, Ed., Alan R. Liss, New York, pp. 155–184.

Bottenstein, J. E. and Sato, G. H. 1979. Growth of a rat neuroblastoma cell line in serum-free supplemented medium. *Proc. Natl. Acad. Sci. USA* **76**, 514–517.

Bottenstein, J. E. and Sato, G. 1980. Fibronectin and polylysine requirement for proliferation of neuroblastoma cells in defined medium. *Exp. Cell Res.* **129**, 361–366.

Bottenstein, J. E., Skaper, S. D., Varon, S. S., and Sato, G. H. 1980. Selective survival of neurons from chick embryo sensory ganglionic dissociates utilizing serum-free supplemented medium. *Exp. Cell Res.* **125**, 183–190.

Bowery, N. G., Brown, D. A., and Marsh, S. 1979. γ-Aminobutyric acid efflux from sympathetic glial cells: effect of "depolarizing" agents. *J. Physiol. (London)* **293**, 75–101.

Bray, D. 1970. Surface movements during the growth of single explanted neurons. *Proc. Natl. Acad. Sci. USA* **65**, 905–910.

Bray, D. 1979. Mechanical tension produced by nerve cells in tissue culture. *J. Cell Sci.* **37**, 391–410.

Bray, D. G. and Hall, Z. W. 1975. Increased extrajunctional ACh sensitivity produced by chronic post-synaptic neuromuscular blockade. *J. Physiol. (London)* **244**, 659–676.

Bray, D., Thomas, C., and Shaw, G. 1978. Growth cone formation in cultures of sensory neurons. *Proc. Natl. Acad. Sci. USA* **75**, 5226–5229.

Bray, D., Wood, P., and Bunge, R. P. 1980. Selective fasciculation of nerve fibers in culture. *Exp. Cell Res.* **130**, 241–245.

Breakefield, X. O. and Nirenberg, M. W. 1974. Selection for neuroblastoma cells that synthesize certain transmitters. *Proc. Natl. Acad. Sci. USA* **71**, 2530–2535.

Brehm, P., Steinbach, J. H., and Kidokoro, Y. 1982. Channel open time of acetylcholine receptors in *Xenopus* muscle cells in dissociated cell culture. *Dev. Biol.* **91,** 93–103.

Brinkley, B. R., Cox, S. M., Pepper, D. A., Wible, L., Brenner, S. L., and Pardue, R. L. 1981. Tubulin assembly sites and the organization of cytoplasmic microtubules in cultured mammalian cells. *J. Cell Biol.* **90,** 554–562.

Brockes, J. P., Fields, K. L., and Raff, M. C. 1979. Studies on cultured rat Schwann cells I. Establishment of purified populations from cultures of peripheral nerve. *Brain Res.* **165,** 105–118.

Brockes, J. P. and Hall, Z. W. 1975. Acetylcholine receptors in normal and denervated rat diaphragm muscle. II. Comparison of junctional and extrajunctional receptors. *Biochemistry* **14,** 2100–2106.

Brockes, J. P. and Hall, Z. W. 1981. Synthesis of acetylcholine receptor by denervated rat diaphragm muscle. *Proc. Natl. Acad. Sci. USA* **72,** 1368–1372.

Brockes, J. P., Lemke, G. E., and Balzer, D. R. 1980. Purification and preliminary characterization of a glial growth factor from bovine pituitary. *J. Biol. Chem.* **255,** 8374–8377.

Brostrom, M. A., Brostrom, C. O., and Wolff, D. J. 1979. Calcium dependence of hormone-stimulated cAMP accumulation in intact glial tumor cells. *J. Biol. Chem.* **254,** 7548–7557.

Brown, D. A. and Scholfield, C. N. 1974. Movements of labelled sodium ions in isolated rat superior cervical ganglia. *J. Physiol. (London)* **242,** 321–351.

Brown, J. C. 1971. Surface glycoprotein characteristic of the differentiated state of neuroblastoma C-1300 cells. *Exp. Cell Res.* **69,** 440–442.

Brown, K. O., Yeh, Y.-C., and Holley, R. W. 1979. Binding, internalization, and degradation of EGF by Balb 3T3 and BP3T3 cells. *J. Cell Physiol.* **100,** 227–238.

Buckley, P. A. and Konigsberg, I. R. 1977. Do myoblasts *in vivo* withdraw from the cell cycle? A reexamination. *Proc. Natl. Acad. Sci. USA* **74,** 2031–2035.

Bueker, E. D. 1948. Implantation of tumors in the hind limb field of the embryonic chick and the developmental response of the lumbosacral nervous system. *Anat. Rec.* **102,** 369–390.

Bulloch, K., Stallcup, W., and Cohn, M. 1977. The derivation and characterization of neuronal cell lines from rat and mouse brain. *Brain Res.* **135,** 25–36.

Bulloch, K., Stallcup, W. and Cohn, M. 1978. A new method for the establishment of neuronal cells from the mouse brain. *Life Sci.* **22,** 495–504.

Bunge, M. B. 1973. Fine structure of nerve fibers and growth cones of isolated sympathetic neurons in culture. *J. Cell Biol.* **56,** 713–735.

Bunge, M. B. 1977. Initial endocytosis of peroxidase or ferritin by growth cones of cultured nerve cells. *J. Neurocytol.* **6,** 407–439.

Bunge, M. B., Williams, A. K., Wood, P. M., Uitto, J., and Jeffrey, J. J. 1980. Comparison of nerve cell and nerve cell plus Schwann cell cultures, with particular emphasis on basal lamina and collagen formation. *J. Cell Biol.* **84,** 184–202.

Bunge, R. P. 1975. Changing uses of nerve tissue culture: 1950–1975. In *The Nervous System,* D. B. Tower, Ed., Vol. 1, Raven Press, New York, pp. 31–41.

Bunge, R. P. and Bunge, M. B. 1981. Evidence that contact with connective tissue matrix is required for normal interaction between Schwann cells and nerve fibers. *J. Cell Biol.* **90,** 943–950.

Bunge, R., Bunge, M. B., Williams, A. K., and Wartels, L. K. Does the dystrophic mouse nerve lesion result from an extracellular matrix abnormality? In *Disorders of the Motor Unit,* D. L. Schotland, Ed., John Wiley & Sons, New York, pp. 23–36.

Bunge, R., Johnson, M., and Ross, C. D. 1978. Nature and nurture in development of the autonomic neuron. *Science* **199,** 1409–1416.

Burden, S. 1977. Acetylcholine receptors at the neuromuscular junction: developmental change in receptor turnover. *Dev. Biol.* **61,** 79–85.

Burden, S. J., Sargent, P. B., and McMahan, U. J. 1979. Acetylcholine receptors in regenerating muscle accumulate at original synaptic sites in the absence of the nerve. *J. Cell Biol.* **82,** 412–425.

Burridge, K. and Bray, D. 1975. Purification and structural analysis of myosins from brain and other non-muscle tissues. *J. Mol. Biol.* **99,** 1–14.

Burry, R. W. 1980. Formation of apparent presynaptic elements in response to polybasic compounds. *Brain Res.* **184,** 85–98.

Buskirk, D. R., Thiery, J.-P., Rutishauser, U., and Edelman, G. M. 1980. Antibodies to a neural adhesion molecule disrupt histogenesis in cultured chick retinae. *Nature* **285,** 488–489.

Cajal, S. R. 1890. A quelle epoque apparaissent les expansions des cellules nerveuses de la moelle epiniere du poulet. *Anatomischer Anzeiger,* Numeros 21 y 22.

Cameron, I. L. and Jeter, J. R. 1971. Relation between cell proliferation and cytodifferentiation in embryonic chick tissues. In *Developmental Aspects of the Cell Cycle,* I. L. Cameron and D. Zimmerman, Eds., Academic Press, New York, pp. 191–222.

Campenot, R. B. 1977. Local control of neurite development by nerve growth factor. *Proc. Natl. Acad. Sci. USA* **74,** 4516–4519.

Cardenas, J. M., Bandman, E., Walker, C., and Strohman, R. 1979. Pyruvate kinase isozymic shifts of differentiating chick myogenic cells *in vivo* and in culture. *Dev. Biol.* **68,** 326–333.

Carmon, Y., Neuman, S., and Yaffe, D. 1978. Synthesis of tropomyosin in myogenic cultures and in RNA-directed cell-free systems: Qualitative changes in polypeptides. *Cell* **14,** 393–401.

Carter, S. B. 1967. Haptotaxis and the mechanism of cell motility. *Nature* **213,** 256–261.

Cassel, D., Wood, P. M., Bunge, R. P., and Glaser, L. 1982. Mitogenicity of brain axolemma membranes and soluble factors for dorsal root ganglion Schwann cells. *J. Cell. Biochem.* **18,** 433–445.

Cathcart, M. K. and Culp, L. A. 1979. Phospholipid composition of substrate adhesion sites of normal, virus-transformed, and revertant murine cells. *Biochemistry* **18,** 1167–1176.

Cejka, J., Peterson, W. D., Belamaric, J., and Kithier, K. 1975. *In vitro* synthesis of β_2-microglobulin by human fetal tissues. *Dev. Biol.* **43,** 200–203.

Ceri, H., Kobiler, D., and Barondes, S. H. 1981. Heparin-inhibitible lectin. *J. Biol. Chem.* **256,** 390–394.

Ceri, H., Shadle, P. J., Kobiler, D., and Barondes, S. H. 1979. Extracellular lectin and its glycosaminoglycan inhibitor in chick muscle cultures. *J. Supramol. Struct.* **11,** 61–67.

Chamley, J. H., Campbell, G. R., and Burnstock, G. 1973. An analysis of the interactions between sympathetic nerve fibers and smooth muscle cells in culture. *Dev. Biol.* **33,** 344–361.

Chang, C. C. and Huang, M. C. 1975. Turnover of junctional and extrajunctional acetylcholine receptors of the rat diaphragm. *Nature* **253,** 643–644.

Changeux, J. P. 1981. The acetylcholine receptor: an "allosteric" membrane protein. *Harvey Lect.* **75,** 85–154.

Chao, N., Young, S. H., and Poo, M. 1981. Localization of cell membrane components by surface diffusion into a "trap." *Biophys. J.* **36,** 139–153.

Charlton, R. R. and Venter, J. C. 1980. Cell cycle-specific changes in β-adrenergic receptor concentrations in C6 glioma cells. *Biochem. Biophys. Res. Comm.* **94,** 1221–1226.

Chen, M. G., Chen, J. S., and Levimont, R. 1978. Sympathetic nerve fibers in growth in the central nervous system of neonatal rodent upon intracerebral NGF injections. *Arch. Ital. Biol.* **116,** 53–84.

Chowrashi, P. K. and Pepe, F. A. 1982. The Z-band: 85,000 dalton amorphin and alpha-actin and their relation to structure. *J. Cell Biol.* **94,** 565–573.

Chun, L. and Patterson, P. H. 1977. Role of nerve growth factor in the development of rat sympathetic neurons *in vitro*. I. Survival, growth and differentiation of catecholamine production. *J. Cell Biol.* **75,** 694–704.

Claude, P., Hawrot, E., Dunis, D. A., and Campenot, R. B. 1982. Binding, internalization and retrograde transport of ^{125}I-NGF in cultured rat sympathetic neurons. *J. Neurosci.* **2,** 431–442.

Cohen, M. W. 1980. Development of an amphibian neuromuscular junction *in vivo* and in culture. *J. Exp. Biol.* **89,** 43–56.

Cohen, M. W. and Weldon, P. B. 1980. Localization of acetylcholine receptors and synaptic ultrastructure at nerve–muscle contacts in culture: dependence on nerve type. *J. Cell Biol.* **86,** 388–401.

Cohen, S. 1958. A nerve growth-promoting protein. In *The Symposium on the Chemical Basis of Development,* W. D. McElroy and B. Glass, Eds., John Hopkins Press, Baltimore, pp. 665–676.

Cohen, S. A. and Fischbach, G. D. 1977. Clusters of acetylcholine receptors located at identified nerve–muscle synapses *in vitro*. *Dev. Biol.* **59,** 24–38.

Cohen, S. and Levi-Montalcini, R. 1956. A nerve growth stimulating factor isolated from snake venom. *Proc. Natl. Acad. Sci. USA* **42,** 571–574.

Cohn, M. 1967. Natural history of the myeloma. *Cold Spring Harbor Symp. Quant. Biol.* **32,** 211–221.

Collier, H. O. J. 1980. Cellular site of opiate dependence. *Nature* **283,** 625–629.

Collins, F. 1978*a*. Axon initiation by ciliary neurons in culture. *Dev. Biol.* **65,** 50–57.

Collins, F. 1978*b*. Induction of neurite outgrowth by a conditioned-medium factor bound to the culture substratum. *Proc. Natl. Acad. Sci. USA* **75,** 5210–5213.

Collins, F. and Dawson, A. 1982. Conditioned medium increases the rate of neurite elongation: separation of this activity from the substratum bound inducer of neurite outgrowth. *J. Neurosci.* **2,** 1005–1010.

Collins, F. and Garrett, J. E. 1980. Elongating nerve fibers are guided by a pathway of material released from embryonic nonneuronal cells. *Proc. Natl. Acad. Sci. USA* **77,** 6226–6228.

Collins, F. and Lee, M. R. 1982. A reversible developmental change in the ability of ciliary ganglion neurons to extend neurites in culture. *J. Neurosci.* **2,** 424–430.

Cook, G. M. W. 1968. Glycoproteins in membranes. *Biol. Rev.* **43,** 363–391.

Connolly, J. L., Green, S. A., and Greene, L. A. 1981. Pit formation and rapid changes in surface morphology of sympathetic neurons in response to NGF. *J. Cell Biol.* **90,** 176–180.

Connolly, J. L., Greene, L. A., Viscarello, R. R., and Riley, W. D. 1979. Rapid, sequential changes in surface morphology of PC12 pheochromocytoma cells in response to NGF. *J. Cell Biol.* **82,** 820–827.

Constantinides, P. G., Jones, P. A., and Gevers, W. 1977. Functional striated muscle cells from non-myoblast precursors following 5-azacytidine treatment. *Nature* **267,** 364–366.

Cowan, M. W. 1979. Selection and control of neurogenesis. In *The Neurosciences Fourth Study Program,* F. O. Schmidt and F. G. Worden, Eds., MIT Press, Cambridge, MA., pp. 59–80.

Cowan, M. W., Martin, A. H., and Wenger, E. 1968. Mitotic patterns in the optic tectum of the chick during normal development and after early removal of the optic vesicle. *J. Exp. Zool.* **169,** 71–92.

Craig, S. W. and Pollard, T. D. 1982. Actin binding proteins. *Trends Biochem. Sci.* **7,** 88–92.

Crain, S. M. 1956. Resting and action potentials of cultured chick embryo spinal ganglion cells. *J. Comp. Neurol.* **104,** 285–330.

Crain, S. M. 1966. Development of "organotypic" bioelectric activities in central nervous tissues during maturation in culture. *Int. Rev. Neurobiol.* **9,** 1–43.

Culp, L. 1975. Topography of substrate-attached glycoproteins from normal and virus-transformed cells. *Exp. Cell Res.* **92,** 467–477.

Culp, L. A., Murray, B. A., and Rollins, B. J. 1979. Fibronectin and proteoglycans as determinants of cell–substratum adhesion. *J. Supramol. Struct.* **11,** 401–427.

Cummins, P. and Perry, S. V. 1974. Chemical and immunochemical characteristics of tropomyosins from striated and smooth muscle. *Biochem. J.* **141,** 43–49.

Dale, H. H. 1935. Pharmacology and nerve-endings. *Proc. R. Soc. Med.* **28,** 319–332.

Damsky, C. H., Knudsen, K. A., Dorio, R. J., and Buck, C. A. 1981. Manipulation of cell–cell and cell–substratum interactions of mouse mammary tumor epithelial cells using broad spectrum antisera. *J. Cell Biol.* **89,** 173–184.

Damsky, C. H., Levy-Benshimol, A., Buck, C. A., and Warren, L. 1979. Effect of tunicamycin on the synthesis, intracellular transport, and shedding of membrane glycoproteins in BHK cells. *Exp. Cell Res.* **119,** 1–13.

Darmon, M., Stallcup, W. B., and Pittman, Q. J. 1982. Induction of neural differentiation by serum deprivation in cultures of the embryonal carcinoma cell line 1003. *Exp. Cell Res.* **138,** 73–78.

Dawson, G. 1979. Complex carbohydrates of cultured neuronal and glial cell lines. In *Complex Carbohydrates of the Nervous Tissue.* R. U. Margolis and R. K. Margolis, Eds., Plenum Press, New York, pp. 291–325.

Denham, S. 1967. A cell proliferation study of the neural retina in the two-day rat. *J. Embryol. Exp. Morphol.* **18,** 53–66.

Dennis, M. J., Ziskind-Conhaim, L., and Harris, A. J. 1981. Development of neuromuscular junctions in rat embryos. *Dev. Biol.* **81,** 266–279.

Desarmenien, M., Feltz, P., and Headley, P. M. 1980. Does glial uptake affect GABA responses? An intracellular study on rat dorsal root ganglion neurons *in vitro. J. Physiol. (London)* **307,** 163–182.

Deschatrette, J., Moore, E. E., Dubois, M., and Weiss, M. 1980. Dedifferentiated variants of a rat hepatoma: reversion analysis. *Cell* **19,** 1043–1051.

Deshmukh, K. and Sawyer, B. D. 1977. Synthesis of collagen by chondrocytes in suspension culture: Modulation by calcium, 3′,5′-cyclic AMP and prostaglandins. *Proc. Natl. Acad. Sci. USA* **74,** 3864–3868.

DeVitry, F., Camier, M., Czernichow, P., Benda, P., Cohen, P., and Tixier-Vidal, A. 1974. Establishment of a clone of mouse hypothalamic neurosecretory cells synthesizing neurophysin and vasopressin. *Proc. Natl. Acad. Sci. USA* **71,** 3575–3579.

DeVitry, F., Delaunoy, J. P., Thibault, J., Buisson, N., Lamando, N., Legault, L., Delasalle, A., and Dupouey, P. 1983. Induction of oligodendrocyte-like properties in a primitive hypothalamic cell line. *EMBO J.* **2,** 199–203.

DeVitry, F., Picart, R., Jacque, C., Legault, L., Dupouey, P., and Tixier-Vidal, A. 1980. Presumptive common precursor for neuronal and glial cell lineages in mouse hypothalamus. *Proc. Natl. Acad. Sci. USA* **77,** 4165–4169.

Devlin, R. B. and Emerson, C. P. 1978. Coordinate regulation of contractile protein synthesis during myoblast differentiation. *Cell* **13,** 599–611.

Devlin, R. B. and Emerson, C. P. 1979. Coordinate accumulation of contractile protein mRNAs during myoblast differentiation. *Dev. Biol.* **69,** 202–216.

Devreotes, P. N. and Fambrough, D. M. 1976. Synthesis of acetylcholine receptors by cultured chick myotubes and denervated mouse extensor digitorum longus muscles. *Proc. Natl. Acad. Sci. USA* **73,** 161.

Dhoot, G. K. and Perry, S. V. 1979. Distribution of polymorphic forms of troponin components and tropomyosin in skeletal muscle. *Nature* **278,** 714–718.

Diamond, J. and Miledi, R. 1962. A study of foetal and new-born rat muscle fibres. *J. Physiol. (London)* **162,** 393–408.

Dichter, M. A., Tischler, A. S., and Greene, L. A. 1977. Nerve growth factor-induced increase in electrical excitability and acetylcholine sensitivity of a rat pheochromocytoma cell line. *Nature* **268,** 501–504.

Dimpfel, W., Huang, R., and Habermann, E. 1977. Gangliosides in nervous tissue cultures and binding of ^{125}I-labelled tetanus toxin, a neuronal marker. *J. Neurochem.* **29,** 329–334.

Doering, J. L. and Fischman, D. A. 1974. The *in vitro* cell fusion of chick muscle without DNA synthesis. *Dev. Biol.* **36,** 225–235.

Doetschman, T. C. 1980. The effects of Con A on cell surface shedding in cell cultures. *J. Cell Sci.* **46,** 221–234.

Doetschman, T. C. and Jewett, J. 1981. The contribution of cell death to medium-released fractions of cell cultures. *In Vitro* **17,** 178–184.

Doljanski, F. and Kapeller, M. 1976. Cell surface shedding—the phenomenon and its possible significance. *J. Theor. Biol.* **62,** 253–270.

Dollenmeier, P., Turner, D. C., and Eppenberger, H. M. 1981. Proliferation and differentiation of chick skeletal muscle cells cultured in a chemically defined medium. *Exp. Cell Res.* **135,** 47–61.

Doyle, D. and Baumann, H. 1979. Turnover of the plasma membrane of mammalian cells. *Life Sci.* **24,** 951–966.

Drachman, D. B. 1967. Is ACh the trophic neuromuscular transmitter? *Arch. Neurol.* Chicago **17,** 206–218.

Dribin, L. B. and Barrett, J. N. 1980. Conditioned medium enhances neuritic outgrowth from rat spinal cord explants. *Dev. Biol.* **74,** 184–195.

Druckrey, H., Preussmann, R., Ivankovic, S., and Schmahl, D. 1967. Organotrope carcinogene Wirkungen bei 65 uerschiedenen *N*-Nitroso-Verbindungen an BD-Ratten. *Z. Krebsforsch.* **69,** 103–210.

Dunn, G. A. 1971. Mutual contact inhibition of extension of chick sensory nerve fibers *in vitro*. *J. Comp. Neurol.* **143,** 491–508.

Dym, H. and Yaffe, D. 1979. Expression of creatine kinase isoenzymes in myogenic cell lines. *Dev. Biol.* **68,** 592–599.

Eagle, H. and Levine, E. M. 1967. Growth regulatory effects of cellular interaction. *Nature* **213,** 1102–1106.

Ebashi, S. and Kodama, A. 1965. A new protein factor promoting aggregation of tropomyosin. *J. Biochem.* (*Tokyo*) **58,** 107–109.

Ebendal, T. 1976. The relative roles of contact inhibition and contact guidance in orientation of axons extending on aligned collagen fibrils *in vitro*. *Exp. Cell Res.* **98,** 159–169.

Eccles, J. C. 1957. The clinical significance of research work on the chemical transmitter substances of the nervous system. *Med. J. Aust.,* **June 1, 1957,** 745–753.

Edde, B., Jeantet, C., and Gros, F. 1981. One β-tubulin subunit accumulates during neurite outgrowth in mouse neuroblastoma cells. *Biochem. Biophys. Res. Comm.* **103,** 1035–1043.

Edmunds, L. N. and Adams, K. J. 1981. Clocked cell cycle clocks. *Science* **211,** 1002–1013.

Edstrom, A., Kanje, M., and Walum, E. 1974. Effects of dibutyryl cyclic AMP and prostaglandin E1 on cultured human glioma cells. *Exp. Cell Res.* **85,** 217–223.

Edwards, C. and Frisch, H. L. 1976. A model for localization of acetylcholine receptors at the muscle endplate. *J. Neurobiol.* **7,** 377–381.

Ehrlich, P. 1979. The effect of multivalency on the specificity of protein and cell interactions. *J. Theor. Biol.* **81,** 123–127.

Ellisman, M. H. and Porter, K. R. 1980. Microtrabecular structure of the axoplasmic matrix. Visualization of cross-linking structures and their distribution. *J. Cell Biol.* **87,** 464–479.

Ellisman, M. H., Rash, J. E., Staehelin, L. A., and Porter, K. R. 1976. Studies of excitable membranes. II. A comparison of specializations at neuromuscular junctions and nonjunctional sarcolemmas of mammalian fast and slow twitch muscle fibers. *J. Cell Biol.* **68,** 752–774.

Elson, H. F. and Ingwall, J. S. 1980. The cell substratum modulates skeletal muscle differentiation. *J. Supramol. Struct.* **14,** 313–328.

Emerson, C. P. 1977. Control of myosin synthesis during myoblast differentiation. In *Pathogenesis of Human Muscular Dystrophies,* L. P. Rowland, Ed., Excerpta Medica, Amsterdam, pp. 799–811.

Emerson, C. P. and Beckner, S. H. 1975. Activation of myosin synthesis in fusing and mononucleated myoblasts. *J. Mol. Biol.* **93,** 431–447.

Emr, S. D., Hall, M. N., and Silhavy, T. J. 1980. A mechanism of protein localization: the signal hypothesis and bacteria. *J. Cell Biol.* **86,** 701–711.

Eng, L. F. 1980. The glial fibrillary acidic (GFA) protein. In *Proteins of the Nervous System,* R. A. Bradshaw and D. M. Schneider, Eds., Raven Press, New York, pp. 85–117.

Engelhardt, J. K., Ishikawa, K., and Katase, D. K. 1980. Low potassium is critical for observing developmental increase in muscle resting potential. *Brain Res.* **190,** 564–568.

Engelman, D. M. and Steitz, T. A. 1981. The spontaneous insertion of proteins into and across membranes: the helical hairpin hypothesis. *Cell* **23,** 411–422.

Eppenberger, H. M., Eppenberger, M., Richterich, R., and Aebi, H. 1964. The ontogeny of creatine kinase isozymes. *Dev. Biol.* **10,** 1–16.

Epstein, M. L. and Gilula, N. B. 1977. A study of communication specificity between cells in culture. *J. Cell Biol.* **75,** 769–787.

Fambrough, D. M. 1979. Control of acetylcholine receptors in skeletal muscle. *Physiol. Rev.* **59,** 165–227.

Fambrough, D. and Rash, J. E. 1971. Development of acetylcholine sensitivity during myogenesis. *Dev. Biol.* **26,** 55–68.

Feldman, E. L., Axelrod, D., Schwartz, M., Heacock, A. M., and Agranoff, B. W. 1981. Studies on the localization of newly added membrane in growing neurites. *J. Neurobiol.* **12,** 591–598.

Fessler, J. H. and Fessler, L. I. 1978. Biosynthesis of procollagen. *Annu. Rev. Biochem.* **47,** 129–162.

Fields, K., Gosling, C., Megson, M., and Stern, P. L. 1975. New cell surface antigens in rat defined by tumors of the nervous system. *Proc. Natl. Acad. Sci. USA* **72,** 1296–1300.

Fischbach, G. D. and Cohen, S. A. 1973. The distribution of acetylcholine sensitivity over uninnervated and innervated muscle fibers grown in cell culture. *Dev. Biol.* **31,** 147–162.

Fischbach, G. D. and Schuetze, S. M. 1980. A post-natal decrease in acetylcholine channel open time at rat end-plates. *J. Physiol. (London)* **303,** 125–137.

Fischman, D. A. 1970. The synthesis and assembly of myofibrils in embryonic muscle. *Curr. Top. Dev. Biol.* **5,** 235–280.

Fiszman, M. Y. and Fuchs, P. 1975. Temperature-sensitive expression of differentiation in transformed myoblasts. *Nature* **254,** 429–431.

Florini, J. R. and Roberts, S. B. 1979. A serum-free medium for the growth of muscle cells in culture. *In Vitro* **15,** 983–992.

Folkman, J. and Moscona, A. 1978. Role of cell shape in growth control. *Nature* **273,** 345–349.

Frank, E. and Fischbach, G. D. 1979. Early events in neuromuscular junction formation *in vitro*. *J. Cell Biol.* **83,** 143–158.

Frankel, F. R. 1976. Organization and energy-dependent growth of microtubules in cells. *Proc. Natl. Acad. Sci. USA* **73,** 2798–2802.

Frazier, W. A., Angeletti, R. H., and Bradshaw, R. A. 1972. Nerve growth factor and insulin. *Science* **176,** 482–488.

Friedlander, M., Beyer, E., and Fischman, D. A. 1978. Muscle development *in vitro* following X-irradiation. *Dev. Biol.* **66,** 457–469.

Friedman, D. L. 1976. Role of cyclic nucleotides in cell growth and differentiation. *Physiol. Rev.* **56,** 652–708.

Fryberg, E. A. Kindle, K. L., and Davidson, N. 1980. The actin genes of *Drosophila:* a dispersed multigene family. *Cell* **19,** 365–378.

Fujii, D. K., Massoglia, S. L., Savion, M., and Gospadorwicz, D. 1982. Neurite outgrowth and protein synthesis of PC12 cells as a function of substratum and NGF. *J. Neurosci.* **2,** 1157–1175.

Fukuda, J., Fischbach, G. D., and Smith, T. G. 1976. A voltage clamp study of the sodium, calcium, and chloride spikes of chick skeletal muscle cells grown in tissue culture. *Dev. Biol.* **49,** 412–424.

Fulton, A. B., Prives, J., Farmer, S., and Penman, S. 1981. Developmental reorganization of the skeletal framework and its surface lamina in fusing muscle cells. *J. Cell Biol.* **91,** 103–112.

Furcht, L. T., Mosher, D. F., and Wendelschafer-Crabb, G. 1978. Immunocytochemical localization of fibronectin (LETS protein) on the surface of L6 myoblasts: light and electron microscopic studies. *Cell* **13,** 263–271.

Furmanski, P., Silverman, D. J., and Lubin, M. 1971. Expression of differentiated functions in mouse neuroblastoma mediated by DBcAMP. *Nature* **233,** 413–415.

Furshpan, E. J. and Potter, D. D. 1968. Low resistance junctions between cells in embryos and tissue culture. *Curr. Top. Dev. Biol.* **3,** 95–127.

Furshpan, E. J., Potter, D. D., and Landis, S. C. 1981. On the transmitter reper-
toire of sympathetic neurons in culture. *Harvey Lect.* **76,** 149–191.

Gagnon, J., Palmiter, R. D., and Walsh, K. A. 1978. Comparison of the NH_2-
terminal sequence of ovalbumin as synthesized *in vitro* and *in vivo. J. Biol.
Chem.* **253,** 7464–7468.

Garrels, J. I. 1979*a*. Two-dimensional gel electrophoresis and computer analysis of proteins synthesized by clonal cell lines. *J. Biol. Chem.* **254,** 7961–
7977.

Garrels, J. I. 1979*b*. Changes in protein synthesis during myogenesis in a clonal
cell line. *Dev. Biol.* **73,** 134–152.

Garrels, J. and Gibson, W. 1976. Identification and characterization of multiple
forms of actin. *Cell* **9,** 793–805.

Garrels, J. I. and Hunter, T. 1979. Post-translational modifications of actins synthesized *in vitro. Biochem. Biophys. Acta* **564,** 517–525.

Garrels, J. and Schubert, D. 1979. Modulation of protein synthesis by nerve
growth factor. *J. Biol. Chem.* **254,** 7978–7985.

Gauthier, G. F., Lowey, S., and Hobbs, A. W. 1978. Fast and slow myosin in
developing muscle fibres. *Nature* **274,** 25–29.

Gaynor, R., Irie, R., Morton, D., and Herschman, H. R. 1980. S100 protein is
present in cultured human malignant melanomas. *Nature* **286,** 400–401.

Geiger, B., Tokyuasu, K. T., Dutton, A. H., and Singer, S. J. 1980. Vinculin, an
intracellular protein localized at specialized sites where microfilament bundles terminate at cell membranes. *Proc. Natl. Acad. Sci. USA* **77,** 4127–
4131.

Gelfant, S. 1981. Cycling–noncycling cell transitions in tissue aging, immunological surveillance, transformation, and tumor growth. *Int. Rev. Cytol.* **70,**
1–25.

Gillard, G. C., Birnbaum, P., Reilly, H. C., Merrilees, M. J., and Flint, M. H. 1979.
The effect of charged synthetic polymers on proteoglycan synthesis and
sequestration in chick embryo fibroblast cultures. *Biochem. Biophys. Acta*
584, 520–528.

Giller, E. L., Neale, J. H., and Bullock, P. N. 1977. Choline acetyltransferase
activity of spinal cord cell cultures is increased by co-culture with muscle.
J. Cell Biol. **74,** 16–29.

Gilman, A. G. and Nirenberg, M. 1971. Effect of catecholamines on the adenosine 3′:5′-cyclic monophosphate concentration of clonal satellite cells of
neurons. *Proc. Natl. Acad. Sci. USA* **68,** 2165–2168.

Gilman, A. G. and Nirenberg, M. 1971. Regulation of adenosine 3′,5′-cyclic
monophosphate metabolism in cultured neuroblastoma cells. *Nature* **234,**
356–358.

Gilula, N. B., Reeves, O. R., and Steinbach, A. 1972. Metabolic coupling, ionic
coupling and cell contacts. *Nature* **235,** 262–265.

Giotta, G. J., Brunson, K. W., and Lotan, R. 1980*a*. The effects of 5-bromodeoxy-
uridine on cyclic AMP levels and cytoskeletal organization in malignant
melanoma cells. *Cell Biol. Int. Rep.* **4,** 105–116.

Giotta, G. J., Heitzmann, J., and Cohn, M. 1980*b*. Properties of two temperature-sensitive Rous sarcoma virus transformed cerebellar cell lines. *Brain Res.* **202,** 445–458.

Goldberg, S. 1977. Unidirectional, bidirectional, and random growth of embryonic optic axons. *Exp. Eye Res.* **25,** 399–404.

Gomer, R. H. and Lazarides, E. 1981. The synthesis and deployment of filamin in chicken skeletal muscle. *Cell* **23,** 524–532.

Goodman, R. and Herschman, H. R. 1978. Nerve growth factor-mediated induction of tyrosine hydroxylase in a clonal pheochromocytoma cell line. *Proc. Natl. Acad. Sci. USA* **75,** 4587–4590.

Gordon, J. R. and Bernfield, M. R. 1980. The basal lamina of the postnatal mammary epithelium contains glycosaminoglycans in a precise ultrastructural organization. *Dev. Biol.* **74,** 118–135.

Granger, B. L. and Lazarides, E. 1978. The existence of an insoluble Z disc scaffold in chicken skeletal muscle. *Cell* **15,** 1253–1263.

Greene, L. A. 1977. Quantitative *in vitro* studies on the nerve growth factor (NGF) requirement of neurons. *Dev. Biol.* **58,**106–113.

Greene, L. A. and Rein, G. 1977. Synthesis, storage, and release of acetylcholine by noradrenergic pheochromocytoma cell line. *Nature* **268,** 349–351.

Greene, L. A. and Shooter, E. M. 1980. Nerve growth factor: biochemistry, synthesis, and mechanism of action. *Annu. Rev. Neurosci.* **3,** 353–402.

Greene, L. A., Shooter, E. M., and Varon, S. 1968. Enzymatic activities of mouse nerve growth factor and its subunits. *Proc. Natl. Acad. Sci. USA* **60,** 1383–1388.

Greene, L. A. and Tischler, A. S. 1976. Establishment of a noradrenergic clonal line of rat adrenal pheochromocytoma cells which respond to nerve growth factor. *Proc. Natl. Acad. Sci. USA* **73,** 2424–2428.

Grinvald, A. and Farber, I. 1981. Optical recording of Ca^{++} action potentials from growth cones of cultured neurons using a laser microbeam. *Science* **212,** 1164–1169.

Gruener, R. and Kidokoro, Y. 1982. Acetylcholine sensitivity of innervated and noninnervated *Xenopus* muscle cells in culture. *Dev. Biol.* **91,** 86–92.

Gruenstein, E. and Rich, A. 1975. Non-identity of muscle and non-muscle actins. *Biochem. Biophys. Res. Commun.* **64,** 472–477.

Gundersen, R. W. and Barrett, J. N. 1979. Neuronal chemotaxis: chick dorsal-root axons turn toward high concentrations of nerve growth factor. *Science* **206,** 1079–1080.

Gundersen, R. W. and Barrett, J. N. 1980. Characterization of the turning response of dorsal root neurites toward nerve growth factor. *J. Cell Biol.* **87,** 546–554.

Guroff, G., Dickens, G., End, D., and Landos, C. 1981. The action of adenosine analogs on PC12 cells. *J. Neurochem.* **37,** 1431–1439.

Guth, L. 1968. "Trophic" influences of nerve on muscle. *Physiol. Rev.* **48,** 645–685.

Hagiwara, S. and Jaffe, L. A. 1979. Electrical properties of egg cell membranes. *Annu. Rev. Biophys. Bioeng.* **8,** 385–416.

Hagiwara, S. and Miyazaki, S. 1977. Ca and Na spikes in egg cell membrane. In

Cellular Neurobiology, Z. Hall, R. Kelly, and C. F. Fox, Eds., Vol. 5, Alan R. Liss, New York, pp. 147–158.

Halegoua, S. and Patrick, J. 1980. Nerve growth factor mediates the phosphorylation of specific proteins. *Cell* **22,** 571–581.

Hall, Z. 1973. Multiple forms of acetylcholinesterase and their distribution in endplate and non-endplate regions of rat diaphragm muscle. *J. Neurobiol.* **4,** 343–362.

Hall, Z. W. and Kelly, R. B. 1971. Enzymatic detachment of endplate acetylcholinesterase from muscle. *Nature [New Biol.]* **232,** 62–63.

Hamburger, V. and Levi-Montalcini, R. 1949. Proliferation, differentiation and degeneration in the spinal ganglia of the chick embryo under normal and experimental conditions. *J. Exp. Zool.* **111,** 457–501.

Hanson, G. R., Iversen, P. L., and Partlow, L. M. 1982*a.* Preparation and partial characterization of highly purified primary cultures of neurons and non-neuronal (glial) cells from embryonic chick cerebral hemispheres and several other regions of the nervous system. *Dev. Brain Res.* **3,** 529–545.

Hanson, G. R., Partlow, L. M., and Iversen, P. L. 1982*b.* Neuronal stimulation of nonneuronal (glial) cell proliferation. *Dev. Brain Res.* **3,** 547–555.

Hardingham, T. 1981. Proteoglycans: their structure, interactions and molecular organization in cartilage. *Biochem. Soc. Trans.* **9,** 489–498.

Harris, A. 1973. Behavior of cultured cells on substrata of variable adhesiveness. *Exp. Cell Res.* **77,** 285–297.

Harris, A. J. 1974. Inductive functions of the nervous system. *Annu. Rev. Physiol.* **36,** 251–305.

Harris, A. J. 1981. Embryonic growth and innervation of skeletal muscles. III. Neural regulation of junctional and extrajunctional acetylcholine receptor clusters. *Philos. Trans. R. Soc. London [Biol.]* **293,** 287–314.

Harris, A. J., Heinemann, S., Schubert, D., and Tarikas, H. 1971. Trophic interaction between cloned tissue culture lines of nerve and muscle. *Nature* **31,** 296–301.

Harris, A. K. 1976. Is cell sorting caused by differences in the work of intercellular adhesion? A critique of the Steinberg hypothesis. *J. Theor. Biol.* **61,** 267–285.

Harris, M. 1971. Mutation rates in cells at different ploidy levels. *J. Cell Physiol.* **78,** 177–184.

Harrison, R. G. 1907. Observations on the living developing nerve fiber. *Proc. Soc. Exp. Biol. Med.* **4,** 140–143.

Harrison, R. G. 1910. The outgrowth of the nerve fiber as a mode of protoplasmic movement. *J. Exp. Zool.* **9,** 787–846.

Harvey, A. M. 1975. Johns Hopkins—the birthplace of tissue culture: the story of Ross G. Harrison, Warren H. Lewis, and George Grey. *John Hopkins Med. J.* **136,** 142–149.

Hastings, K. and Emerson, C. 1982. cDNA clone analysis of six co-regulated mRNAs encoding skeletal muscle contractile proteins. *Proc. Natl. Acad. Sci. USA* **79,** 1553–1557.

Hata, R., Yoshifumi, N., Nagai, Y., and Tsukada, Y. 1980. Biosynthesis of interstitial types of collagen by albumin-producing rat liver parenchymal cell (hepatocyte) clones in culture. *Biochemistry* **19,** 169–176.

Hatanaka, H. 1981. NGF-mediated stimulation of tyrosine hydroxylase activity in a clonal rat pheochromocytoma cell line. *Brain Res.* **222,** 225–233.

Hauschka, S. D. and Konigsberg, I. R. 1966. The influence of collagen on the development of muscle clones. *Proc. Natl. Acad. Sci. USA* **55,** 119–126.

Hauschka, S. D. and White, N. K. 1972. Studies of myogenesis *in vitro.* In *Research in Muscle Development and the Muscle Spindle.* B. Q. Banker, R. J. Przybylski, J. P. van der Meulen, and M. Victor, Eds., Excerpta Medica, Amsterdam, pp. 53–71.

Hausman, R. E. and Moscona, A. A. 1975. Purification and characterization of the retina-specific cell-aggregating factor. *Proc. Natl. Acad. Sci. USA* **72,** 916–920.

Hausman, R. E. and Moscona, A. A. 1976. Isolation of retina-specific cell-aggregating factor from membranes of embryonic neural retina tissue. *Proc. Natl. Acad. Sci. USA* **73,** 3594–3598.

Hausman, R. E. and Moscona, A. A. 1979. Immunologic detection of retina cognin on the surface of embryonic cells. *Exp. Cell Res.* **119,** 191–204.

Hawrot, E. 1980. Cultured sympathetic neurons: Effects of cell-derived and synthetic substrata on survival and development. *Devel. Biol.* **74,** 136–151.

Hay, E. D. 1978. Role of basement membranes in development and differentiation. In *Biology and Chemistry of Basement Membranes,* N. A. Kefalides, Ed., Academic Press, New York, pp. 119–136.

Heidemann, S. R., Landers, J. M., and Hamborg, M. A. 1981. Polarity orientation of axonal microtubules. *J. Cell Biol.* **91,** 661–665.

Helfand, S., Riopello, R. J., and Wessells, N. K. 1978. Non-equivalence of conditioned medium and NGF for sympathetic, parasympathetic, and sensory neurons. *Exp. Cell Res.* **113,** 39–45.

Helfand, S. L., Smith, G. A., and Wessells, W. K. 1976. Survival and development in culture of dissociated parasympathetic neurons from ciliary ganglia. *Dev. Biol.* **50,** 541–547.

Hellerquest, C. G. 1979. Intercellular adhesion as a function of the cell cycle traverse. *J. Cell Biol.* **82,** 682–687.

Henkart, P., Humphreys, S., and Humphreys, T. 1973. Characterizations of sponge aggregation factor. A unique proteoglycan complex. *Biochemistry* **12,** 3045–3052.

Henn, F. A., Haljamae, H., and Hamberger, A. 1972. Glial cell function: active control of extracellular K^+ concentration. *Brain Res.* **43,** 437–443.

Herman, B. A. and Fernandez, S. M. 1978. Changes in membrane dynamics associated with myogenic cell fusion. *J. Cell. Physiol.* **94,** 253–264.

Herrup, K. and Thoenen, H. 1979. Properties of the nerve growth factor receptor of a clonal line of rat pheochromocytoma (PC12) cells. *Exp. Cell Res.* **121,** 71–78.

Herschman, H. R., Levine, L., and DeVellis, J. 1971. Appearance of a brain-specific antigen (S100 protein) in the developing rat brain. *J. Neurochem.* **18,** 629–633.

Heumann, R., Ocalan, M., Kachel, V., and Hamprecht, B. 1979. Clonal hybrid cell lines expressing cholinergic and adrenergic properties. *Proc. Natl. Acad. Sci. USA* **76,** 4674–4677.

Heumann, R., Schwab, M., and Thoenen, H. 1981*a*. A second messenger required for nerve growth factor biological activity? *Nature* **292,** 838–840.

Heumann, R., Villegas, J., and Herzfeld, D. W. 1981*b*. Acetylcholine synthesis in the Schwann cell and axon in the giant nerve fiber of the squid. *J. Neurochem.* **36,** 765–768.

Heuser, J. E. and Salpeter, S. R. 1979. Organization of acetylcholine receptors in quick-frozen, deep-etched, and rotary-replicated *Torpedo* postsynaptic membrane. *J. Cell Biol.* **82,** 150–173.

Higgins, D., Lacovitti, L., Joh, T. H., and Burton, H. 1981. The immunocytochemical localization of tyrosine hydroxylase within rat sympathetic neurons that release acetylcholine in culture. *J. Neurosci.* **1,** 126–131.

Hill, C. E., Hendry, I. A., and Bonyhady, R. E. 1981. Avian parasympathetic neurotrophic factors: age-related increases and lack of regional specificity. *Dev. Biol.* **85,** 258–261.

Hirata, F. and Axelrod, J. 1980. Phospholipid methylation and biological signal transmission. *Science* **209,** 1082–1090.

Hokfelt, T., Lundeberg, J. M., Schultzberg, M., Johansson, O., Ljungdahl, A., and Rehfeld, J. 1980. Coexistence of peptides and putative transmitters in neurons. In *Neural Peptides and Neuronal Communication.* E. Costa and M. Trabucchi, Eds., Raven Press, New York, pp. 1–23.

Holley, R. W., Armour, R., Baldwin, J. H., Brown, K. D., and Yeh, Y. C. 1977. Density-dependent regulation of growth of BSC-1 cells in cell culture: control of growth by serum factors. *Proc. Natl. Acad. Sci. USA* **74,** 5046–5050.

Holtzer, H., Biehl, J., Yeoh, G., Meganathan, R., and Kaji, A. 1975. Effect of oncogenic virus on muscle differentiation. *Proc. Natl. Acad. Sci. USA* **72,** 4051–4055.

Horwitz, A. F., Wight, A., and Knudsen, K. 1979. A role for lipid in myoblast fusion. *Biochem. Biophys. Res. Commun.* **86,** 514–521.

Howard, A. and Pelc, S.R. 1953. Synthesis of deoxyribonucleic acid in normal and irradiated cells and its relation to chromosome breakage. *Heredity (London)* [Suppl.] **6,** 261–273.

Hsie, A. W. and Puck, T. T. 1971. Morphological transformation of Chinese hamster cells by dibutyryl adenosine 3',5'-monophosphate and testosterone. *Proc. Natl. Acad. Sci. USA* **68,** 358–361.

Hsieh, P. and Sueoka, N. 1980. Antisera inhibiting mammalian cell spreading and possible cell-surface antigens involved. *J. Cell. Biol.* **86,** 866–873.

Huesgen, A. and Gerisch, G. 1975. Solubilized contact sites from cell membranes of *Dictyostelium discoideum. FEBS Lett.* **56,** 46–49.

Hunter, T. and Garrels, J. I. 1977. Characterization of the mRNAs for α, β, and γ actin. *Cell* **12,** 767–781.

Hutchison, H. T., Werrbach, K., Vance, C., and Haber, B. 1974. Uptake of neurotransmitters by clonal lines of astrocytoma and neuroblastoma in culture. I. Transport of γ-aminobutyric acid. *Brain Res.* **66,** 265–274.

Huxley, H. E. 1963. Electron microscope studies on the structure of natural and synthetic protein filaments from striated muscle. *J. Mol. Biol.* **7,** 281–308.

Huxley, H. E. 1982. The mechanism of force production in muscle. In *Disorders of the Motor Unit,* D. L. Schotland, Ed., John Wiley & Sons, New York, pp. 1–14.

Hynes, R. O., Martin, G. S., Shearer, M., Critchley, D. R., and Epstein, C. J. 1976. Viral transformation of rat myoblasts: effects on fusion and surface properties. *Dev. Biol.* **48,** 35–46.

Imamoto, K. and Leblond, C. P. 1978. Radioautographic investigation of gliogenesis in the corpus callosum of young rats. II. Origin of microglial cells. *J. Comp. Neurol.* **180,** 139–164.

Inestrosa, N. C., Silberstein, L., and Hall, Z. W. 1982. Association of the synaptic form of acetylcholinesterase with extracellular matrix in cultured mouse muscle cells. *Cell* **29,** 71–79.

Isenberg, G. and Small, J. V. 1978. Filamentous actin, 100A filaments and microtubules in neuroblastoma cells: their distribution in relation to sites of movement and neuronal transport. *Cytobiologie* **16,** 326–337.

Ishii, K., Mizumoto, T., Hanaoka, M., and Matsumoto, A. 1977. Differential susceptibility to immune cytolysis of 3T3 cells in monolayer and in suspension. *Exp. Cell Res.* **105,** 237–244.

Ishikawa, H. R., Bischoff, R., and Holtzer, H. 1969. Formation of arrowhead complexes with heavy meromyosin in a variety of cell types. *J. Cell Biol.* **43,** 312–328.

Isobe, T., Ishioka, N., and Okuyama, T. 1981. Structural relation of two S100 proteins in bovine brain. Subunit composition of S100a protein. *Eur. J. Biochem.* **115,** 469–474.

Ito, Y. and Miledi, R. 1977. The effect of calcium ionophores on acetylcholine release from Schwann cells. *Proc. R. Soc. Lond. [Biol.]* **196,** 51–58.

Iversen, L. L. and Kelly, J. S. 1975. Uptake and metabolism of γ-aminobutyric acid by neurones and glial cells. *Biochem. Pharmacol.* **24,** 933–938.

Jacobson, M. 1978. *Developmental Neurobiology,* Plenum Press, New York.

Jameson, L. and Caplow, M. 1981. Modification of microtubule steady-state dynamics by phosphorylation of the microtubule-associated proteins. *Proc. Natl. Acad. Sci. USA* **78,** 3413–3417.

Jessell, T. M., Siegel, R. E., and Fischbach, G. D. 1979. Induction of acetylcholine receptors on cultured skeletal muscle by a factor extracted from brain and spinal cord. *Proc. Natl. Acad. Sci. USA* **76,** 5397–5401.

Jetten, A. M., Jetten, M. E. R., and Sherman, M. I. 1979. Analyses of cell surface and secreted proteins in primary cultures of mouse extraembryonic membranes. *Dev. Biol.* **70,** 89–104.

Jilek, F. and Hormann, H. 1979. Fibronectin (cold-insoluble globulin) VI. Influence of heparin and hyaluronic acid on the binding of native collagen. *Hoppe Seylers Z. Physiol. Chem.* **360,** 597–603.

Jockusch, H. and Jockusch, B. M. 1980. Structural organization of the Z-line protein, α-actinin, in developing skeletal muscle cells. *Dev. Biol.* **75,** 231–238.

John, P. C. L., Ed. 1981. *The Cell Cycle. Society for Experimental Biology Seminar Series 10.* Cambridge University Press, Cambridge.

Johnson, G. S., Friedman, R. M., and Pastan, I. 1971. Restoration of several morphological characteristics of normal fibroblasts in sarcoma cells treated with adenosine-3′ : 5′-cyclic monophosphate and its derivatives. *Proc. Natl. Acad. Sci. USA* **68,** 425–429.

Johnson, G. S., Morgan, W. D., and Pastan, I. 1972. Regulation of cell motility by cyclic AMP. *Nature* **235,** 54–56.

Johnson, J. D. and Epel, D. 1975. Relationship between release of surface proteins and metabolic activation of sea urchin eggs at fertilization. *Proc. Natl. Acad. Sci. USA* **72,** 4474–4478.

Johnson, M. I., Ross, C. D., Meyers, M., Spitznagel, E. L., and Bunge, R. P. 1980*a.* Morphological and biochemical studies in the development of cholinergic properties in cultured sympathetic neurons I. Correlative changes in choline acetyltransferase and synaptic vesicle cytochemistry. *J. Cell Biol.* **84,** 680–691.

Johnson, M. I., Ross, C. D., and Bunge, R. P. 1980*b.* Morphological and biochemical studies on the development of cholinergic properties in cultured sympathetic neurons II. Dependence on postnatal age. *J. Cell Biol.* **84,** 692–704.

Jones, P. H. 1977. Isolation and culture of myogenic cells from regenerating rat skeletal muscle. *Exp. Cell Res.* **107,** 111–117.

Jones, R. and Vrbova, G. 1974. Two factors responsible for the development of denervation hypersensitivity. *J. Physiol. (London)* **236,** 517–538.

Jumblatt, J. E., Schlup, V., and Burger, M. 1980. Cell–cell recognition: specific binding to *Microcona* sponge aggregation factor to homotypic cells and the role of calcium ions. *Biochem. J.* **19,** 1038–1042.

Kalderon, N. 1979. Migration of Schwann cells and wrapping of neurites *in vitro:* A function of protease activity (plasmin) in the growth medium. *Proc. Natl. Acad. Sci. USA* **76,** 5992–5996.

Kalderon, N., Epstein, M. L., and Gilula, N. B. 1977. Cell-to-cell communication and myogenesis. *J. Cell Biol.* **75,** 788–806.

Kalderon, N. and Gilula, N. B. 1979. Membrane events involved in myoblast fusion. *J. Cell Biol.* **81,** 411–425.

Karlin, A., Cox, R., Dwork, A., Kaldany, R., Kao, P., Lebel, P., Yodh, N., and Holtzman, E. 1984. The arrangement and functions of the chains of the acetylcholine receptor of *Torpedo* electric tissue. *Cold Spring Harbor Symp. Quant. Biol.* **48,** 1–8.

Karnovsky, M. J., Unanue, E. R., and Leventhal, M. 1972. Ligand induced movement of lymphocyte membrane macromolecules. II. Mapping of surface moieties. *J. Exp. Med.* **136,** 907–930.

Karp, G. 1979. *Cell Biology.* McGraw-Hill, New York.

Katcoff, D., Nudel, U., Zevin-Sonkin, D., Carmon, Y., Shani, M., Lehrach, H., Frischant, A. M., and Yaffe, D. 1980. Construction of recombinant plasmids containing rat muscle actin and myosin light chain DNA sequences. *Proc. Natl. Acad. Sci. USA* **77,** 960–964.

Katz, B. and Miledi, R. 1959. Spontaneous subthreshold activity at denervated amphibian endplates. *J. Physiol. (London)* **146,** 44–45P.

Katz, M. J., Lasek, R. J. and Nauta, H. J. W. 1980. Ontogeny of substrate pathways and the origin of the neural circuit pattern. *Neuroscience* **5,** 821–833.

Kaufman, S. J. and Parks, C. M. 1977. Loss of growth control and differentiation in the fu-1 variant of the LB line of rat myoblasts. *Proc. Natl. Acad. Sci. USA* **74,** 3888–3892.

Kavinsky, C. J. and Garber, B. B. 1979. Fibronectin associated with the glial component of embryonic brain cell cultures. *J. Supramol. Struct.* **11,** 269–281.

Keller, J. M. and Nameroff, M. 1974. Induction of creatine phosphokinase in cultures of chick skeletal myoblasts without concomitant cell fusion. *Differentiation* **2,** 19–23.

Keller, L. R. and Emerson, C. P. 1980. Synthesis of adult myosin light chains by embryonic muscle cultures. *Proc. Natl. Acad. Sci. USA* **77,** 1020–1024.

Kelly, A. M. 1978. Satellite cells and myofiber growth in rat soleus and extensor digitorum longus muscles. *Dev. Biol.* **65,** 1–10.

Kelly, A. M. and Zacks, S. I. 1969. The fine structure of motor endplate morphogenesis. *J. Cell Biol.* **42,** 154–169.

Ketley, J. N., Orkin, R. W., and Martin, G. R. 1976. Collagen in developing chick muscle *in vivo* and *in vitro. Exp. Cell Res.* **99,** 261–268.

Kidokoro, Y. 1975a. Developmental changes of membrane electrical properties in a rat skeletal muscle cell line. *J. Physiol. (London)* **244,** 129–143.

Kidokoro, Y. 1975b. Sodium and calcium components of the action potential in a developing skeletal muscle cell line. *J. Physiol. (London)* **244,** 145–159.

Kidokoro, Y. 1980. Developmental changes of spontaneous synaptic potential properties in the rat neuromuscular contact formed in culture. *Dev. Biol.* **78,** 231–241.

Kidokoro, Y. 1981. Electrophysiological properties of developing skeletal muscle cells in culture. In *Excitable Cells in Tissue Culture,* P. Nelson and M. Lieberman, Eds., Plenum Press, New York, pp. 319–339.

Kidokoro, Y., Anderson, M. J., and Gruener, R. 1980. Changes in synaptic potential properties during acetylcholine receptor accumulation and neurospecific interaction in *Xenopus* nerve–muscle cell culture. *Dev. Biol.* **78,** 464–483.

Kidokoro, Y. and Gruener, R. 1982. Distribution and density of α-bungarotoxin binding sites on innervated and noninnervated *Xenopus* muscle cells in culture. *Dev. Biol.* **91,** 78–85.

Kidokoro, Y., Heinemann, S., Schubert, D., Brandt, B. L., and Klier, F. G. 1975. Synapse formation and neurotrophic effects of muscle cell lines. *Cold Spring Harbor Symp. Quant. Biol.* **40,** 373–388.

Kidokoro Y. and Yeh, E. 1982. Initial synaptic transmission at the growth cone in *Xenopus* nerve–muscle cultures. *Proc. Natl. Acad. Sci. USA* **79,** 6727–6731.

Kim, U., Baumler, A., Carruthers, C., and Bielat, K. 1975. Immunological escape mechanism in spontaneously metastasizing mammary tumors. *Proc. Natl. Acad. Sci. USA* **72,** 1012–1016.

Kimes, B., Tarikas, H., and Schubert, D. 1974. Neurotransmitter synthesis by two clonal nerve cell lines: changes with culture growth and morphological differentiation. *Brain Res.* **79,** 291–295.

Kimhi, Y., Palfrey, C., Spector, I., Barak, Y., and Littauer, U. Z. 1976. Maturation of neuroblastoma cells in the presence of dimethylsulfoxide. *Proc. Natl. Acad. Sci. USA* **73,** 462–466.

Kinoshita, Y., Maikta, A., and Takeuchi, T. 1982. Bromodeoxyuridine-induced molecular species conversion of sialic acids of gangliosides and the alteration of cellular phenotypic expression in B16 melanoma cells. *J. Biochem. (Tokyo)* **92,** 801–808.

Klebe, R. J. 1974. Isolation of a collagen-dependent cell attachment factor. *Nature* **250,** 248–251.

Kleinman, H. K., Klebe, R. J., and Martin, G. R. 1981. Role of collagenous matrices in the adhesions and growth of cells. *J. Cell Biol.* **88,** 473–485.

Kleinman, H. K., Martin, G. R., and Fishman, P. H. 1979. Ganglioside inhibition of fibronectin-mediated cell adhesion to collagen. *Proc. Natl. Acad. Sci. USA* **76,** 3367–3371.

Kleinman, H. K., McGoodwin, E. B., Martin, G. R., Klebe, R. J., Fietzek, P. P., and Woolley, D. E. 1978. Localization of the binding site for cell attachment in the $\alpha1(I)$ chain of collagen. *J. Biol. Chem.* **253,** 5642–5646.

Klier, F. G., Schubert, D., and Heinemann, S. 1977. The ultrastructural differentiation of the clonal myogenic cell line L6 in normal and high K^+ medium. *Dev. Biol.* **57,** 440–449.

Konigsberg, I. R. 1963. Clonal analysis of myogenesis. *Science* **140,** 1273–1284.

Konigsberg, I. R., McElvain, N., Tootle, M., and Herrmann, H. 1960. The dissociability of DNA synthesis from the development of multinuclearity of muscle cells in cultures. *J. Biophys. Biochem. Cytol.* **8,** 333–343.

Konigsberg, I. R., Sollmann, P. A., and Mixter, L. O. 1978. The duration of the terminal G1 of fusing myoblasts. *Dev. Biol.* **63,** 11–26.

Knudsen, K. A. and Horwitz, A. F. 1978. Differential inhibition of myoblast fusion. *Dev. Biol.* **66,** 294–307.

Krayanek, S. and Goldberg, S. 1981. Oriented extracellular channels and axonal guidance in the embryonic chick retina. *Dev. Biol.* **84,** 41–50.

Kreis, T. E. and Birchmeier, W. 1980. Stress fiber sarcomeres of fibroblasts are contractile. *Cell* **22,** 555–561.

Krystosek, A. and Seeds, N. 1981. Plasminogen activator release at the neuronal growth cone. *Science* **213,** 1532–1534.

Kudlow, J. E., Rae, P. A., Gutmann, N. S., Schimmer, B. P., and Burrow, G. N. 1980. Regulation of ODC activity by corticotropin in adrenocortical tumor cell clones: roles of cAMP and cAMP dependent protein kinase. *Proc. Natl. Acad. Sci. USA* **77,** 2767–2771.

Kuffler, S. W. 1943. Specific excitability of the endplate region in normal and denervated muscle. *J. Neurophysiol.* **6,** 99–110.

Kuffler, S. W. and Nicholls, J. G. 1966. The physiology of neuroglial cells. *Ergeb. Physiol.* **57,** 1–90.

Kuhl, U., Timpl, R., and Von Der Mark, K. 1982. Synthesis of Type IV collagen and laminin in cultures of skeletal muscle cells and their assembly on the surface of myotubes. *Dev. Biol.* **93,** 344–354.

Kuhn, L. C. and Kraehenbuhl, J. P. 1979. Role of secretory component, a secreted glycoprotein, in the specific uptake of IgA dimer by epithelial cells. *J. Biol. Chem.* **254,** 11072–11081.

Kurkinen, M. and Alitalo, K. 1979. Fibronectin and procollagen produced by a clonal line of Schwann cells. *FEBS Lett.* **102,** 64–68.

Labourdette, G. and Marks, A. 1975. Synthesis of S100 protein in monolayer cultures of rat glial cells. *Eur. J. Biochem.* **58,** 73–79.

LaCorbiere, M. and Schubert, D. 1981. The role of cAMP in the induction of ornithine decarboxylase by nerve and epidermal growth factors. *J. Neurosci. Res.* **6,** 211–216.

Lage-Davila, A., Krust, B., Hofmann-Clerc, F., Torpier, G., and Montagnier, L. 1979. Bromodeoxyuridine-induced reversion of transformed characteristics in BHK21 cells: changes at the plasma membrane level. *J. Cell. Physiol.* **100,** 95–107.

Lajtha, L. G. 1968. On the concept of the cell cycle. *J. Cell. Comp. Physiol.* **62,** 143–145.

Lallermayer, G. and Neupert, W. 1976. In *Genetics and Biogenesis of Chloroplasts and Mitochondria.* T. Bucher, Ed. Elsevier, Amsterdam, pp. 806–820.

LaMont, J. T., Thornton-Gammon, M., and Isselbacher, K. J. 1977. Cell-surface glycosyltransferases in cultured fibroblasts: increased activity and release during serum stimulation of growth. *Proc. Natl. Acad. Sci. USA* **74,** 1086–1090.

Lander, A. D., Fujii, D. K., Gospodarowicz, D., and Reichardt, L. F. 1982. Characterization of a factor that promotes neurite outgrowth: evidence linking activity to a heparan sulfate proteoglycan. *J. Cell Biol.* **94,** 574–585.

Landreth, G. and Shooter, E. 1980. NGF receptor on PC12 cells: ligand-induced conversion from low to high affinity states. *Proc. Natl. Acad. Sci. USA* **77,** 4751–4755.

Lasek, R. J. and Tytell, M. A. 1981. Macromolecular transfer from glia to axon. *J. Exp. Biol.* **95,** 153–165.

Laug, W. E., Jones, P. A., Nye, C. A., and Benedict, W. F. 1976. The effect of cAMP and prostaglandins on the fibrinolytic activity of mouse neuroblastoma cells. *Biochem. Biophys. Res. Comm.* **68,** 114–119.

Layman, D. L., McGoodwin, E. B., and Martin, G. R. 1971. The nature of the collagen synthesized by cultured human fibroblasts. *Proc. Natl. Acad. Sci. USA* **68,** 454–458.

Lazarides, E. 1981. Intermediate filaments—chemical heterogeneity in differentiation. *Cell* **23,** 649–650.

Lazarides, E. and Revel, J. P. 1980. The molecular basis of cell movement. *Sci. Am.* **241,** 100–113.

Lazarus, K. J., Bradshaw, R. A., West, N. R., and Bunge, R. P. 1976. Adaptive survival of rat sympathetic neurons cultured without supporting cells or exogenous NGF. *Brain Res.* **113,** 159–164.

LeBeau, J., LaCorbiere, M., and Schubert, D. 1984. Adherons induce neurite outgrowth in cultured cells. In preparation.

LeDouarin, N. M., Smith, J., and LeLievre, C. S. 1981. From the neural crest to the ganglia of the peripheral nervous system. *Annu. Rev. Physiol.* **43,** 653–671.

LeDouarin, N. M. and Teillet, M. A. 1974. Experimental analysis of the migration and differentiation of neuroblasts of the autonomic nervous system and of neuroectodermal mesenchymal derivatives, using a biological cell marking technique. *Dev. Biol.* **41,** 162–184.

Lentz, T. L. 1967. Fine structure of nerves in the regenerating limb of the newt *Triturus. Am. J. Anat.* **121,** 647–669.

Lester, H. A. 1977. The response to acetylcholine. *Sci. Am.* **236**(2), 107–116.

Letourneau, P. C. 1975*a*. Possible roles of cell-to-substratum adhesion in neuronal morphogenesis. *Dev. Biol.* **44,** 77–91.

Letourneau, P. C. 1975*b*. Cell-to-substratum adhesion and guidance of axonal elongation. *Dev. Biol.* **44,** 92–101.

Letourneau, P. C. 1978. Chemotactic response of nerve fiber elongation to NGF. *Dev. Biol.* **66,** 183–196.

Letourneau, P. C. 1979. Cell–substratum adhesion of neurite growth cones, and its role in neurite elongation. *Exp. Cell Res.* **124,** 127–138.

Letourneau, P. C. 1981. Immunocytochemical evidence for colocalization in neurite growth cones of actin and myosin and their relationship to cell–substratum adhesions. *Dev. Biol.* **85,** 113–122.

Letourneau, P. C. 1982. Analysis of microtubule number and length in cytoskeletons of cultured chick sensory neurons. *J. Neurosci.* **2,** 806–814.

Letourneau, P. C. and Wessells, N. K. 1974. Migratory cell locomotion versus nerve axon elongation. Differences based on the effects of lanthanum ion. *J. Cell Biol.* **61,** 56–69.

Levi, A., Shechter, Y., Neurfeld, E. J., and Schlessinger, J. 1980. Mobility, clustering, and transport of NGF in embryonal sensory cells and in a sympathetic neuronal cell line. *Proc. Natl. Acad. Sci. USA* **77,** 3469–3473.

Levi-Montalcini, R. 1964. Growth control of nerve cells by a protein factor and its antiserum. *Science* **143,** 105–110.

Levi-Montalcini, R. 1976. The NGF: Its role in growth, differentiation and function of the sympathetic adrenergic neuron. *Prog. Brain Res.* **45,** 235–258.

Levi-Montalcini, R. and Booker, B. 1960. Destruction of the sympathetic ganglia in mammals by an antiserum to a nerve-growth protein. *Proc. Natl. Acad. Sci. USA* **46,** 384–391.

Levi-Montalcini, R. and Calissano, P. 1979. The nerve growth factor. *Sci. Am.* **240,** 68–77.

Levi-Montalcini, R. and Hamburger, V. 1951. Selective growth stimulating effects of mouse sarcoma on the sensory and sympathetic nervous system of the chick embryo. *J. Exp. Zool.* **116,** 321–361.

Levi-Montalcini, R., Meyer, H., and Hamburger, V. 1954. *In-vitro* experiments on the effects of mouse sarcomas on the sensory and sympathetic ganglia of the chick embryo. *Cancer Res.* **14,** 49–57.

Levin, E. G. and Loskutoff, D. J. 1980. Serum-mediated suppression of cell-associated plasminogen activator activity in cultured endothelial cells. *Cell* **22,** 701–707.

Levitt, P., Cooper, M. L., and Rakic, P. 1981. Coexistence of neuronal and glial precursor cells in the cerebral ventricular zone of the fetal monkey: An ultrastructural immunoperoxidase analysis. *J. Neurosci.* **1,** 27–39.

Libby, P. and Goldberg, A. L. 1981. Comparison of the control of pathways for degradation of the acetylcholine receptor and average protein in cultured muscle cells. *J. Cell. Physiol.* **107,** 185–194.

Liepins, A. and Hillman, A. J. 1981. Shedding of tumor cell surface membranes. *Cell Biol. Int. Rep.* **5,** 15–26.

Lilien, J. E. 1968. Specific enhancement of cell aggregation *in vitro. Dev. Biol.* **17,** 657–678.

Lilien, J. E. and Moscona, A. A. 1967. Cell aggregation: its enhancement by a supernatant from cultures of homologous cells. *Science* **157,** 70–72.

Lipton, B. H. 1977*a*. A fine-structural analysis of normal and multinucleate cells in myogenic cultures. *Dev. Biol.* **60,** 26–47.

Lipton, B. H. 1977*b*. Collagen synthesis by normal and bromodeoxyuridine-modulated cells in myogenic culture. *Dev. Biol.* **61,** 153–165.

Lipton, B. H. and Schultz, E. 1979. Developmental fate of skeletal muscle satellite cells. *Science* **205,** 1292–1294.

Liskay, R. M. 1977. Absence of a measureable G2 phase in two Chinese hamster cell lines. *Proc. Natl. Acad. Sci. USA* **74,** 1622–1625.

Loewenstein, W. R. 1981. Junctional intercellular communication: The cell-to-cell channel. *Physiol. Rev.* **61,** 829–911.

Lofberg, J., Ahlfors, K., and Fallstrom, C. 1980. Neural crest cell migration in relation to extracellular matrix organization in the embryonic axolotl trunk. *Dev. Biol.* **75,** 148–167.

Lømo, T. and Rosenthal, J. 1972. Control of ACh sensitivity by muscle activity in the rat. *J. Physiol.* (*London*) **221,** 493–510.

Longo, A. M. and Penhoet, E. E. 1974. Nerve growth factor in rat glioma cells. *Proc. Natl. Acad. Sci. USA* **71,** 2347–2349.

Lough, J. and Bischoff, R. 1977. Differentiation of creatine phosphokinase during myogenesis: quantitative fractionation of isozymes. *Dev. Biol.* **57,** 330–344.

Luckenbill-Edds, L., Van Horn, C., and Greene, L. A. 1979. Fine structure of initial outgrowth of processes induced in a pheochromocytoma cell line (PC12) by nerve growth factor. *J. Neurocytol.* **8,** 493–511.

Lundberg, J. M. 1981. Evidence for coexistence of VIP and acetylcholine in neurons of cat exocrine glands—morphological, biochemical and functional studies. *Acta Physiol. Scand.* [*Suppl.*] **496,** 1–57.

Lyman, G. H., Preisler, H. D., and Papahadjopoulos, D. 1976. Membrane action of DMSO and other chemical inducers of Friend leukaemic cell differentiation. *Nature* **262,** 360–363.

MacCallum, D. K., Lillie, J. H., Scaletta, L. J., Occhino, J. C., Frederick, W. G., and Ledbetter, S. R. 1982. Bovine corneal epithelium *in vitro*. Elaboration and organization of basement membrane. *Exp. Cell Res.* **139,** 1–13.

MacDonnell, P. C., Nagaiah, K., Lakshmanan, J., and Guroff, G. 1977. NGF increases activity of ornithine decarboxylase in superior cervical ganglia of young rats. *Proc. Natl. Acad. Sci. USA* **74,** 4681–4684.

Markelonis, G. J., Oh, T. H., Eddefrawl, M. E., and Guth, L. 1982. Sciatin: A myotrophic protein increases the number of acetylcholine receptors and receptor clusters in cultured skeletal muscle. *Dev. Biol.* **89,** 353–361.

Martin, D. L. and Shain, W. 1979. High affinity transport of taurine and β-alanine and low affinity transport of GABA by a single transport system in cultured glioma cells. *J. Biol. Chem.* **254,** 7076–7084.

Martin, D. W., Tomkins, G. M., and Bresler, M. A. 1969. Control of specific gene expression examined in synchronized mammalian cells. *Proc. Natl. Acad. Sci. USA* **63,** 842–849.

Max, S. R., Rohrer, H., Otten, U., and Thoenen, H. 1978. NGF-mediated induction of tyrosine hydroxylase in rat superior cervical ganglia *in vitro*. *J. Biol. Chem.* **253,** 8013–8015.

McCarthy, K. D. and Partlow, L. M. 1976. Neuronal stimulation of [³H]thymidine incorporation by primary cultures of highly purified non-neuronal cells. *Brain Res.* **114,** 415–426.

McClay, D. R. and Moscona, A. A. 1974. Purification of the specific cell-aggregating factor from embryonic neural retina cells. *Exp. Cell Res.* **87,** 438–443.

McCormick, P. J., Keys, B. J., Pucci, C., and Millis, A. J. T. 1979. Human fibroblast-conditioned medium contains a 100K dalton glucose-regulated cell surface protein. *Cell* **18,** 173–182.

McMahon, D. 1974. Chemical messengers in development: a hypothesis. *Science* **185,** 1012–1021.

McMahon, U. J. 1984. Molecular components of synaptic basal lamina that cause aggregation of ACh receptors on regenerating myofibers. *Cold Spring Harbor Symp. Quant. Biol.* **48,** 653–665.

McMorris, F. A. 1977. Norepinephrine induces glial-specific enzyme activity in cultured glioma cells. *Proc. Natl. Acad. Sci. USA* **74,** 4501–4504.

Meiri, H., Spira, M. E., and Parnas, I. 1981. Membrane conductance and action potential of a regenerating axonal tip. *Science* **211,** 709–712.

Michler, A. and Sakmann, B. 1980. Receptor stability and channel conversion in the subsynaptic membrane of the developing mammalian neuromuscular junction. *Dev. Biol.* **80,** 1–17.

Miledi, R. and Slater, C. R. 1968. Electrophysiology and electron microscopy of rat neuromuscular junctions after nerve degeneration. *Proc. R. Soc. Lond.* [*Biol.*] **169,** 289–306.

Millar, T. J. and Unsicker, K. 1982. Ultrastructure of growth cones formed by isolated rat adrenal medullary chromaffin cells *in vitro* after treatment with nerve growth factor. *Dev. Brain Res.* **2,** 577–582.

Miller, R. A. and Ruddle, F. H. 1974. Enucleated neuroblastoma cells form neurites when treated with dibutyryl cyclic AMP. *J. Cell Biol.* **63,** 295–299.

Milstein, C., Brownlee, G. G., Harrison, T. M., and Mathews, M. B. 1972. A possible precursor of immunoglobulin light chains. *Nature [New Biol.]* **239,** 117–120.

Minchin, M. C. W. and Iversen, L. L. 1974. Release of [³H]gamma aminobutyric acid from glial cells in rat dorsal root ganglia. *J. Neurochem.* **23,** 533–540.

Minna, J., Glazer, D., and Nirenberg, M. 1972. Genetic dissection of neural properties using somatic cell hybrids. *Nature [New Biol.]* **235,** 225–231.

Mirsky, R. 1980. Cell-type-specific markers in nervous system cultures. *Trends Neurosci.* **3,** 190–192.

Mirsky, R., Winter, J., Abney, E. R., Pruss, R. M., Gavrilovic, J., and Raff, M. C. 1980. Myelin-specific proteins and glycolipids in rat Schwann cells and oligodendrocytes in culture. *J. Cell Biol.* **84,** 483–494.

Mintz, B. and Baker, W. W. 1967. Normal mammalian muscle differentiation and gene control of isocitrate dehydrogenase. *Proc. Natl. Acad. Sci. USA* **58,** 592–598.

Miyake, M. 1978. The development of action potential mechanism in a mouse neuronal cell line *in vitro. Brain Res.* **143,** 349–354.

Moody-Corbett, F. and Cohen, M. W. 1981. Localization of cholinesterase at sites of high ACh receptor density on embryonic amphibian muscle cells cultured without nerve. *J. Neurosci.* **1,** 596–605.

Moody-Corbett, F. and Cohen, M. W. 1982. Influence of nerve on the formation and survival of ACh receptor and cholinesterase patches on embryonic *Xenopus* muscle cells in culture. *J. Neurosci.* **2,** 633–646.

Moore, B. W. and McGregor, D. 1965. Chromatographic and electrophoretic fractionation of soluble proteins of brain and liver. *J. Biol Chem.* **240,** 1647–1653.

Moore, E. 1976. Cell to substratum adhesion-promoting activity released by normal and virus transformed cells in culture. *J. Cell Biol.* **70,** 634–647.

Morgan, J. L. and Seeds, N. W. 1975. Tubulin constancy during morphological differentiation of mouse neuroblastoma cells. *J. Cell Biol.* **67,** 136–145.

Morris, G. E. and Cole, R. J. 1979. Calcium and the control of muscle specific creatine kinase accumulation during skeletal muscle differentiation *in vitro. Dev. Biol.* **69,** 146–158.

Morrison, R. S. and DeVellis, J. 1981. Growth of purified astrocytes in a chemically defined medium. *Proc. Natl. Acad. Sci. USA* **78,** 7205–7209.

Moscona, A. 1961. Rotation-mediated histogenetic aggregation of dissociated cells. *Exp. Cell Res.* **22,** 455–475.

Moscona, A. and Moscona, H. 1952. The dissociation and aggregation of cells from organ rudiments of the early chick embryo. *J. Anat.* **86,** 287–301.

Mosesson, M. W. and Umfleet, R. A. 1980. The cold-insoluble globulin of human plasma. *J. Biol. Chem.* **245,** 5728–5732.

Mosher, D. F. and Wing, D. A. 1976. Synthesis and secretion of alpha-2-macroglobulin by cultured human fibroblasts. *J. Exp. Med.* **143,** 462–467.

Moss, P. S., Honeycutt, N., Pawson, T., and Martin, G. S. 1979. Viral transformation of chick myogenic cells. *Exp. Cell Res.* **123,** 95–105.

Murphy, R. A., Oger, J., Saide, J. D., Blanchard, M. H., Arnason, B., Hogan, C., Pantazis, N. J., and Young, M. 1977. Secretion of nerve growth factor by central nervous system glioma cells in culture. *J. Cell Biol.* **72,** 769–773.

Nadal-Ginard, B. 1978. Commitment, fusion and biochemical differentiation of a myogenic cell line in the absence of DNA synthesis. *Cell* **15,** 855–864.

Napias, C., Olsen, R. W., and Schubert, D. 1980. GABA and picrotoxin in receptors in clonal nerve cells. *Nature* **283,** 298–299.

Narumi, S. and Maki, S. 1978. Stimulatory effects of substance P on neurite extension and cyclic AMP levels in cultured neuroblastoma cells. *J. Neurochem.* **30,** 1321–1326.

Neher, E. and Sakmann, B. 1976. Noise analysis of drug induced voltage clamp currents in denervated frog muscle fibers. *J. Physiol.* **258,** 705–736.

Nelson, P. G. and Lieberman, M., Eds. 1981. *Excitable Cells in Tissue Culture.* Plenum Press, New York.

Nelson, P. G. and Peacock, J. H. 1972. Acetylcholine responses in L cells. *Science* **177,** 1005–1007.

Newmark, P. 1979. Pathway to secretion. *Nature* **281,** 629–630.

Nichols, R. A., Chandler, C. E., and Shooter, E. M. 1981. Enucleation of PC12 cells. *Soc. Neurosci. Abstr.* **7,** 695.

Nikodijevic, B., Nikodijevic, O., Yu, M., Pollard, H., and Guroff, G. 1975. The effect of NGF on cyclic AMP levels in superior cervical ganglia of the rat. *Proc. Natl. Acad. Sci. USA* **72,** 4769–4771.

Nishi, R. and Berg, D. K. 1979. Survival and development of ciliary ganglion neurons grown alone in cell culture. *Nature* (*London*) **277,** 232–234.

Nishi, R. and Berg, D. K. 1981. Effects of high K^+ concentrations in the growth and development of ciliary ganglion neurons in cell culture. *Dev. Biol.* **87,** 301–307.

Noda, M., Takahashi, H., Tanabe, T., Toyosato, M., Kiloytani, S., Hiroshe, T., Asai, M., Takashima, H., Ihayana, S., Miyata, T., and Numa, S. 1983. Primary structures of β and γ subunit precursors of *Torpedo californica* acetylcholine receptor derived from cDNA sequences. *Nature* **301,** 251–255.

Nolan, J. C., Ridge, S., Oronsky, A. L., Slakey, L. L., and Kerwar, S. S. 1978. Synthesis of a collagenase inhibitor by smooth muscle cells in culture. *Biochem. Biophys. Res. Commun.* **83,** 1183–1190.

Norling, B., Glimelius, B., Westermark, B., and Wasteson, A. 1978. A chondroitin sulphate proteoglycan from human cultured glial cells aggregates with hyaluronic acid. *Biochem. Biophys. Res. Comm.* **84,** 914–921.

Norr, S. C. 1973. *In vitro* analysis of sympathetic neuron differentiation from chick neural crest cells. *Dev. Biol.* **34,** 16–39.

Nowak, T. P., Kobiler, D., Roel, L. E., and Barondes, S. 1977. Developmentally regulated lectin from embryonic chick pectoral muscle. *J. Biol. Chem.* **252,** 6026–6030.

Numa, S., Noda, M., Takahashi, H., Tanabe, T., Toyosato, M., Furutani, Y., and

Kikyotani, S. 1984. Molecular structure of *Torpedo californica* acetylcholine receptor. *Cold Spring Harbor Symp. Quant. Biol.* **48,** 57–69.

Obata, K. 1981. Requirements for growth of neurites from embryonic chick cerebellum in culture. *Dev. Brain Res.* **1,** 444–449.

Obata, K. and Tanaka, H. 1980. Conditioned medium promotes neurite growth from both central and peripheral neurons. *Neurosci. Lett.* **16,** 27–33.

Obinata, T., Masaki, T., and Takano, H. 1980. Types of myosin light chains present during the development of fast skeletal muscle in chick embryo. *J. Biochem. (Tokyo)* **87,** 81–88.

O'Day, W. T. and Young, R. W. 1978. Rhythmic daily shedding of outer-segment membranes by visual cells in the goldfish. *J. Cell Biol.* **76,** 593–604.

O'Donnell-Tormey, J. and Quigley, J. P. 1981. Inhibition of plasminogen activator release from transformed chicken fibroblasts by a protease inhibitor. *Cell* **27,** 85–95.

Oey, J. 1976. Noradrenaline-refractoriness in cultured glioma cells. *Exp. Brain Res.* **25,** 359–368.

O'Farrell, P. H. 1975. High resolution two-dimensional electrophoresis of proteins. *J. Biol. Chem.* **250,** 4007–4021.

Oh, E., Pierschbacher, M., and Ruoslahti, E. 1981. Deposition of plasma fibronectin in tissues. *Proc. Natl. Acad. Sci. USA* **78,** 3218–3221.

Okada, T. S. 1976. Transdifferentiation of cells of specialized eye tissues in cell culture. In *Tests of Teratogenicity in Vitro,* J. Ebert and M. Marois, Eds., North Holland, Amsterdam, pp. 91–110.

Olden, K., Parent, J. B., and White, S. L. 1982. Carbohydrate moieties of glycoproteins: a re-evaluation of their function. *Biochim. Biophys. Acta* **650,** 209–232.

Olender, E. J. and Stach, R. W. 1980. Sequestration of ^{125}I-labeled β-NGF by sympathetic neurons. *J. Biol. Chem.* **255,** 9338–9343.

O'Neill, J. P., Schroder, C. H., and Hsie, A. W. 1975. Hydrolysis of butyryl derivatives of adenosine cyclic 3':5'-monophosphate by Chinese hamster ovary cell extracts and characterization of products. *J. Biol. Chem.* **250,** 990–995.

O'Neill, M. C. and Stockdale, F. E. 1972. Differentiation without cell division in cultured skeletal muscle. *Dev. Biol.* **29,** 410–418.

O'Neill, M. C. and Strohman, R. C. 1969. Changes in DNA polymerase activity associated with cell fusion in cultures of embryonic muscle. *J. Cell. Physiol.* **73,** 61–68.

Orgel, L. E. 1973. Aging of clones of mammalian cells. *Nature* **243,** 441–445.

Orkland, R. K., Nicholls, J. G., and Kuffler, S. W. 1966. Effect of nerve impulses on the membrane potential of glial cells in the central nervous system of *Amphibia. J. Neurophysiol.* **29,** 788–806.

Ortiz, J. R., Yamada, T., and Hsie, A. W. 1973. Induction of the stellate configuration in cultured iris epithelial cells by adenosine and compounds related to adenosine 3':5'-cyclic manophosphate. *Proc. Natl. Acad. Sci. USA* **70,** 2286–2290.

Osborn, M. and Weber, K. 1977. The display of microtubules in transformed cells. *Cell* **12,** 561–571.

Osborne, N. N. 1979. Is Dale's Principle Valid? *Trends Neurosci.* **2,** 73–75.

Ostlund, R. and Pastan, I. 1975. Fibroblast tubulin. *Biochemistry* **14,** 4064–4068.

Otsuka, H. and Moskowitz, M. 1975. Difference in transport of leucine in attached and suspended 3T3 cells. *J. Cell. Physiol.* **85,** 665–674.

Otten, U., Schwab, M., Gagnon, C., and Thoenen, H. 1977. Selective induction of tyrosine hydroxylase and dopamine β-hydroxylase by nerve growth factor. Comparison between adrenal medulla and sympathetic ganglia of adult and newborn rats. *Brain Res.* **133,** 291–303.

Palade, G. E. 1975. Intracellular aspects of the process of protein synthesis. *Science* **189,** 347–358.

Palfrey, C., Kimhi, Y., Littauer, U. Z., Reuben, R., and Marks, P. 1977. Induction of differentiation in mouse neuroblastoma cells by hexamethylene bisacetamide. *Biochem. Biophys. Res. Commun.* **76,** 937–942.

Pappas, G. D., Peterson, E. R., Masurovsky, E. B., and Crain, S. M. 1971. Electron microscopy of the *in vitro* development of mammalian motor endplates. *Ann. NY Acad. Sci.* **183,** 33–45.

Pappenheimer, A. M. and Gill, D. M. 1973. Diphtheria. *Science* **182,** 353–358.

Paravicini, U., Stoeckel, K., and Thoenen, H. 1975. Biological importance of retrograde axonal transport of nerve growth factor in adrenergic neurons. *Brain Res* **84,** 279–291.

Partlow, L. M. and Larrabee, M. G. 1971. Effects of nerve growth factor, embryo age and metabolic inhibitors on growth of fibres and on synthesis of RNA and protein in embryonic sympathetic ganglia. *J. Neurochem.* **18,** 2101–2118.

Pastan, I. H., Johnson, G. S., and Anderson, W. B. 1975. Role of cyclic nucleotides in growth control. *Annu. Rev. Biochem.* **44,** 491–522.

Patel, J. and Marangos, P. J. 1982. Modulation of protein phosphorylation by the S-100 protein. *Biochim. Biophys. Res. Commun.* **109,** 1089–1093.

Patterson, P. H. 1978. Environmental determination of autonomic neurotransmitter functions. *Annu. Rev. Neurosci.* **1,** 1–17.

Pearlstein, E. 1976. Plasma membrane glycoprotein which mediates adhesion of fibroblasts to collagen. *Nature* **262,** 497–500.

Pearlstein, E., Gold, L. I., and Garcia-Pardo, A. 1980. Fibronectin: a review of its structure and biological activity. *Mol. Cell. Biochem.* **29,** 103–128.

Peng, H. B. and Cheng, P.-C. 1982. Formation of postsynaptic specializations induced by latex beads in cultured muscle cells. *J. Neurosci.* **2,** 1760–1774.

Peng, H. B., Cheng, P.-C., and Luther, P. W. 1981. Formation of AChR clusters induced by positively charged latex beads. *Nature* **292,** 831–834.

Perriard, J. 1979. Developmental regulation of creatine kinase isoenzymes in myogenic cell cultures from chicken. *J. Biol. Chem.* **254,** 7036–7041.

Peters, A., Palay, S. L., and Webster, H. de F. 1976. The fine structure of the nervous system. *The Neurons and Supporting Cells,* W. B. Saunders, Philadelphia, pp. 250–254.

Peterson, J. A. and Rubin, H. 1969. The exchange of phospholipids between cultured chick embryo fibroblasts and their growth medium. *Exp. Cell Res.* **58,** 365–378.

Pfeiffer, S. E., Herschman, H. R., Lightbody, J., and Sato, G. 1970. Synthesis by a clonal line of rat glial cells of a protein unique to the nervous system. *J. Cell. Physiol.* **75,** 329–340.

Pfenninger, K. H. and Johnson, M. P. 1981. NGF stimulates phospholipid methylation in growing neurites. *Proc. Natl. Acad. Sci. USA* **78,** 7797–7800.

Pierschbacher, M. and Ruoslahti, E. 1981. Deposition of plasma fibronectin in tissues. *Proc. Natl. Acad. Sci. USA* **78,** 3218–3221.

Pintar, J. E. 1978. Distribution and synthesis of glycosaminoglycans during quail neural crest morphogenesis. *Dev. Biol.* **67,** 444–464.

Pockett, S. 1977. Clustering of acetylcholine receptors during neuromuscular synapse formation in rat and chick. Ph.D. Thesis, University of Otago, Dunedin, New Zealand.

Podleski, T. R., Axelrod, D., Ravdin, P., Greenberg, I., Johnson, M. M., and Salpeter, M. M. 1978. Nerve extract induces increase and redistribution of acetylcholine receptors on cloned muscle cells. *Proc. Natl. Acad. Sci. USA* **75,** 2035–2039.

Podleski, T. R. and Greenberg, I. 1980. Distribution and activity of endogenous lectin during myogenesis as measured with antilectin antibody. *Proc. Natl. Acad. Sci. USA* **77,** 1054–1058.

Pollard, T. D. 1981. Cytoplasmic contractile proteins. *J. Cell Biol.* **91,** 1565–1655.

Porter, K. R. 1976. Motility in cells. In *Cell Motility,* R. Goldman, T. Pollard, and J. Rosenbaum, Eds., Vol. 3, Cold Spring Harbor, N.Y., pp. 1–28.

Poste, G., Greenham, L. W., Mallucci, L., Reeve, P., and Alexander, D. J. 1973. The study of cellular "microexudates" by ellipsometry and their relationship to the cell coat. *Exp. Cell Res.* **78,** 303–313.

Prasad, K. N., Gilmer, K., and Kumar, S. 1973. Morphologically "differentiated" mouse neuroblastoma cells induced by noncyclic AMP agents: levels of cyclic AMP, nucleic acid and protein. *Proc. Soc. Exp. Biol. Med.* **143,** 1168–1171.

Prasad, K. N. and Hsie, A. W. 1971. Morphologic differentiation of mouse neuroblastoma cells induced *in vitro* by dibutyryl adenosine 3':5'-cyclic monophosphate. *Nature* [*New Biol.*] **233,** 141–142.

Prasad, K. N. and Mandal, B. 1972. Catechol-o-methyl transferase activity in DBcAMP, prostaglandin, and X-ray induced differentiated neuroblastoma cell culture. *Exp. Cell Res.* **74,** 532–534.

Prasad, K. N. and Sheppard, J. R. 1972. Neuroblastoma cell culture: membrane changes during cyclic AMP-induced differentiation. *Proc. Soc. Exp. Biol. Med.* **141,** 240–243.

Prasad, K. N. and Sinha, P. K. 1976. Effect of sodium butyrate on mammalian cells in culture: a review. *In Vitro* **12,** 125–132.

Pratt, R. M., Larsen, M. A., and Johnston, M. C. 1975. Migration of cranial neural crest cells in a cell-free hyaluronate matrix. *Dev. Biol.* **44,** 298–305.

Preisler, H. D., Housman, D., Scher, W., and Friend, D. 1973. Effects of 5-bromo-2-deoxyuridine on production of globin messenger RNA in dimethyl sulfoxide-stimulated Friend leukemia cells. *Proc. Natl. Acad. Sci. USA* **70,** 2956–2959.

Prives, J., Fulton, A. B., Penman, S., Daniels, M. P., and Christan, C. N. 1982.

Interaction of the cytoskeletal framework with ACh receptor on the surface of embryonic muscle cells in culture. *J. Cell Biol.* **9,** 231–236.

Prives, J. and Shinitzky, M. 1977. Increased membrane fluidity precedes fusion of muscle cells. *Nature* **268,** 761–763.

Raff, M. C. 1979. The control of microtubule assembly *in vivo. Int. Rev. Cytol.* **59,** 1–96.

Raff, M. C., Fields, K. L., Hakomori, S., Mirsky, R., Pruss, R. M., and Winter, J. 1979. Cell-type specific markers for distinguishing and studying neurons and the major classes of glial cells in culture. *Brain Res.* **174,** 283–308.

Raff, M. C., Hornby-Smith, A., and Brockes, J. P. 1978a. Cyclic AMP as a mitogenic signal for cultured rat Schwann cells. *Nature* **273,** 672–673.

Raff, M. C., Mirsky, R., Fields, K. L., Lisak, R. P., Dorfman, S. H., Silberberg, D. H., Gregson, N. A., Leibowitz, S., and Kennedy, M. 1978b. Galactocerebroside is a specific cell surface antigenic marker for oligodendrocytes in culture. *Nature* **274,** 813–816.

Raju, T. R., Bignami, A., and Dahl, D. 1980. Glial fibrillary acidic protein in monolayer cultures of C6 glioma cells: effect of aging and dibutyryl cyclic AMP. *Brain Res.* **200,** 225–230.

Rash, J. E. and Fambrough, D. 1973. Ultrastructural and electrophysiological correlates of cell coupling and cytoplasmic fusions during myogenesis *in vitro. Dev. Biol.* **30,** 166–186.

Rash, J. E. and Staehelin, L. A. 1974. Freeze–cleave demonstration of gap junctions between skeletal myogenic cells *in vivo. Dev. Biol.* **36,** 455–461.

Rasmussen, H. and Goodman, D. B. P. 1977. Relationships between calcium and cyclic nucleotides in cell activation. *Physiol. Rev.* **57,** 421–509.

Reeves, R. and Cserjesi, P. 1979. Sodium butyrate induces new gene expression in Friend erythroleukemic cells. *J. Biol. Chem.* **254,** 4283–4290.

Reff, M. E. and Davidson, R. L. 1979. Deoxycytidine reverses the suppression of pigmentation caused by 5-BrdUrd without changing the distribution of 5-BrdUrd in DNA. *J. Biol. Chem.* **254,** 6869–6872.

Reichardt, L. F. and Patterson, P. H. 1977. Neurotransmitter synthesis and uptake by isolated sympathetic neurones in microcultures. *Nature* **270,** 147–151.

Rein, D., Gruenstein, E., and Lessard, J. 1980. Actin and myosin synthesis during differentiation of neuroblastoma cells. *J. Neurochem.* **34,** 1459–1469.

Reiness, C. G. and Weinberg, C. B. 1981. Metabolic stabilization of acetylcholine receptors at newly formed neuromuscular junctions in rat. *Dev. Biol.* **84,** 247–254.

Reyero, C. and Waksman, B. H. 1980. Cell cycle dependent expression of cell surface antigens in murine T lymphoma cells. *Exp. Cell Res.* **130,** 265–274.

Reznik, M. 1976. Origin of the myogenic cell in the adult striated muscle of mammals. *Differentiation* **7,** 65–73.

Richler, C. and Yaffe, D. 1970. The *in vitro* cultivation and differentiation capacities of myogenic cell lines. *Dev. Biol.* **23,** 1–22.

Rieger, F., Koenig, J., and Vigny, M. 1980. Spontaneous contractile activity and the presence of the 16S form of acetylcholinesterase in rat muscle cells in culture: reversible suppressive action of tetrodotoxin. *Dev. Biol.* **76,** 358–365.

Ritchie, A. K. and Fambrough, D. M. 1975. Electrophysiological properties of the membrane and acetylcholine receptor in developing rat and chick myotubes. *J. Gen. Physiol.* **66,** 327–355.

Robinson, J. M. and Karnovsky, M. J. 1980. Specializations in filopodial membranes at points of attachment to substrate. *J. Cell Biol.* **87,** 562–568.

Robison, G. A., Butcher, R. W., and Sutherland, E. W. 1968. Cyclic AMP. *Annu. Rev. Biochem.* **37,** 149–174.

Robison, G. A., Butcher, R. W., and Sutherland, E. W. 1971. *Cyclic AMP,* Academic Press, New York.

Rogers, J., Ng, S. K. C., Coulter, M. B., and Sanwal, B. D. 1975. Inhibition of myogenesis in a rat myoblast line by 5-bromodeoxyuridine. *Nature* **256,** 438–439.

Rohrer, H., Schafer, T., Korsching, S., and Thoenen, H. 1982. Internalization of NGF by pheochromocytoma PC12 cells: absence of transfer to the nucleus. *J. Neurosci.* **2,** 687–697.

Roisen, F. J. and Murphy, R. A. 1973. Neurite development *in vitro.* II. The role of microfilaments and microtubules in dibutyryl adenosine 3′,5′-cyclic monophosphate and NGF stimulated maturation. *J. Neurobiol.* **4,** 397–412.

Roisen, F. J., Murphy, R. A., and Braden, W. G. 1972a. Neurite development *in vitro.* I. The effects of adenosine 3′,5′-cyclic monophosphate (cyclic AMP). *J. Neurobiol.* **3,** 347–368.

Roisen, F. J., Murphy, R. A., Pichichero, M. E., and Braden, W. G. 1972b. Cyclic adenosine monophosphate stimulation of axonal elongation. *Science* **175,** 73–74.

Ronne, H., Anundir, H., Rask, L., and Peterson, P. A. 1979. Nerve growth factor binds to serum alpha-2-macroglobulin. *Biochem. Biophys. Res. Commun.* **87,** 330–336.

Rosenberg, M. D. 1960. Microexudates from cells grown in tissue culture. *Biophys. J.* **1,** 137–159.

Rotundo, R. L. and Fambrough, D. M. 1980a. Synthesis, transport, and fate of acetylcholinesterase in cultured chick embryo muscle cell. *Cell* **22,** 583–594.

Rotundo, R. L. and Fambrough, D. M. 1980b. Secretion of acetylcholinesterase: relation to acetylcholine receptor metabolism. *Cell* **22,** 595–602.

Rous, P. and Jones, F. S. 1916. A method for obtaining suspensions of living cells from tissues. *J. Exp. Med.* **23,** 549–555.

Roy, R. K., Sreter, F. A., and Sarkar, S. 1979. Changes in tropomyosin subunits and myosin light chains during development of chicken and rabbit striated muscles. *Dev. Biol.* **69,** 15–30.

Rozengurt, E. 1980. Stimulation of DNA synthesis in quiescent cultured cells: exogenous agents, internal signals, and early events. *Curr. Top. Cell. Regul.* **17,** 59–88.

Rubenstein, P., Ruppert, T., and Sandra, A. 1982. Selective isoactin release from cultured embryonic skeletal muscle. *J. Cell Biol.* **92,** 164–169.

Rubin, H. 1980. Is somatic mutation the major mechanism of malignant transformation? *JNCI* **64,** 995–1000.

Rudy, B., Kirschenbaum, B., and Greene, L. A. 1982. NGF-induced increase in saxitoxin binding to rat PC12 pheochromocytoma cells. *J. Neurosci.* **2,** 1405–1411.

Rumsby, G. and Puck, T. T. 1982. Ornithine decarboxylase induction and the cytoskeleton in normal and transformed cells. *J Cell. Physiol.* **111,** 133–139.

Ruoslahti, E., Vaheri, A., Kuusela, P., and Linder, E. 1973. Fibroblast surface antigen: a new serum protein. *Biochem. Biophys. Acta* **322,** 352–358.

Rutishauser, U. and Edelman, G. M. 1980. Effects of fasciculation on the outgrowth of neurites from spinal ganglia in culture. *J. Cell Biol.* **87,** 370–378.

Rutishauser, U., Gall, W. E., and Edelman, G. M. 1978*b*. Adhesion among neural cells of the chick embryo IV. Role of the cell surface molecule CAM in the formation of neurite bundles in cultures of spinal ganglia. *J. Cell Biol.* **79,** 382–393.

Rutishauser, U., Thiery, J.-P., Brackenbury, R., and Edelman, G. M. 1978*a*. Adhesion among neural cells of the chick embryo III. Relationship of the surface molecule CAM to cell adhesion and the development of histotypic patterns. *J. Cell Biol.* **79,** 371–381.

Saad, A. D., Soh, B. M., and Moscona, A. A. 1981. Modulation of cortisol receptors in embryonic retina cells by changes in cell–cell contacts: correlations with induction of glutamine synthetase. *Biochem. Biophys. Res. Commun.* **98,** 701–708.

Sabatini, D. D., Kreibich, G., Morimoto, T., and Adesnik, M. 1982. Mechanisms for the incorporation of proteins in membranes and organelles. *J. Cell Biol.* **92,** 1–22.

Said, S. I. and Rosenberg, R. N. 1976. Vasoactive intestinal polypeptide: abundant immunoreactivity in neural cell lines and normal nervous tissue. *Science* **192,** 907–908.

Salmons, S. and Sreter, F. A. 1976. Significance of impulse activity in the transformation of skeletal muscle type. *Nature* **263,** 30–34.

Salzer, J. L. and Bunge, R. P. 1980. Studies of Schwann cell proliferation I. An analysis of tissue culture proliferation during development, Wallerian degeneration, and direct injury. *J. Cell Biol.* **84,** 739–752.

Salzer, J. L., Bunge, R. P., and Glaser, L. 1980*a*. Studies of Schwann cell proliferation III. Evidence of the surface localization of the neurite mitogen. *J. Cell Biol.* **84,** 767–778.

Salzer, J. L., Williams, A. K., Glaser, L., and Bunge, R. 1980*b*. Studies of Schwann cell proliferation II. Characterization of the stimulation and specificity of the response to a neurite membrane fraction. *J. Cell Biol.* **84,** 753–766.

Sanes, J. R. and Hall, Z. W. 1979. Antibodies that bind specifically to synaptic sites on muscle fiber basal lamina. *J. Cell. Biol.* **83,** 357–370.

Sanes, J. R., Marshall, L. M., and McMahan, U. J. 1978. Reinnervation of muscle fiber basal lamina after removal of myofibers. *J. Cell Biol.* **78,** 176–198.

Sanford, K. K., Earle, W. R., and Likely, G. D. 1948. Growth *in vitro* of single isolated cells. *JNCI* **9,** 229–246.

Sato, G. 1981. A working hypothesis for cell culture. *Horm. Cell Regul.* **5,** 1–13.

Sato, G., Ed. 1974. *Tissue Culture of the Nervous System,* Plenum Press, New York.

Schachner, M. 1974. Nervous system antigen, a glial cell-specific antigenic component of the surface membrane. *Proc. Natl. Acad. Sci. USA* **71,** 1795–1799.

Schaffner, A. E. and Daniels, M. P. 1982. Conditioned medium from cultures of embryonic neurons contains a high molecular weight factor which increases acetylcholine receptor aggregation on cultured myotubes. *J. Neurochem.* **2,** 623–632.

Schechter, A. L. and Bothwell, M. A. 1981. Nerve growth factor receptors on PC12 cells: Evidence for two receptor classes with differing cytoskeletal association. *Cell* **24,** 867–874.

Schimke, R. T., Ed. 1982. *Gene Amplification,* Cold Spring Harbor Laboratories, Cold Spring Harbor, New York.

Schliwa, M. 1981. Proteins associated with cytoplasmic actin. *Cell* **25,** 507–590.

Schliwa, M. and van Blerkom, J. 1981. Structural interaction of cytoskeletal components. *J. Cell Biol.* **90,** 222–235.

Schliwa, M., van Blerkom, J., and Porter, K. R. 1981. Stabilization of the cytoplasmic ground substance in detergent opened cells and a structural and biochemical analysis of its composition. *Proc. Natl. Acad. Sci. USA* **78,** 4329–4333.

Schloss, J. A. and Goldman, R. D. 1979. Isolation of a high molecular weight actin-binding protein from baby hamster kidney (BHK-21) cells. *Proc. Natl. Acad. Sci. USA* **76,** 4484–4488.

Schmechel, D. E. and Rakic, P. 1979. Arrested proliferation of radial glial cells during midgestation in rhesus monkey. *Nature* **277,** 303–305.

Schousboe, A., Drejer, J., and Divac, I. 1980. Regional heterogeneity in astroglial cells: implications of neuron–glia interactions. *Trends Neurosci.* **3,** 13–14.

Schrier, B. K. and Thompson, E. J. 1974. On the role of glial cells in the mammalian nervous system. *J. Biol. Chem.* **249,** 1769–1780.

Schroder, C. H. and Hsie, A. W. 1973. Morphological transformation of enucleated Chinese hamster ovary cells by dibutyryl adenosine 3′,5′-monophosphate and hormones. *Nature* **246,** 58–60.

Schubert, D. 1973. Protein secretion by clonal glial and neuronal cell lines. *Brain Res.* **56,** 387–391.

Schubert, D. 1974. Induced differentiation of clonal rat nerve and glia cells. *Neurobiology* **4,** 376–387.

Schubert, D. 1975. The uptake of GABA by clonal nerve and glia. *Brain Res.* **84,** 87–98.

Schubert, D. 1976. Proteins secreted by clonal cell lines. Changes in metabolism with culture growth. *Exp. Cell Res.* **102,** 329–340.

Schubert, D. 1977. The substrate attached material synthesized by clonal cell lines of nerve, glia, and muscle. *Brain Res.* **132,** 337–346.

Schubert, D. 1978. NGF-induced alterations in protein secretion and substrate-attached material of a clonal nerve cell line. *Brain Res.* **155,** 196–200.

Schubert, D. 1979. Early events after the interaction of NGF with sympathetic nerve cells. *Trends Neurosci.* **2,** 17–21.

Schubert, D., Carlisle, W., and Look, C. 1975. Putative neurotransmitters in clonal cell lines. *Nature* **254,** 341–343.

Schubert, D., Harris, A. J., Heinemann, S., Kidokoro, Y., Patrick, J., and Steinbach, J. H. 1974*a*. Differentiation and interaction of clonal cell lines of nerve and muscle. In *Tissue Culture of the Nervous System,* G. Sato, Ed., Plenum Press, New York, pp. 55–86.

Schubert, D., Heinemann, S., Carlisle, W., Tarikas, H., Kimes, B., Patrick, J., Steinbach, J. H., Culp, W., and Brandt, B. L. 1974*b*. Clonal cell lines from the rat central nervous system. *Nature* **249,** 224–227.

Schubert, D., Heinemann, S., and Kidokoro, Y. 1977. Cholinergic metabolism and synapse formation by a rat nerve cell line. *Proc. Natl. Acad. Sci. USA* **74,** 2579–2583.

Schubert, D., Humphreys, S., Baroni, C., and Cohn, M. 1969. *In vivo* differentiation of a mouse neuroblastoma. *Proc. Natl. Acad. Sci. USA* **64,** 316–320.

Schubert, D. and Jacob, F. 1970. Bromodeoxyuridine induced differentiation of a neuroblastoma. *Proc. Natl. Acad. Sci. USA* **67,** 247–254.

Schubert, D. and Klier, F. G. 1977. Storage and release of acetylcholine by a clonal cell line. *Proc. Natl. Acad. Sci. USA* **74,** 5184–5188.

Schubert, D. and LaCorbiere, M. 1977. Cell type transitions involving muscle cells. In *Pathogenesis of Human Muscular Dystrophies,* L. P. Rowland, Ed., John Wiley & Sons, New York, pp. 812–822.

Schubert, D. and LaCorbiere, M. 1980*a*. Role of a 16S glycoprotein complex in cellular adhesion. *Proc. Natl. Acad. Sci. USA* **77,** 4137–4141.

Schubert, D. and LaCorbiere, M. 1980*b*. Altered collagen and glycosaminoglycan secretion by a skeletal muscle myoblast variant. *J. Biol. Chem.* **255,** 11557–11563.

Schubert, D. and LaCorbiere, M. 1980*c*. A role of secreted glycosaminoglycans in cell–substratum adhesion. *J. Biol. Chem.* **255,** 11564–11569.

Schubert, D. and LaCorbiere, M. 1982*a*. The specificity of extracellular glycoprotein complexes in mediating cellular adhesion. *J. Neurosci.* **2,** 82–89.

Schubert, D. and LaCorbiere, M. 1982*b*. Properties of extracellular adhesion-mediating particles in a myoblast clone and its adhesion-deficient variant. *J. Cell Biol.* **94,** 108–114.

Schubert, D., LaCorbiere, M., Klier, F. G., and Birdwell, C. 1984. The structure and function of adherons in skeletal muscle. *Cold Spring Harbor Symp. Quant. Biol.* **48,** 539–549.

Schubert, D., LaCorbiere, M., Klier, F. G., and Birdwell, C. 1983. A role for adherons in neural retina cell adhesion. *J. Cell Biol.* **96,** 990–999.

Schubert, D., LaCorbiere, M., Klier, F. G., and Steinbach, J. H. 1980. The modulation of neurotransmitter synthesis by steroid hormones and insulin. *Brain Res.* **190,** 67–79.

Schubert, D., LaCorbiere, M., Whitlock, C., and Stallcup, W. 1978. Alterations in the surface properties of cells responsive to NGF. *Nature* **273,** 718–723.

Schubert, D., Munro, A., and Ohno, S. 1968. Immunoglobulin biosynthesis I. A myeloma variant secreting light chain only. *J. Mol. Biol.* **38,** 253–265.

Schubert, D., Tarikas, H., Harris, A. J., and Heinemann, S. 1971. Induction of acetylcholinesterase activity in a mouse neuroblastoma. *Nature [New Biol.]* **233,** 79–80.

Schubert, D., Tarikas, H., Humphreys, S., Heinemann, S., and Patrick, J. 1973. Protein synthesis and secretion in a myogenic cell line. *Dev. Biol.* **33,** 18–37.

Schubert, D., Tarikas, H., and LaCorbiere, M. 1976. Neurotransmitter regulation of adenosine 3′,5′-monophosphate in clonal nerve, glia, and muscle cell lines. *Science* **192,** 471–472.

Schubert, D. and Whitlock, C. 1977. The alteration of cellular adhesion by nerve growth factor. *Proc. Natl. Acad. Sci. USA* **74,** 4055–4059.

Schuetze, S. M., Frank, E. F., and Fischbach, G. D. 1978. Channel open time and metabolic stability of synaptic and extrasynaptic acetylcholine receptors on cultured chick myotubules. *Proc. Natl. Acad. Sci. USA* **75,** 520–523.

Schwab, I. A. and Luger, O. 1980. Reinitiation of DNA synthesis in postmitotic nuclei of myotubes by virus-mediated fusion with embryonic fibroblasts. *Differentiation* **16,** 93–99.

Schwartz, R. J. and Rothblum, K. N. 1981. Gene switching in myogenesis: differential expression of the chicken actin multigene family. *Biochemistry* **20,** 4122–4129.

Scordilis, S. and Adelstein, R. S. 1976. Myoblast myosin—phosphorylation is a prerequisite for actin activation. *Nature* **268,** 558–560.

Seeds, N. W., Gilman, A. G., Amano, T., and Nirenberg, M. 1970. Regulation of axon formation by clonal lines of a neural tumor. *Proc. Natl. Acad. Sci. USA* **66,** 160–167.

Sekiguchi, K. and Hakomori, S. 1980. Functional domain structure of fibronectin. *Proc. Natl. Acad. Sci. USA* **77,** 2661–2665.

Sellstrom, A. and Hamberger, A. 1977. Potassium-stimulated γ-aminobutyric acid release from neurons and glia. *Brain Res.* **119,** 189–198.

Senger, D. R. and Hynes, R. O. 1978. C3 component of complement secreted by established cell lines. *Cell* **15,** 375–384.

Shainberg, A., Yagil, G., and Yaffe, D. 1969. Control of myogenesis *in vitro* by Ca^{2+} concentration in nutritional medium. *Exp. Cell Res.* **58,** 163–167.

Shani, M., Zevin-Sonkin, D., Saxel, O., Carmon, Y., Katcoff, D., Nudei, U., and Yaffe, D. 1981. The correlation between the synthesis of skeletal muscle actin, myosin heavy chain, and myosin light chain and the accumulation of corresponding mRNA sequences during myogenesis. *Dev. Biol.* **86,** 483–492.

Shapiro, D. L. 1973. Morphological and biochemical alterations in fetal rat brain cells cultured in the presence of monobutyryl cyclic AMP. *Nature* **241,** 203–205.

Sharp, G. A., Osborn, M., and Weber, K. 1981. Ultrastructure of multiple microtubule initiation sites in mouse neuroblastoma cells. *J. Cell Sci.* **47,** 1–24.

Shaw, G. and Bray, D. 1977. Movement and extension of isolated growth cones. *Exp. Cell Res.* **104,** 55–62.

Sheppard, J. R. and Prasad, K. N. 1973. Cyclic AMP levels and the morphological differentiation of mouse neuroblastoma cells. *Life Sci.* **12,** 431–439.

Shimada, Y. 1971. Electron microscope observations on the fusion of chick myoblasts *in vitro. J. Cell Biol.* **48,** 128–142.

Shimada, Y. and Fischman, D. A. 1975. Scanning electron microscopy of nerve–muscle contacts in embryonic cell culture. *Dev. Biol.* **43,** 42–61.

Sidman, R. L., Miale, I. L., and Feder, N. 1959. Cell proliferation and migration in the primitive ependymal zone. *Exp. Neurol.* **1,** 322–333.

Sidman, R. L. and Rakic, P. 1973. Neuronal migration, with special reference to the developing human brain. A review. *Brain Res.* **62,** 1–35.

Silberstein, L., Inestrosa, N. C., and Hall, Z. W. 1982. Aneural muscle cell cultures make synaptic basal lamina components. *Nature* **295,** 143–145.

Siminovitch, L. 1976. On the nature of heritable variation in cultured somatic cells. *Cell* **7,** 1–11.

Singer, M. 1974. Neurotrophic control of limb regeneration in the newt. *Ann. NY Acad Sci.* **228,** 308–322.

Singer, M., Nordlander, R. H., and Egar, M. 1979. Axonal guidance during embryogenesis and regeneration in the spinal cord of the newt: the blueprint hypothesis of neuronal pathway patterning. *J. Comp. Neurol.* **185,** 1–22.

Skaper, S. D., Bottenstein, J. E., and Varon, S. 1979. Effects of NGF on cyclic AMP levels on embryonic chick dorsal root ganglia following factor deprivation. *J. Neurochem.* **32,** 1845–1851.

Smith, J. A. and Martin, L. 1973. Do cells cycle? *Proc. Natl. Acad. Sci. USA* **70,** 1263–1267.

Solomon, F. 1979. Detailed neurite morphologies of sister neuroblastoma cells are related. *Cell* **16,** 165–169.

Solomon, F. 1980. Neuroblastoma cells recapitulate their detailed neurite morphologies after reversible microtubule disassembly. *Cell* **21,** 333–338.

Solomon, F. 1981. Specification of cell morphology by endogenous determinants. *J. Cell Biol.* **90,** 547–553.

Spalding, J., Kajiwara, K., and Mueller, G. 1966. The metabolism of basic proteins in Hela cell nuclei. *Proc. Natl. Acad. Sci. USA* **56,** 1535–1542.

Spector, I. and Prives, J. M. 1977. Development of electrophysiological and biochemical membrane properties during differentiation of embryonic skeletal muscle in culture. *Proc. Natl. Acad. Sci. USA* **74,** 5166–5170.

Spencer, P. S. and Weinberg, H. J. 1978. Axonal specification of Schwann cell expression and myelination. In *Physiology and Pathology of Axons* (S. G. Waxman, Ed.) Raven Press, New York, pp. 389–406.

Sperry, R. W. 1963. Chemoaffinity in the orderly growth of nerve fiber patterns and connections. *Proc. Natl. Acad. Sci. USA* **50,** 703–710.

Spiegelman, B. M., Lopata, M. A., and Kirschner, M. W. 1979a. Multiple sites for the initiation of microtubule assembly in mammalian cells. *Cell* **16,** 239–252.

Spiegelman, B. M., Lopata, M. A., and Kirschner, M. W. 1979b. Aggregation of microtubule initiating sites preceding neurite outgrowth in mouse neuroblastoma cells. *Cell* **16,** 253–263.

Spitzer, N. C. 1979. Ion channels in development. *Annu. Rev. Neurosci.* **2,** 363–397.

Spitzer, N. C. 1981. Development of membrane properties in vertebrates. *Trends Neurosci.* **4,** 169–172.

Spitzer, N. C. and Lamborghini, J. E. 1976. The development of the action potential mechanism of amphibian neurons isolated in culture. *Proc. Natl. Acad. Sci. USA* **73,** 1641–1645.

Stallcup, W. B. 1977a. Comparative pharmacology of voltage-dependent sodium channels. *Brain Res.* **135,** 37–53.

Stallcup, W. B. 1977b. Specificity of adhesion between cloned neuronal cell lines. *Brain Res.* **126,** 475–486.

Stallcup, W. B. 1979. Sodium and calcium fluxes in a clonal nerve cell line. *J. Physiol. (London)* **286,** 525–540.

Stallcup, W. B. 1981. The NG2 antigen, a putative lineage marker: Immunofluorescence localization in primary cultures of rat brain. *Dev. Biol.* **83,** 154–165.

Stallcup, W. B. and Cohn, M. 1976. Correlation of surface antigens and cell type in cloned cell lines from the rat central nervous system. *Exp. Cell Res.* **98,** 285–297.

Starostina, M. V., Malup, T. K., and Sviridov, S. M. 1981. Studies on the interaction of Ca$^+$ ions with some fractions of the neurospecific S100 protein. *J. Neurochem.* **36,** 1904–1915.

Steinbach, J. H. 1974. Role of muscle activity in nerve–muscle interaction *in vitro. Nature* **248,** 70–71.

Steinbach, J. H. 1975. Acetylcholine responses in clonal myogenic cells *in vitro. J. Physiol. (London)* **247,** 393–405.

Steinbach, J. H. 1981. Developmental changes in acetylcholine receptor aggregates at rat skeletal neuromuscular junctions. *Dev. Biol.* **84,** 267–276.

Steinbach, J. H., Harris, A. J., Patrick, J., Schubert, D., and Heinemann, S. F. 1973. Nerve–muscle interaction *in vitro.* Role of acetylcholine. *J. Gen. Physiol.* **62,** 255–270.

Steinbach, J. H. and Schubert D. 1975. Multiple modes of dibutyryl cyclic AMP induced process formation by clonal nerve and glial cells. *Exp. Cell. Res.* **91,** 449–453.

Steinberg, M. S. 1970. Does differential adhesion govern self-assembly processes in histogenesis? Equilibrium configurations and the emergence of a hierarchy among populations of embryonic cells. *J. Exp. Zool.* **173,** 395–433.

Stenevi, U., Bjerre, B., Bjorklund, A., and Mobley, W. 1974. Effects of localized intracerebral regenerative growth of lesioned central noradrenergic neurons. *Brain Res.* **69,** 217–223.

Stieg, P. E., Kimelberg, H. K., Mazurkiewicz, J. E., and Banker, G. A. 1980. Distribution of glial fibrillary acidic protein and fibronectin in primary astroglial cultures from rat brain. *Brain Res.* **199,** 493–500.

Stockdale, F. E. and Holtzer, H. 1961. DNA synthesis and myogenesis. *Exp. cell Res.* **24,** 508–520.

Stockdale, F. E., Oazaki, K., Nameroff, M., and Holtzer, H. 1964. 5-Bromodeoxyuridine: effect on myogenesis *in vitro. Science* **146,** 533–535.

Stockdale, F. E., Raman, N., and Baden, H. 1981. Myosin light chains and the developmental origin of fast muscle. *Proc. Natl. Acad. Sci. USA* **78,** 931–935.

Stocklet, J. C. 1981. Calmodulins. *Biochem. Pharmacol.* **30,** 1723–1739.

Storrie, B., Puck, T. T., and Wenger, L. 1978. The role of butyrate in the reverse transformation reaction in mammalian cells. *J. Cell Physiol.* **94,** 69–76.

Strassman, R. J., Letourneau, P. C., and Wessells, N. K. 1973. Elongation of axons in an agar matrix that does not support cell locomotion. *Exp. Cell Res.* **81,** 482–487.

Sturgeon, C. M. 1979. Carcinoembryonic antigen and related glycoproteins as tumor markers. *Trends Biochem. Sci.* **4**(2) 121–123.

Sueoka, N., Imada, M., Tomozawa, Y., Droms, K., Chow, T., and Leighton, T. 1982. Neuronal–glial differentiation of a stem cell line from a rat neurotumor RT4-branch differentiation. In *Stability and Switching in Cellular Differentiation,* R. M. Clayton and D. Truman, Eds., Plenum Press, New York, pp. 165–174.

Sulston, J. E. 1984. Neuronal cell lineages in the nematode *C. elegans. Cold Spring Harbor Symp. Quant. Biol.* **48,** 443–452.

Sutter, A., Riopelle, R. J., Harris-Warrick, R. M., and Shooter, E. M. 1979. Nerve growth factor receptors. *J. Biol. Chem.* **254,** 5972–5982.

Sweadner, K. J. 1981. Environmentally regulated expression of soluble extracellular proteins of sympathetic neurons. *J. Biol. Chem.* **256,** 4063–4070.

Sytkowski, A. J., Vogel, Z., and Nirenberg, M. W. 1973. Development of acetylcholine receptor clusters on cultured muscle cells. *Proc. Natl. Acad. Sci. USA* **70,** 270–274.

Szent-Gyorgyi, A. 1948. *Nature of Life, a Study on Muscle,* Academic Press, New York.

Takahashi, K., Miyazaki, S., and Kidokoro, Y. 1971. Development of excitability on embryonic muscle cell membranes in certain tunicates. *Science* **171,** 415–418.

Tarikas, H. and Schubert, D. 1974. Regulation of adenylate kinase and creatine kinase activities in myogenic cells. *Proc. Natl. Acad. Sci. USA* **71,** 2377–2381.

Tash, J. S., Means, A. R., Brinkley, B. R., Dedman, J. R., and Cox, S. M. 1980. Cyclic nucleotides and Ca^{2+} ion regulation of microtubule initiation and elongation. In *Microtubules and Microtubule Inhibitors,* M. DeBrabander and J. DeMey, Eds., Elsevier-North Holland, Amsterdam, pp. 269–279.

Thiery, J.-P., Brackenbury, R., Rutishauser, U., and Edelman, G. M. 1977. Adhesion among neural cells of the chick embryo II. Purification and characterization of a cell adhesion molecule from neural retina. *J. Biol. Chem.* **252,** 6841–6845.

Thoenen, H. and Barde, Y. A. 1980. Physiology of nerve growth factor. *Physiol. Rev.* **60,** 1284–1335.

Thoenen, H., Schafer, T., Heumann, R., and Schwab, M. 1981. NGF as a retrograde macromolecular messenger between effector cells and innervating neurons. *Horm. Cell Regul.* **5,** 15–34.

Thomas, K. A. and Bradshaw, R. A. 1980. Nerve growth factor. In *Proteins of the Nervous System* (2nd ed.), R. Bradshaw and D. M. Schneider, Eds., Raven Press, New York, pp. 213–230.

Todaro, G. J., Fryling, C., and DeLarco, J. E. 1980. Transforming growth factors produced by certain human tumor cells: polypeptides that interact with EGF receptors. *Proc. Natl. Acad. Sci. USA* **77,** 5258–5262.

Tomozawa, Y. and Sueoka, N. 1978. *In vitro* segregation of different cell lines with neuronal and glial properties from a stem cell line of rat neurotumor RT4. *Proc. Natl. Acad. Sci. USA* **75,** 6305–6309.

Townes, P. L. and Holtfreter, J. 1955. Directed movements and selective adhesion of embryonic amphibian cells. *J. Exp. Zool.* **128,** 53–120.

Trinkaus, J. P. 1980. Formation of protrusions of the cell surface during tissue cell movement. Tumor cell surfaces and malignancy. *Prog. Clin. Biol. Res.* **41,** 887–906.

Truding, R. and Morell, P. 1977. Effect of $N^6,O^{2'}$-dibutyryl adenosine $3':5'$-monophosphate on the release of surface proteins by murine neuroblastoma cells. *J. Biol. Chem.* **252,** 4850–4854.

Truding, R., Shelanski, M. L., Daniels, M. P., and Morell, P. 1974. Comparison of surface membranes isolated from cultured murine neuroblastoma cells in the differentiated or undifferentiated state. *J. Biol. Chem.* **249,** 3973–3982.

Truding, R., Shelanski, M. L., and Morell, P. 1975. Glycoproteins released into the culture medium of differentiating murine neuroblastoma cells. *J. Biol. Chem.* **250,** 9348–9354.

Turley, E. A. 1980. The control of adrenocortical cytodifferentiation by extracellular matrix. *Differentiation* **17,** 93–103.

Unsicker, K., Krisch, B., Otten, U., and Thoenen, H. 1978. NGF-induced fiber outgrowth from isolated rat adrenal chromaffin cell: impairment by glucocorticoids. *Proc. Natl. Acad. Sci. USA* **75,** 3498–3502.

Unwin, P. N. T. and Zampighi, G. 1980. Structure of the junction between communicating cells. *Nature* **283,** 545–548.

Vanderkerckhove, J. and Weber, K. 1978. Mammalian cytoplasmic actins are the products of at least two genes and differ in primary structure in at least 25 identified positions from skeletal muscle actins. *Proc. Natl. Acad. Sci. USA* **75,** 1106–1110.

Varani, J., Wass, J., Piontek, G., and Ward, P. 1981. Chemotactic factor-induced adherence of tumor cells. *Cell Biol. Int. Rep.* **5,** 525–530.

Varon, S., Nomura, J., and Shooter, E. M. 1967. Subunit structure of a high-molecular weight form of NGF from mouse submaxillary gland. *Proc. Natl. Acad. Sci. USA* **57,** 1782–1789.

Vigny, M., Gisiger, V., and Massoulie, J. 1978. "Nonspecific" cholinesterase and acetylcholinesterase in rat tissues. *Proc. Natl. Acad. Sci. USA* **75,** 2588–2592.

Virtanen, I., Lehto, V., Lehtonen, E., Vartio, T., Stenman, S., Kurhi, P., Wager, O., Small, J., Dahl, D., and Badley, R. 1981. Expression of intermediate filaments in cultured cells. *J. Cell Sci.* **50,** 45–63.

Vitteta, E. S. and Uhr, J. W. 1972. Cell surface immunoglobulins: V. Release from murine splenic lymphocytes. *J. Exp. Med.* **136,** 676–696.

Wahl, G. M., Vitto, L., Padgett, R. A., and Stark, G. R. 1982. Single-copy and amplified CAD genes in Syrian hamster chromosomes localized by a highly sensitive method for *in situ* hybridization. *Mol. Cell. Biol.* **2,** 308–319.

Warren, L. and Glick, M. C. 1968. Membranes of animal cells. II. The metabolism and turnover of the surface membrane. *J. Cell Biol.* **37,** 729–746.

Watt, F. A. and Harris, H. 1980. Microtubule-organizing centres in mammalian cells in culture. *J. Cell Sci.* **44,** 103–121.

Weatherbee, J. A. 1981. Membranes and cell movement: interactions of membranes with the proteins of the cytoskeleton. *Int. Rev. Cytol.* [*Suppl.*] **12,** 113–176.

Webster, H. F. 1975. Development of peripheral myelinated and unmyelinated nerve fibers. In *Peripheral Neuropathology,* P. J. Dyck, P. K. Thomas, and E. H. Lambert, Eds., Vol. 1, W. B. Saunders, Philadelphia, pp. 37–61.

Weeds, A. G., Trentham, D. R., Kean, C. J. C., and Buller, A. J. 1974. Myosin from cross-reinnervated cat muscles. *Nature* **247,** 135–139.

Weinberg, C. B. and Hall, Z. W. 1979. Antibodies from patients with myasthenia gravis recognize determinants unique to extrajunctional acetylcholine receptors. *Proc. Natl. Acad. Sci. USA* **76,** 504–508.

Weiss, L. and Lachmann, P. J. 1964. The origin of an antigenic zone surrounding HeLa cells cultured on glass. *Exp. Cell Res.* **36,** 86–91.

Weiss, L., Poste, G., MacKearnin, A., and Willett, K. 1975. Growth of mammalian cells on substrates coated with cellular microexudates. *J. Cell Biol.* **64,** 135–145.

Weiss, P. 1941. Nerve patterns: the mechanics of nerve growth. *Growth* **5,** 163–203.

Weiss, P. 1955. Neurogenesis. In *Analysis of Development,* B. Willier, P. Weiss, and V. Hamburger, Eds., W. B. Saunders, Philadelphia, pp. 346–392.

Weiss, P. 1961. Ruling principles in cell locomotion and cell aggregation. *Exp. Cell Res.* [*Suppl.*] **8,** 260–281.

Weiss, P. and Taylor, A. C. 1960. Reconstitution of complete organs from single-cell suspensions of chick embryos in advanced stages of differentiation. *Proc. Natl. Acad. Sci. USA* **46,** 1177–1185.

Weldon, P. R. and Cohen, M. W. 1979. Development of synaptic ultrastructure at neuromuscular contacts in an amphibian cell culture system. *J. Neurocytol.* **8,** 239–259.

Wessells, N. K., Letourneau, P. C., Nuttall, R. P., Luduena-Anderson, M., and Geiduschek, J. M. 1980. Responses to cell contacts between growth cones, neurites, and ganglionic non-neuronal cells. *J. Neurocytol.* **9,** 647–664.

West, G. J., Uki, J., Stahn, R., and Herschman, H. R. 1977. Neurochemical properties of cell lines from *N*-ethyl-*N*-nitrosourea induced rat tumors. *Brain Res.* **130,** 387–392.

Whalen, R. G., Butler-Browne, G. S., and Gros, F. 1976. Protein synthesis and actin heterogeneity in calf muscle cells in culture. *Proc. Natl. Acad. Sci. USA* **73,** 2018–2022.

Whalen, R. G., Butler-Browne, G. S., and Gros, F. 1978. Identification of a novel form of myosin light chain present in embryonic muscle tissue and cultured muscle cells. *J. Mol. Biol.* **126,** 415–431.

Whalen, R. G., Schwartz, K., Bouveret, P., Sell, S. M., and Gros, F. 1979. Contractile protein isozymes in muscle development: identification of an embryonic form of myosin heavy chain. *Proc. Natl. Acad. Sci. USA* **76,** 5197–5201.

Whalen, R. G., Sell, S. M., Butler-Browne, G. S., Schwartz, K., Bouveret, P., and Pinset-Harstrom, I. 1981. Three myosin heavy-chain isozymes appear sequentially in rat muscle development. *Nature* **292,** 805–809.

Wheater, P., Burkitt, G. H., and Daniels, A. 1979. *Functional Histology.* Churchill Livingstone, New York.

Wickner, W. 1980. Assembly of proteins into membranes. *Science* **210,** 861–868.

Wilkinson, J. M. 1978. The components of troponin from chicken fast skeletal muscle. *Biochem. J.* **169,** 229–238.

Willingham, M. C. 1976. Cyclic AMP and cell behavior in cultured cells. *Int. Rev. Cytol.* **44,** 319–363.

Willingham, M. C. and Pastan, I. 1975. Cyclic AMP and cell morphology in cultured fibroblasts. *J. Cell Biol.* **67,** 146–159.

Wilson, B. W., Nieberg, P. S., Walter, C. R., Linkhart, T. A., and Fry, D. M. 1973. Production and release of acetylcholinesterase by cultured chick embryo muscle. *Dev. Biol.* **33,** 285–299.

Wilson, S. S., Baetge, E. E., and Stallcup, W. B. 1981. Antisera specific for cell lines with mixed neuronal and glial properties. *Dev. Biol.* **83,** 145–153.

Wolfe, R. A., Wu, R., and Sato, G. 1980. Epidermal growth factor-induced down regulation of receptor does not occur in HeLa cells grown in defined medium. *Proc. Natl. Acad. Sci. USA* **77,** 2735–2739.

Wolff, J. and Hope-Cook, G. 1975. Charge effects in the activation of adenylate cyclase. *J. Biol. Chem.* **250,** 6897–6903.

Wolosewick, J. J. and Porter, K. R. 1979. Microtrabecular lattice of the cytoplasmic ground substance. Artifact or reality? *J. Cell Biol.* **82,** 114–139.

Wood, P. M. and Bunge, R. P. 1975. Evidence that sensory axons are mitogenic for Schwann cells. *Nature* **256,** 662–664.

Worton, R. G., Duff, C., and Campbell, C. E. 1980. Marker segregation without chromosome loss at the *emt* locus in Chinese hamster cell hybrids. *Somatic Cell Genet.* **6,** 199–213.

Wrathall, J. R., Oliver, C., Silagi, S., and Essner, E. 1973. Suppression of pigmentation in mouse melanoma cells by 5-bromodeoxyuridine. *J. Cell Biol.* **57,** 406–423.

Yablonka, Z. and Yaffe, D. 1977. Synthesis of myosin light chains and accumulation of translatable mRNA coding for light chain-like polypeptides in differentiating muscle cultures. *Differentiation* **8,** 133–143.

Yaffe, D. 1968. Retention of differentiation potentialities during prolonged cultivation of myogenic cells. *Proc. Natl. Acad. Sci. USA* **61,** 477–483.

Yaffe, D. and Feldman, M. 1965. The formation of hybrid multinucleated muscle fibers from myoblasts of different genetic origin. *Dev. Biol.* **11,** 300–317.

Yamada, K. M. and Olden, K. 1978. Fibronectin–adhesive glycoprotein of cell surface and blood. *Nature* **275,** 179–184.

Yamada, K., Spooner, B. S., and Wessells, N. K. 1970. Axon growth: roles of microfilaments and microtubules. *Proc. Natl. Acad. Sci. USA* **66,** 1206–1212.

Yamada, K. M., Spooner, B. S., and Wessells, N. K. 1971. Ultrastructure and function of growth cones and axons of cultured nerve cells. *J. Cell Biol.* **49,** 614–635.

Yamada, K. M., Yamada, S. S., and Pastan, I. 1976. Cell surface protein partially restores morphology, adhesiveness, and contact inhibition of movement to transformed fibroblasts. *Proc. Natl. Acad. Sci. USA* **73,** 1217–1221.

Yaoi, Y. and Kanaseki, T. 1972. Role of microexudate carpet in cell division. *Nature* **237,** 283–285.

Yarden, Y., Schreiber, A. B., and Schlessinger, J. 1982. A nonmitogenic analogue of epidermal growth factor induces early responses mediated by epidermal growth factor. *J. Cell Biol.* **92,** 687–693.

Yu, A. and Cohen, E. P. 1974. Studies on the effect of specific antisera on the metabolism of cellular antigens. II. The synthesis and degradation of TL antigens of mouse cells in the presence of TL antiserum. *J. Immunol.* **112,** 1296–1307.

Yu, M. W., Nikodijevic, B., Lakshmanan, J., Rowe, V., MacDonnell, P., and Guroff, G. 1975. Nerve growth factor and the activity of tyrosine hydroxylase in organ cultures of rat superior cervical ganglia. *J. Neurochem.* **28,** 835–842.

Zalin, R. 1979. The cell cycle, myoblast differentiation and prostaglandin as a developmental signal. *Dev. Biol.* **71,** 274–288.

Zalin, R. and Montague, W. 1974. Changes in adenylate cyclase, cyclic AMP, and protein kinase levels in chick myoblasts, and their relationship to differentiation. *Cell* **2,** 103–108.

Ziller, C., Dupin, E., Brazeau, P., Paulin, D., and LeDouarin, N. M. 1983. Early segregation of a neuronal precursor cell line in the neural crest as revealed by culture in a chemically defined medium. *Cell* **32,** 627–638.

Zisapel, N. and Littauer, U. Z. 1979. Expression of external-surface membrane proteins in differentiated and undifferentiated mouse neuroblastoma cells. *Eur. J. Biochem.* **95,** 51–59.

Ziskind, L. and Harris, A. J. 1979. Reinnervation of adult muscle in organ culture restores tetrodotoxin sensitivity in absence of electrical activity. *Dev. Biol.* **69,** 388–399.

Zomzely-Neurath, C. E. and Walker, W. A. 1980. Nervous system-specific proteins 14-3-2 protein, neuron-specific enolase, and S-100 protein. In *Proteins of the Nervous System,* 2nd ed., R. A. Bradshaw and D. M. Schneider, Eds., Raven Press, New York, pp. 1–57.

Index